PHYSIOLOGY OF CELL EXPANSION
DURING PLANT GROWTH

edited by

DANIEL J. COSGROVE AND DANIEL P. KNIEVEL

PROCEEDINGS OF THE SECOND ANNUAL PENN STATE
SYMPOSIUM IN PLANT PHYSIOLOGY

(MAY 21-23, 1987)

THE PENNSYLVANIA STATE UNIVERSITY

Sponsored by the Intercollege Graduate
Program in Plant Physiology, The Pennsylvania
State University, University Park, PA 16802

Published by:
American Society of Plant Physiologists
15501-A Monona Drive
Rockville, Maryland 20855

Library of Congress Cataloging in Publication Data

Main entry under title:

    Symposium in Plant Physiology (2nd: 1987: The Pennsylvania State University)
        Physiology of Cell Expansion During Plant Growth

Symposium proceedings.
Includes bibliographies and index.

    "Proceedings sponsored by the Intercollege Graduate Program in Plant Physiology" [*et al.*]
    1. Plant Cell Expansion--Congresses. 2. Biochemistry and Physiology--Congresses. 3. Water and Solute Transport--Congresses.

    I. Cosgrove, Daniel J., 1952- II. Knievel, Daniel P., 1943- III. The Pennsylvania State University, Intercollege Graduate Program in Plant Physiology. V. Title.

Library of Congress Catalog Card Number: 87-072933
ISBN 0-943088-11-9

EDITORS' INTRODUCTION

One remarkable aspect of plant growth is that individual cells undergo a period of prolonged cell enlargement during their development; plant cells may enlarge in volume by one hundred-fold or more. This contrasts markedly with the growth of bacterial or animal cells, in which cells typically only double in size before dividing or ceasing growth. Prolonged cell expansion in plants is uniquely important because it gives rise to the shape, form and size of plants. For experimentalists this is an attractive subject because it is controlled internally by various hormones or growth substances, and is externally influenced by light, water availability, gravity and other environmental stimuli. Thus there are many probes and prods which may be used to examine growth. For the same reasons it is of practical interest; namely, there exists a large, and relatively untapped, potential to modify plant growth for agricultural purposes. Moreover, some environmental conditions such as drought and salinity have pronounced and detrimental effects on crop yield through reductions in cell expansion.

A symposium, held at The Pennsylvania State University, 21-23 May 1987, focused on the topic of cell expansion during plant growth. Twenty leading researchers were invited to present their ideas and results in oral presentations, ranging from the molecular level to whole plant level. In addition, 28 posters were contributed. This volume represents the proceedings of this symposium. Two informal workshops were also held: one on the current state of the acid-growth hypothesis, another on new physical and mechanical techniques for measuring wall extensibility. Both workshops stimulated considerable discussion and debate, but they are not recorded here.

We had three aims in putting together this volume. First, we wanted an up-to-date statement on the current ideas and working hypotheses in this field. Second, we wanted to cover a wide spectrum of topics that touch on the subject of cell expansion, and particularly those that employ new methods. Cell expansion can be studied at many levels, including molecular, genetic, physical, chemical, cellular, ecological and agricultural levels. However, it is unusual for molecular biologists and water-relations experts to discuss their work and exchange ideas. By juxtaposing these different approaches, we hope to stimulate new thinking and promote crossover of ideas, methods, and problems. Finally, we wanted this volume to be timely and affordable. The cooperation of the authors and publication by the American Society of Plant Physiogists made this possible.

This volume begins with an overview of plant cell expansion by Peter M. Ray. In his article, Ray outlines the main assumptions and principles that underlie most of our studies of cell growth. This is followed by four articles dealing with the mechanics and biochemistry of wall expansion. Robert E. Cleland discusses the mechanism of wall loosening in plants. Nicholas C. Carpita reviews the biochemical composition and structure of primary walls. Karen J. Biggs and Stephen C. Fry discuss their ideas on the mechanism and importance of wall rigidification by phenolic cross-links. Debra B. Folsom and R. Malcolm Brown, Jr. provide a cytological view of growth during gravitropism. To provide a point of comparison and contrast with our ideas on plant growth, Arthur L. Koch reviews the mechanism of bacterial cell expansion.

Another set of three articles discusses water and solute transport during growth. Daniel J. Cosgrove emphasizes the importance of wall relaxation as a linking process between wall expansion and water uptake. Roger M. Spanswick reviews the role of proton-pumping ATPases in growth, and John S. Boyer discusses the hydraulics of growth, particularly with respect to the effects and consequences of water deficits.

Molecular and genetic studies of cell expansion are presented next. Nevins et al. show how antibodies directed against wall proteins may provide a new way to examine the molecular basis for wall loosening. Athanasios Theologis reviews auxin-regulated gene expression and its possible linkage with cell expansion. The regulation of nuclear enzymes is discussed by S. J. Roux. The usefulness of genetic mutants in the study of growth is pointed out by Jake MacMillan and Bernard O. Phinney, who discuss gibberellin action on stem elongation, and by Koornneef and his coworkers who review gibberellin and photoreceptor mutants.

A final group of articles deal with specific growth responses or growing systems. Theodore C. Hsiao and J. Jing illustrate the differing responses of leaf and root growth to water deficits. Dicot leaf growth is summarized by E. Van Volkenburgh, while Hans Kende reviews the control of internodal growth of deepwater rice. Michael L. Evans and Karl-H. Hasenstein present a model for intracelluar regulation of growth by auxin and gravity via the calcium/calmodulin system and the inositol phospholipid system. At the multicellular level, U. Kutschera reviews evidence for the epidermis as the primary controlling layer in organ growth.

Following the invited review articles, twenty five poster abstracts are presented.

We wish to thank the authors for their diligent efforts and their timeliness, and Ms. Bobbi Barnhart-Roth for her preparation of the manuscripts for camera-ready reproduction. Special thanks go to Jack C. Shannon for his dedication and work in planning and supporting this Second Annual Penn State Symposium in Plant Physiology.

The symposium and these proceedings would not have been possible without the financial support of Penn State's College of Science, College of Agriculture, and Graduate School, and from the U.S. Departments of Agriculture and of Energy. We gratefully acknowledge gifts in support of the symposium from: Ciba-Geigy Corp., Dekalb-Pfizer Genetics, E.I. du Pont de Nemours and Co., Hershey Foods Corp., Monsanto Agricultural Co., and Union Carbide Agricultural Products Co., Inc.

October 1987   Daniel J. Cosgrove
                                                     Daniel P. Knievel

A NOTE ON SYMBOLS, TERMS, AND ABBREVIATIONS

   In a volume of this sort, it is inevitable that special terms and symbols will reflect common use. Where ambiguities, synonyms or differing conventions are found in the literature, they will also rear their ugly heads here to confuse the unwary. In some cases, the editors were able to resolve these problems satisfactorily; in other cases, the problems were too deep set to resolve without doing violence to the opinions and conventions of the authors. So, careful readers will note differences in certain symbols and conventions in the various chapters. This nonuniformity in style was allowed to persist in this volume, in large part because it reflects the nonuniform usage in the current literature. A few terms merit special comment.

   Osmotic pressure is consistently given the symbol π. It is always a positive quantity. *Osmotic potential* is consistently given the symbol ψ_s. Numerically it is the negative value of osmotic pressure.

   Turgor pressure is sometimes symbolized by P and sometimes by ψ_p. They are equivalent numerically.

   The term *extensibility* is used in this volume in several different ways. In some cases it is used to represent the proportionality constant between growth rate and turgor pressure in excess of the yield threshold; for example, in the equation,

$$\text{Growth Rate} = \phi \ (P - Y)$$

ϕ is frequently termed extensibility. In other cases extensibility is used to denote some mechanical parameter of walls, measured by stress/strain analysis. There are still other meanings in the literature for the word extensibility. These various uses are not synonymous, so the reader must be careful to discern how the word is used.

   Two symbols have been used in this volume to denote the proportionality constant between growth rate and turgor pressure in excess of the yield threshold, as in the above equation. In most case, ϕ is used when growth rate is measured as a relative growth rate (i.e., a percent change in volume per unit time). In Cleland's article, growth rate is measured as an absolute quantity (change in volume per unit time), and the symbol used there is **m**. Thus, ϕ and **m** have different units, but are related terms. However, as Hsiao and Jing point out (p. 183), an alteration in the spacial pattern of growth may lead to some misleading conclusions when only the absolute growth rate is measured.

   In some cases, ϕ is simply called wall extensibility, in other cases it is called, more specifically, a *wall yield coefficient*.

   Certain abbreviations appear in some of the articles or poster abstracts without identification. The more common abbreviations are listed below. Other abbreviations may be found by consulting the January issue of Plant Physiology.

ABA, abscisic acid	P, turgor pressure
GA, gibberellin	UDP, uridine diphosphate
IAA, indole-3-acetic acid	Y, yield threshold

FOREWARD

With this volume, the American Society of Plant Physiologists continues its series of publications on timely topics in plant physiology. Publication of proceedings devoted to focus areas, such as the present one on the physiology of cell expansion during plant growth, is designed to share information from the symposia with other plant scientists. It is the wish of the Publications Committee and the Executive Committee to make these publications as useful as possible and, to this end, suggestions will always be welcome.

The ASSP Publications Committee
Jack C. Shannon, Chairman
Lowell D. Owens
Machi F. Dilworth

ASPP Symposia Publications from previous years are:

1982: CRASSULACEAN ACID METABOLISM, Eds., I.P. Ting and M. Gibbs.

1983: BIOSYNTHESIS AND FUNCTION OF PLANT LIPIDS, Eds., W.W. Thomson, J.B. Mudd and M. Gibbs.

1984: STRUCTURE, FUNCTION AND BIOSYNTHESIS OF PLANT CELL WALLS, Eds., W.M. Dugger and S. Bartnicki-Garcia.

1985: REGULATION OF CARBON PARTITIONNING IN PHOTOSYNTHETIC TISSUE, Eds., R.L. Heath and J. Preiss.

1985: INORGANIC CARBON UPTAKE BY AQUATIC PHOTOSYNTHETIC ORGANISMS, Eds., W.J. Lucas and J.A. Berry.

1985: EXPLOITATION OF PHYSIOLOGICAL AND GENETIC VARIABILITY TO ENHANCE CROP PRODUCTIVITY, Eds., J.E. Harper, L.E. Schrader and R.W. Howell.

1986: MOLECULAR BIOLOGY OF SEED STORAGE PROTEINS AND LECTINS, Eds., L.M. Shannon and M.J. Chrispeels.

1986: REGULATION OF CARBON AND NITROGEN REDUCTION AND UTILIZATION IN MAIZE, Eds., J.C. Shannon, D.P. Knievel, and C.D. Boyer.

1987: PLANT SENSCENCE: ITS BIOCHEMISTRY AND PHYSIOLOGY, Eds., W.W. Thomson, E.A. Nothnagel and R.C. Huffaker.

CONTENTS

Editors' Introduction...iii

Symbols, Terms, and Abbreviations..v

Foreward...vi

Contents..vii

CHAPTERS

- Principles of Plant Cell Growth..1
  Peter M. Ray

- The Mechanism of Wall Loosening and Wall Extension 18
  Robert E. Cleland

- The Biochemistry of "Growing" Cell Walls................................... 28
  Nicholas C. Carpita

- Phenolic Cross-linking in the Cell Wall... 46
  Karen J. Biggs and Stephen C. Fry

- Changes in Cellulose Microfibril Orientation During Differential
  Growth in Oat Coleoptiles.. 58
  Debra B. Folsom and R. Malcolm Brown, Jr.

- The Bacteria's Solution to the Problem of Growing Within
  Its Own Pressurized Space.. 74
  Arthur L. Koch

- Linkage of Wall Extension with Water and Solute Uptake.................... 88
  Daniel J. Cosgrove

- Properties of Proton-Pumping ATPases and Their Involvement
  in Growth...101
  Roger M. Spanswick

- Hydraulics, Wall Extensibility and Wall Proteins 109
  John S. Boyer

- Inhibition of Elongation by Antibodies Specific for
  Wall Proteins 122
  D. J. Nevins, R. Hatfield, T. Hoson and M. Inouhe

- Possible Link Between Auxin Regulated Gene Expression,
  H$^+$ Secretion, and Cell Elongation: A Hypothesis 133
  Athanasios Theologis

- Regulation of Nuclear Enzymes in Plants 145
  S. J. Roux

- Biochemical Genetics and the Regulation of Stem Elongation
  by Gibberellins..156
  Jake MacMillan and Bernard O. Phinney

- The Use of Gibberellin and Photomorphogenetic Mutants
  for Growth Studies 172
  M. Koornneef, P. Adamse, G. W. M. Barendse,
  C. M. Karssen and R. E. Kendrick

- Leaf and Root Expansive Growth in Response to Water Deficits 180
  Theodore C. Hsiao and Jiahai Jing

- Regulation of Dicotyledonous Leaf Growth.......................193
  E. Van Volkenburgh

- Stimulus-Response Coupling in the Action of Auxin
  and Gravity on Roots 202
  Michael L. Evans and Karl-H. Hasenstein

- Cooperation Between Outer and Inner Tissues in
  Auxin-Mediated Plant Organ Growth.............................215
  U. Kutschera

- Studies on Internodal Growth Using Deepwater Rice.........227
  Hans Kende

POSTER ABSTRACTS

- Apoplastic Solutes and Lowered Water Potential in
  Elongating Sugarcane Leaves...239
  F. C. Meinzer and P. H. Moore

- Does Leaf Turgor Determine Leaf Expansion Rates in
  Dry or Saline Soils? 241
  Rana Munns

- Leaf Wall Yield Threshold as Determined by Vapor Pressure
  Psychrometery for Field-Grown Soybeans................................243
  H. C. Randall and T. R. Sinclair

- Sensitivity of Vessel Diameter to Climatic Conditions in
  Bur Oak (*Quercus macrocarpa*) 245
  Deborah Woodcock

- Extrusion Pores: Localized Deformation of Cotyledonary
  Cell Walls During Imbibition..249
  Stephen C. Spaeth

- Solute Supply and Stem Elongation: Effect of Cotyledon
  Removal on Growth, Solute and Turgor Pressures 252
  Judy Gougler Schmalstig and Daniel J. Cosgrove

- The Biophysical Mechanism of Reduced Growth in *Pisum
  sativum* Seedlings Under Saline Conditions..............................255
  Cynde Margritz and Daniel J. Cosgrove

- Rates of Cell Wall Accumulation in Higher Plants:
  Measurement with Interference Microscopy 258
  Tobias I. Baskin, M. Syndonia Bret-Harte and Paul B. Green

- Xyloglucan Fucosyltransferase Activity is Localized in
  Golgi Dictyosomes 262
  A. Camirand, D. A. Brummel and G. A. Maclachlan

- Properties of Cell Wall-Associated Enzymes in the Bound
  and Free State..265
  G. Nagahashi, S.-I. Tu and P. M. Barnett

- Autolysis of Potato Tuber Cell Walls 267
  *Ken Sasaki, Gerald Nagahashi, Pauline Barnett, and
  Landis Doner*

- Changes in Wall Metabolism of Tobacco Suspension-Cultured
  Cells Upon Adaption to Osmotic Stress 270
  Naim Mohammad Iraki and Nicholas C. Carpita

- Solubility of Polysaccharides Synthesized *In Vitro* by Golgi
  Apparatus and Plasma Membrane of *Zea mays*
  Coleoptiles 273
  David M. Gibeaut and Nicholas C. Carpita

- Cell Wall Formation by Isolated Protoplasts and Plasmolyzed
  Cells From Carrot Suspension Cultures 276
  E. M. Shea and N. C. Carpita

- Root-cap Net in Corn..280
  F. C. Guinel, B. G. Pickard, J. E. Varner and M. E. McCully

- β-Glucan Synthase from Red Beet Root: CHAPS Solubilization,
  Functional Reconstitution and Group Specific Chemical
  Modification..284
  *Theresa L. Mason, Margaret E. Sloan, Panayotis Rodis,
  and Bruce P. Wasserman*

- Production of Monoclonal Antibodies to Cell Wall Peroxidase
  and Cellulase..289
  Sung-Ha Kim, M. E. Terry and S. J. Roux

- Antibodies Against Salt-Extractable Pea Cell Wall Proteins
  and Their Effect on Growth 292
  Melissa A. Melan and Daniel J. Cosgrove

- Physiological Evidence for Auxin Compartmentation...........................295
  David J. Parrish

- Calcium Antagonist Inhibits Auxin-Stimulated Growth
  and Wall Synthesis...298
  D. A. Brummell and G. A. Maclachlan

- The Relationship Between Accumulation of Auxin into,
  and Growth of, Coleoptile Cells..301
  M. J. Vesper, C. L. Kuss and J. M. Maxson

- Growth Regulation in Amphibious Buttercups..................................304
  R. F. Horton, C. Briand and S. Hubler

- Photocontrol of Leaf Unrolling 307
  Neil Viner and Harry Smith

- Effect of Low Fluence Red Light on Growth and NAA
  Binding in Maize Mesocotyl..310
  Alan M. Jones and Shubo Zhou

- The Effect of S-3307 on the IAA Levels in Elongating
    Stem of Alaska Pea 313
    R. H. Hamilton and D. M. Law

- Index..315

Physiology of Cell Expansion During Plant Growth, *D.J. Cosgrove and D.P. Knievel* Eds.,
Copyright ©1987, The American Society of Plant Physiologists

PRINCIPLES OF PLANT CELL EXPANSION

PETER M. RAY

Department of Biological Sciences
Stanford University, Stanford, CA 94305

INTRODUCTION

Permanent growth in size of a plant cell requires that its cell wall undergo a permanent expansion. This is called an irreversible extension, to distinguish it from the reversible (elastic) extension that must exist in the wall of any cell when the cell possesses a turgor pressure. Irreversible wall extension, and thus growth in size of a plant cell, cannot occur simply from the synthesis and accumulation of cytoplasmic components such as proteins, unless there are components whose synthesis causes, or is coupled to processes that cause, irreversible wall extension. Synthesis of cytoplasmic components without wall extension would just force an equal volume of water out of the cell, not yield an increase in cell size. Similarly, cell division (and biosynthetic processes such as DNA replication that are necessary for it) cannot of itself cause any growth in size of the cell or of the plant, unless the division is accompanied by irreversible wall extension and thus enlargement of the daughter cells (or of the mother cell prior to its division).

A cell may grow throughout its surface, or throughout its length in the case of a cell that is elongating. This is *diffuse growth*; irreversible expansion is occurring in all parts of the cell's wall, or of its side walls in the case of a diffusely elongating cell. Some types of cells instead show *localized growth*, due to irreversible extension occurring only in a local area or patch of cell wall. The most common instance is tip growth, in which a cell elongates into a tubular shape by wall expansion restricted entirely to the tip of the tube: examples are root hairs, pollen tubes, and fungal hyphae.

Dating from the middle of the last century, the classical explanation of wall growth was that biosynthetic incorporation of new wall polymers into the interior of the existing wall increases its surface area. This was called "growth by intussusception." The theory could be questioned because the structure of many walls suggests they are built up by depositing a succession of layers at the inner surface of the then-existing wall, a process called apposition. Apposition of a new surface wall layer cannot cause a wall to extend, but would tend instead to stiffen it. For reasons to be mentioned later, the predominant view now is that wall biosynthesis does not cause wall extension during growth. Although this view may be misleading, in order to appreciate the issues considered hereafter it is important to realize from the outset that one cannot just assume biosynthesis to be the basis of wall expansion, as people unfamiliar with the principles in this field often do.

1

Growth in size of a cell requires water uptake, most of the volume of any growing cell being water, whether its volume is increasing mainly through an increase in amount of cytoplasm (as in most meristematic cells) or in vacuolar volume (as in most non-dividing, growing cells). This water uptake comprises the hydraulic aspect of cell growth.

Irreversible wall extension requires turgor pressure, and does not occur in a wilted (turgor-less) cell even though such a cell may be metabolically and biosynthetically quite active. This reflects the existence of a mechanical aspect of plant cell growth. The capacity of a growing cell's wall to mechanically extend has often been referred to as plasticity. However, the physical meaning of this term denotes a specific type of irreversible extension process, from which irreversible wall extension seems to differ significantly.

Wall extension, and the cell enlargement it leads to, ceases rather quickly if energy metabolism is inhibited. The rate of the mechanical aspect of growth is evidently under the control of the cell, via some kind of metabolism-dependent input from the cell into its wall. This comprises the biochemical aspect of cell growth. The cell's action that leads to a capacity of its wall to undergo irreversible extension is called wall loosening, a neutral term chosen so as not to prejudge the biophysical nature of the phenomenon. A tip-growing cell can grow in its characteristic localized manner presumably by exerting localized wall loosening.

The following sections sketch some of the basic principles that have been perceived, and alternatives that still need consideration, regarding the mechanisms of the hydraulic, mechanical, and biochemical aspects of cell growth, and their regulation. These matters are covered more extensively in the remaining chapters of this volume.

HYDRAULIC ASPECT OF GROWTH

As far as we can tell, the water uptake that is required for cell enlargement is entirely passive, i.e., osmotic. For it to occur, a difference in water potential ($\Delta\psi$) must exist between a growing cell and its surroundings. Things are more complicated for cells that are part of a growing tissue, whose water uptake involves transport from the outside, or from its vascular system, through several to many cell layers. These complications are neglected in the present discussion, which is restricted to a single cell's requirements.

The rate of water uptake during cell growth is defined by the elementary dynamic equation for osmotic transport,

$$Q = K (\Delta\psi) = K (\Delta\pi - P) \qquad (1)$$

where Q is the rate of uptake, K is the cell membrane's hydraulic conductance (conductivity times membrane area), $\Delta\pi$ is the difference in osmotic pressure between cell contents and surroundings, and P is the cell's turgor pressure, the term in brackets on the right being of course the water potential difference between the cell and its surroundings (the cell's surroundings are assumed to be at atmospheric pressure).

If a cell is not growing (and not losing water by evaporation, or to other cells), Q has to be zero, so $\Delta\psi$ must become zero, hence P must become equal to $\Delta\pi$ (the osmotic equilibrium value of turgor pressure). In order for water uptake and cell enlargement to begin, P must somehow drop below this equilibrium value, so that a $\Delta\psi$ will arise. The process of wall loosening, which would cause irreversible wall extension if the cell were already increasing in volume, will instead, in these circumstances, cause a diminution or relaxation of the force, or stress, borne by the cell wall as a reaction to the cell's turgor pressure. This stress relaxation will cause a decrease in P itself, since P exists only by virtue of the equal and opposite reaction force exerted by the wall on the cell's contents. As a result of this drop in P, a $\Delta\psi$ has arisen, which will drive water uptake, cell volume increase, and hence wall extension. These processes will continue for as long as wall loosening continues to cause the molecular changes that result in stress relaxation (see Cosgrove's article in this volume).

Lockhart's Law. The growing cell faces conflicting roles of turgor pressure in the hydraulic and mechanical aspects of growth: P promotes irreversible wall extension, but as just noted, P opposes water uptake. As recognized originally by Lockhart (9), the cell must compromise these conflicting dependences, in effect adjusting its turgor pressure (via the stress relaxation phenomenon) until the rate of water uptake given by eq. (1) equals the rate of increase in volume of the cell wall chamber due to irreversible wall extension. This principle can be called Lockhart's Law. Lockhart deduced an expression for it by assuming that the rate of irreversible extension (E, in terms of volume increase) depends on P in the following manner:

$$E = \phi(P - Y) \qquad (\text{for } P > Y; \text{ for } P < Y, E = 0) \qquad (2)$$

where Y is a "yield threshold" or value of P below which no irreversible extension occurs at all, and ϕ is a "yield coefficient", or rate coefficient that governs how fast the wall extends for a unit increase in P, above Y (Fig. 1, line A). If the above expressions for E and Q are set equal to each other, as they must be during steady growth to satisfy Lockhart's Law, the result can be solved for P to find the value, P_s, that satisfies these requirements:

$$P_s = \frac{K \Delta\pi + \phi Y}{K + \phi} \qquad (3)$$

P_s is the value to which P must adjust, by stress relaxation, during steady growth. This expression can be substituted for P in either of the rate equations (1 or 2) to give the growth rate in terms of independent parameters (P_s itself being a dependent variable):

$$Q \text{ (or E)} = \frac{K \phi (\Delta\pi - Y)}{K + \phi} \qquad (4)$$

Equation (4) has two limiting cases. If the conductance for water (K) is small compared to the yield coefficient (ϕ), the equation reduces to eq. (1), with Y substituted for P. This is conductance- or water uptake-limited cell enlargement. In this situation, the rate of cell enlargement can be substantially

3

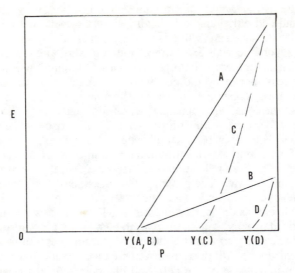

FIG. 1. Schematic dependence of cell wall extension rate (E) on turgor pressure (P): A and B, according to eq. (2) for steady-state growth with (A) or without (B) auxin; C and D, according to results of Green et al. (4, 5) for immediate extension rate with (C) or without (D) auxin. Y symbols indicate apparent yield thresholds for the different curves. The slope of lines A and B equals ø in eq. (2).

raised by increasing the conductance for water. If K is large compared to φ, on the other hand, eq. (4) reduces to eq. (2), with $\Delta\pi$ substituted for P. This is yield coefficient-limited cell enlargement. In this situation a negligible $\Delta\psi$ is all that is required to support the hydraulic aspect of cell growth, so as can be seen from eq. (3) P_s rises virtually to the osmotic equilibrium value ($\Delta\pi$), and the growth rate cannot be increased by increasing K, but can be by increasing the value of φ. In the intermediate situation, where K and φ are of the same magnitude, the rate of growth can be increased by increasing either parameter, but to only a limited extent if the other is not also raised. It appears that growing plant cells (except for unusual cases of extremely rapid or "explosive" cell enlargement) generally approach more closely to yield coefficient-limited than to conductance-limited cell enlargement. This issue is considered in the articles by Cosgrove and by Boyer in this volume.

Osmoregulation. Equation (4) shows that increasing the value of $\Delta\pi$, by raising a growing cell's solute concentration, would tend to increase its growth rate whether it were conductance- or yield coefficient-limited. This is a possible basis for regulatory effects on growth, such as hormone actions, but very few instances of it seem to be on record. In all cases of growth stimulation by auxin that have been studied in this respect, the cells' solute concentration actually declines rather than rises, after auxin is given.

Water uptake during growth tends to dilute, in proportion to cell volume, a cell's solute concentration, reducing $\Delta\pi$ and therefore growth rate (which would fall to zero if $\Delta\pi$ fell to the same value as Y). To offset this, substantial solute uptake normally accompanies growth, maintaining a relatively constant solute concentration as the cell expands. Hormonal stimulation of

4

growth is accompanied by stimulated solute uptake, which in cases that have been studied largely counteracts the dilution of cell solutes by the water uptake induced by the hormone. In response to water stress situations, a substantial rise in growing cells' solute concentration can occur, allowing cell growth to continue when it otherwise could not, for lack of sufficient turgor pressure. Examples of these phenomena are presented in the articles by Boyer and by Hsiao in this volume.

MECHANICAL ASPECT OF CELL GROWTH

As noted above, a tip-growing cell can achieve an elongated shape by restricting wall loosening to a localized patch at the tip of the cell. The shape that a diffusely growing cell develops during its growth is determined, on the other hand, by structural features of its cell wall that influence the directionality of the mechanical aspect of growth. Oriented deposition of the microfibrillar component of the wall (cellulose, in most plants), illustrated in Folsom and Brown's article in this volume, stiffens the wall in the direction of microfibril orientation, reducing irreversible extension in that direction compared with others. The direction of microfibril deposition during wall synthesis is controlled by cytoplasmic microtubules located adjacent to the cell's plasma membrane, but how they do this is still not adequately explained. This 3-dimensional aspect of cell expansion is also under hormonal control, notably by ethylene (2).

The evidence that irreversible extension depends on turgor pressure in the manner given above in eq. (2) comes primarily from osmotic experiments, in which growing cells' P is forced to vary by keeping them in media of different solute concentration, thus changing the $\Delta\pi$ term in eq. (3). Experiments of this type, performed on coleoptile or stem segments whose growth is auxin-inducible, or on leaf disks whose growth is light-inducible (cf. Van Volkenburgh's article in this volume), indicate that these stimuli do not affect the value of Y in eq. (2), but increase the wall's yield coefficient, $\emptyset$ (lines A and B, Fig. 1).

A new approach to this matter is to measure the stress relaxation that occurs, as mentioned previously, in a growing cell if it is unable to take up water and increase in volume. Cosgrove (cf. his article in this book) has used a pressure probe to follow the relaxation in turgor pressure that occurs after the cells of a growing tissue are taken out of contact with their water source. P declines gradually, over a period of about 1-3 hr, to the same value as the "yield threshold" for growth (Y, eq. 2) that is indicated by osmotic experiments. This is termed *in vivo* stress relaxation (to distinguish it from artificial stress relaxation measurements on isolated walls).

The exponential time course observed for *in vivo* stress relaxation conforms with a linear dependence of relaxation rate on P above the yield threshold, i.e., the relaxation has the same form as eq. (2). Auxin stimulates stress relaxation in auxin-sensitive tissue. By taking into account the role that wall elasticity plays in stress relaxation, an apparent value of ϕ can be calculated from an *in vivo* stress relaxation's time course. This gives values that agree

5

with those obtained from osmotic experiments on comparable tissue, and shows a similarly large auxin effect on ϕ.

Equation (2) has a form equivalent to viscoplastic flow. In this type of deformation, irreversible extension occurs only if forces are imposed that exceed a certain yield threshold (possibly representing a breaking strength for bonds between structural elements in the material), and the flow rate rises more or less linearly as the force is raised above that threshold (flow apparently being retarded by viscous friction between molecular components that have to move relative to one another).

Artificial extension tests of several kinds, discussed in Cleland's article in this volume, have been made on walls from growing cells, to try to elucidate the biophysics of irreversible wall extension and hormonal effects on it. These tests give various kinds of biophysical information which it is impractical to try to summarize here; besides Cleland's article, a recent review by Taiz (14) can be consulted for a detailed look at the subject. The mechanical behavior of cell walls and the auxin effects on it, revealed by most of these tests, seem to differ in important ways from the apparently viscoplastic extension that walls undergo during growth, as represented by eq. (2). This is not surprising, because eq. (2) describes a steady state of wall extension during growth, which depends, as previously noted, on some kind of metabolic wall loosening action exerted by the cell. The parameters in eq. (2), although formally resembling physical coefficients, do not necessarily relate to the purely physical process of extension that results from this wall loosening. We can appreciate this better if we roughly depict the situation as:

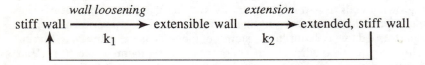

Artificial extension tests, and the decline in growth rate after metabolic inhibition, both indicate that the extensible state created by wall loosening decays as a result of extension ("strain hardening"), and/or by a parallel, time-dependent process. These two types of decay are simplistically represented, in this scheme, as being governed collectively by one rate coefficient (k_2). Now during steady growth the actual rates of the first and second steps must be equal (otherwise, the rate of extension would be changing). Somewhat as in Lockhart's Law, we can visualize two limiting cases of rate control in this scheme.

If k_2 is relatively small (i.e., a small fractional decline in extensibility per min), the wall loosening process governed by k_1 will gradually convert a "stiff" into an "extensible" wall, and within the time scale over which this extensible state persists the growth rate will be determined by physical parameters such as viscosities and yield stresses that govern the true physical extension process, and that should be artificially measurable. We may call this a physical extensibility-limited growth rate. The steady-state situation differs, however, from that in Lockhart's Law, where different potentially rate-controlling processes are in parallel. Here, because of the series relationship between the steps, the steady-state growth rate will be determined by the

6

coefficient k_1 for wall loosening, even though the immediate wall extension process does not depend on (is not "limited" by) simultaneous wall loosening.

If, instead, k_2 is large (e.g., a large fractional decline in extensibility per sec), the extension and extensibility decay processes will proceed virtually to completion as fast as extensibility is generated, so the rate of wall extension even in the short term will depend on the rate of wall loosening. We may call this a wall loosening-limited growth rate. Its most extreme case is a chemorheological extension, wherein a covalently crosslinked molecular network, incapable of physical flow, extends under a load as a result of a chemical reaction breaking bonds within or between the molecules. An apparent example is the extension of the bacterial cell wall during growth, which is considered by Koch in this volume.

In growing plant cells so far studied the situation seems to be closer to the wall loosening-limited, than to the physical extensibility-limited, case. For example, the fact that the large temperature coefficient for steady growth, attributed to its biochemical aspect (cf. Cleland's article in this volume), affects the immediate growth rate without an appreciable time lag (13), indicates an essentially ϕ wall loosening-limited growth rate. Therefore, the coefficients ϕ and Y in eq. (2) are not really physical parameters, but factors that at least partly reflect the dependence of the wall-loosening process on P. The thinking behind artificial extension experiments nevertheless still seems often to assume that measurable physical coefficients should be governing the overall rate of growth. What artificial extension measurements should really relate to is the instantaneous dependence of a cell's growth rate on turgor-generated stress in its wall, the true physical extension process in growth.

Characteristics of the true physical extension process. To determine these, the rate of irreversible extension should be followed before, and just after, step decreases or increases in a cell's P. Irreversible extension must be distinguished from elastic changes in cell size that result from the change in P, a distinction that may be difficult to make while P is still changing. Measurements of this kind by Green et al. on growing *Nitella* cells (3) and on oat and rye coleoptiles (4, 5) indicated that immediate wall extension depends much more steeply on turgor pressure than called for by the value of ϕ in eq. (2), and this dependence seems (at least for coleoptiles) to be curvilinear (lines C and D, Fig. 1) rather than linear as in eq. (2). The form of eq. (2) for steady growth apparently results from biological adjustment of the extension rate upward after a turgor step-down, and downward after a step-up. These adjustments seemed to be due to shifts in the yield point of the physical extension process, as in the difference between curves C and D in Fig. 1.

Possibly similar growth rate adjustments after step changes in P in maize leaves are described by Hsiao in this volume. These phenomena apparently reflect the operation of steps 1 and 2 of the above scheme at different rates, when the overall wall extension process is not at a steady state immediately after P has been changed. They may also, however, involve biological up- or down-regulation of the wall-loosening process in response to changes in P, as the analysis by Green et al. (4, 5) suggested. Much remains to be learned about the principles of the mechanical aspect of cell growth.

BIOCHEMICAL ASPECT OF CELL GROWTH

About wall loosening there are really two major questions: 1. What molecular change(s) in the wall's polymers is (are) responsible for converting a "stiff" into an "extensible" wall? 2. How does the cell cause such change(s)? The main kinds of possible processes that might be considered are (a) breakage of covalent bonds within wall polymer backbones or of crosslinks between them; (b) modification of wall polymers or their environment, weakening or breaking noncovalent bonding associations between them such as hydrogen or ionic bonds; and (c) biosynthetic insertion of new polymers. Basic to these possibilities, is whether the plant cell wall, like bacterial wall peptidoglycan, is effectively a covalently crosslinked network structure that can extend only by splitting covalent bonds, as against being held together to a significant extent by noncovalent associations that can allow polymers to be displaced and the wall to extend without covalent bond splitting.

The Albersheim group's model (8) for the structure of the dicot primary wall depicted most of the matrix polymers (pectins, hemicelluloses) as glycosidically linked to one another. This model, often construed as a structure requiring covalent bond breakage for its extension, has heavily influenced thinking about wall loosening, even though considerable general information on wall polymers is contrary to its representation of most of the wall's matrix polymers as being glycosidically interlinked. For example, L. Talbott in my laboratory has been able to separate the bulk of the pea cell wall's major matrix polymers from one another in a procedure that should not break glycosidic bonds. Although other types of cross links have been discovered (see articles by Carpita and by Biggs and Fry in this volume), a covalent network has not been demonstrated, and some workers think that noncovalent associations between wall polymers, as depicted in models given in Carpita's article in this volume, are the primary basis for wall integrity. Thus a cell's options for causing wall loosening do not in general seem to be restricted to covalent bond splitting reactions, as often assumed.

Experimental approaches to wall loosening. The principal approach available is to turn on growth by an appropriate stimulus (hormone, light, etc.) in an isolated tissue system, and look for concomitantly stimulated metabolic changes that affect, or are capable of affecting, the cell wall. Some nice experiments of this type have been done with gibberellin (GA) on lettuce hypocotyl segments and with light on leaf disks (see Van Volkenburgh's article in this volume), but most of our thinking about these questions is based on extensive auxin experiments with stem and coleoptile segments. An alternative "*in vitro*" approach will be to look for biochemical changes in isolated walls under conditions in which they extend irreversibly in a manner that apparently depends on chemical reactions (see Cleland's article in this volume).

Several kinds of possibly relevant *in vivo* auxin effects have been detected: (a) Auxin stimulates cell wall synthesis and enhances the activity of Golgi-localized polysaccharide synthases involved in it. (b) Auxin induces degradation of non-cellulosic cell wall glucan in grass coleoptiles, and an apparently degradative modification of xyloglucan in legume stem tissues (pea, bean; for the nature of these polymers, see Carpita's article in this volume). (c) Auxin induces energy-dependent extrusion of H^+ ions from the cell into the cell

8

wall (discussed by Spanswick in this volume), which by acidifying the cell wall can cause wall loosening, as shown in Cleland's article. This last effect is the basis of the "acid growth" theory, that H^+ ions extruded from the cell are the "second messenger" that couples wall loosening to primary auxin action.

Virtually all of the auxin effects listed above have been rejected as causes of wall loosening, based on what appears to be critical experimental evidence. For example, because auxin stimulates wall synthesis by a smaller factor than it stimulates growth, and this stimulation of wall synthesis (although rather rapid) develops more slowly than the stimulation of growth, it has been concluded that wall synthesis is not the basis of wall loosening. This conclusion has been reinforced by the finding that isolated wall specimens, in which no synthesis can be occuring, will extend irreversibly if held under a load at a low pH (see Cleland's article in this volume), showing that wall synthesis is not "necessary" for irreversible extension. Auxin-induced non-cellulosic glucan breakdown in coleoptiles seems to begin later than growth induction, does not seem to occur at all if the coleoptile segments are given a supply of sugar, and in other respects deviates from correlating closely with wall extension. The "acid growth" theory has been rejected because of several observations, such as that auxin stimulation of H^+ extrusion does not lead to a low enough cell wall pH to account for the auxin stimulation of growth.

These disproofs generally depend on at least some measurements or treatments whose validity or appropriateness can be questioned. More fundamentally, the validity of most such disproofs depends crucially on a usually unmentioned (and probably often unrecognized) assumption that *only one* mechanism is responsible for wall loosening, or for controlling its rate, in the system under study. Considering the biochemical and structural complexity of the plant cell wall, as brought out in the articles by Carpita and by Biggs and Fry in this volume, it actually seems more likely that cells have a variety of mechanisms available to them for causing wall loosening, and for controlling it. Such multiplicity is biologically plausible, because it would allow cells alternative ways of accomplishing growth and making it responsive to internal signals, as well as of adapting it to the variety of environmental fluctuations and stresses that a plant must tolerate. It seems best, therefore, to view the kinds of results on this topic summarized briefly in the preceding list as revealing a variety of processes that probably contribute to wall loosening or its control. No one of these processes probably provides a complete answer to the questions posed at the start of this section. If one finds a model system (such as an artificial wall extension experiment) for which "a" mechanism can be established, one should beware of assuming that this will explain the growth of a normal cell.

In this article it is impractical to undertake a more detailed discussion of most of the specific mechanisms (or potential mechanisms) of wall loosening and its control that available information suggests; some are covered elsewhere in this volume. However, after introducing another principle about wall loosening, wall synthesis will be specifically discussed, since after some decades of relative neglect, interest in its role is reviving, and it is not really covered in other articles in this volume.

Primary versus accessory processes. What we shall call a primary process in cell growth is one that causes (or directly contributes to) wall loosening. An accessory process is one that normally accompanies wall loosening, and that is necessary for wall loosening and growth to continue over the longer term, but is not required for the actual wall loosening events.

To illustrate this distinction, suppose (as currently often assumed) that wall loosening is due to enzymatic splitting of covalent bonds, the rate of which is promoted by auxin-stimulated H^+ extrusion, because the wall enzyme involved has an acidic pH optimum. The primary process is the bond-splitting enzymatic reaction, and its control by wall pH. An obligatory accessory process would be synthesis and transport, into the cell wall, of the responsible enzyme, to maintain its activity against decline over time due both to dilution by wall extension and wall synthesis, and to any inherent instability of the enzyme. Another accessory process would be solute uptake or production, since the rate of wall loosening depends on turgor pressure (and thus on the cell's π) as we saw earlier.

For growth to be stimulated over the long term, both primary and accessory processes may have to be stimulated. For example, in the preceding model it seems likely that for auxin-stimulated growth to be sustained over any substantial amount of cell enlargement, expression of the gene for the "wall-loosening enzyme" would also have to be increased by auxin. Over a suitably long time scale an accessory process can be rate-controlling for the overall rate of growth. In terms of the foregoing model, the growth rate of slowly growing tissue culture cells that do not significantly extrude H^+ ions could be controlled simply by the rate of synthesis and secretion of the wall-loosening enzyme.

Roles of wall synthesis. To compensate for the thinning of the wall caused by extension during growth, wall synthesis is needed to maintain the wall's thickness and strength, in order that the cell not become osmotically and mechanically unstable. This function does not presuppose any role of wall deposition in causing wall extension, and can be served perfectly well by apposition of new wall layers. However, it is effectively an accessory process for growth since if it does not occur at an adequate rate, the cell will eventually burst and thus not continue to grow. (Tip-growing cells such as root hairs and pollen tubes will burst within a few seconds of inhibiting their wall synthesis by hypoxia, a long-known phenomenon called "plasmoptysis".) One might expect, therefore, a mechanism for stimulating wall synthesis when growth is hormonally stimulated; this could be all the significance there is to the widely observed auxin stimulation of cell wall synthesis. Formation of covalent crosslinks between wall polymers may be an accessory process with a similar function.

Recently, however, the view has arisen that wall synthesis is also needed to maintain, in effect, the substrate supply for wall loosening, in order to allow irreversible extension to continue over the long term. This supposes that the wall polymers or inter-polymeric bonds present at any given time have only a limited capacity to be modified in the manner that causes wall loosening. When this capacity is used up by continued operation of the wall loosening action, wall extension would come to a halt if new polymers with their additional capacity to be loosened had not been added. This function would represent a second kind of accessory process.

This concept is tenable only if new polymers can be introduced *between* existing polymers in the wall. If not, for example if the new polymers were added by apposition, they would be mechanically in parallel with the preexisting, load-bearing wall layers, so any loosening action on the new material could not relax stress in the older layers nor allow them and the wall as a whole to extend. However, if new molecules can become intercalated, in the mechanical sense, between existing, load-bearing wall polymers of the wall, this intercalation would relax the load that is being borne by the polymers that are already in place; in other words it would constitute a primary process in wall loosening.

Electron microscope autoradiography after biosynthetic incorporation of tritium-labeled sugars into the thick outer epidermal walls of growing oat coleoptile and pea stem segments (11) showed that new cellulose is added to these walls entirely by apposition. However, part of the newly synthesized wall matrix material becomes incorporated internally into the wall, mostly into its inner layers near the plasma membrane, but at least some going all the way to its outer surface. Evidence was obtained that new polysaccharides are also incorporated into the interior of the much thinner walls of internal parenchyma cells (cortex, pith). The observations on epidermal walls conflict with the classical theory of growth by intussusception, from which a uniform incorporation of new material throughout the wall thickness would be expected. However, they make tenable the idea that wall synthesis might serve to maintain a supply of substrate for wall loosening, and also contribute directly to wall loosening, especially in view of the concept (discussed elsewhere in this volume) that the inner wall layers are mechanically the most important ones for wall extension.

Evidence that wall matrix polymer synthesis is important for wall extension comes from inhibition of coleoptile segment growth by galactose (7), and of pea segment growth by Golgi-inhibiting antibiotics like monensin, as discussed in Kutschera's article in this volume. Kutschera also gives results suggesting a more rapid auxin stimulation of wall matrix synthesis in pea stem epidermal cells than has been found in whole tissue segments, implying a possible role of auxin-stimulated wall synthesis in the initiation of auxin-stimulated elongation. Stimulation of total wall synthesis is not actually required, however, for wall synthesis to cause wall loosening, because loosening could be achieved by just diverting, from apposition to internal incorporation, a proportion of the large amount of new wall material that a young plant cell is typically making (as much as 50% of its total biosynthetic activity). Autoradiographic results (11) indicated that auxin treatment does increase the proportion of new matrix material that becomes internally incorporated, as against being deposited by apposition, in oat coleoptile and pea stem epidermal cells.

The key to promoting internal incorporation and resultant wall loosening would be to *retard* the rate at which newly exported wall polymer molecules associate with one another or with the wall surface, allowing them time to diffuse through interstices in the wall matrix, rather than becoming part of a new layer of material at the surface. New polymer molecules that diffuse into the wall interior could integrate into the wall structure, and relax stress in it, by exchanging bonding relationships with pairs of load-bearing, noncovalently associated polymers that are already part of the structure. One means by which cells could slow down association of new wall polymers could be to retard the formation of covalent crosslinks between them (see the Biggs and Fry article).

11

Another could be low wall pH, which by repressing dissociation of carboxyl groups of pectins and glucuronoxylans could retard formation of intermolecular ionic bonds involving divalent cations such as Ca^{2+}, or between pectins and hydroxyproline-rich wall protein ("extensin") chains (see Carpita's article, in this volume). Low wall pH could thus play a role in wall loosening different from the covalent bond-splitting reactions usually assumed.

HORMONE AND GENE ACTION IN CELL EXPANSION

Any hormone or gene whose action influences the growth rate of plant cells must act at least indirectly (not necessarily a primary action) by affecting the rate of irreversible wall extension. Over a long enough time scale, growth-inhibitory influences such as that of ABA could in principle result from an effect on any of the components of the growth mechanism considered above. Growth-stimulating effects like those of auxin and GA, on the other hand, would have to involve the rate of wall loosening (via either a primary or an accessory process) if cell expansion is yield coefficient-limited, which was suggested above as the usual situation.

The key question usually asked about a hormone effect is of course whether the primary action of the hormone that leads to the display is to act as an effector of some enzyme or transport process, or to regulate the expression of genes whose products participate in the display. In regard to auxin action this issue has been extensively debated since it was recognized that the inhibition of auxin-induced growth by RNA and protein synthesis inhibitors is not valid evidence that the hormone acts via gene expression, and realized that the auxin stimulation of growth rate begins so rapidly (ca. 10 min) as to strain the credibility of a gene activity induction mechanism. By now, auxin effects attributed to effector action, and auxin induction of gene expression, are both experimentally established. In view of the multiple possibilities for primary and accessory effects on growth rate explained previously, these kinds of effects should not necessarily be viewed as mutually exclusive alternatives (as they often are), but complementary phenomena that may go well beyond the control of growth rate and help explain the variety of other developmental and metabolic processes influenced by this hormone.

The same question can be asked about growth stimulation by GA, but has less often been addressed. The much-studied GA induction of gene activity for α-amylase in grass aleurone cells might suggest that GA stimulation of elongation is also based on gene expression. The GA effect on elongation rate occurs much more rapidly than α–amylase induction, but considerably slower than the auxin growth effect, over a time scale (ca. 40 min) that might readily allow it to be attributed to some kind of gene activity induction. Since most of the work yet done on hormone actions that lead to growth has involved auxin, only auxin action will be considered here.

Action at the transport level. The previously mentioned auxin stimulation of H^+ extrusion, reviewed in Spanswick's article in this volume, is the principal auxin effect widely assumed to be an effector action, viz., stimulation of a plasma membrane H^+-ATPase. However, an *in vitro* auxin effect on such a system that convincingly explains the auxin stimulation of *in vivo* H^+ extrusion has still not been demonstrated.

The auxin induction of proton extrusion (and growth) has an absolute latent period of 7-10 min (or more, in some systems) that conflicts with a direct action on a plasmalemma ATPase: fusicoccin, which apparently does induce H^+ extrusion by activating a plasmalemma ATPase, induces H^+ extrusion (and cell enlargement) with a lag of less than 1 min, and analogously rapid effects are exerted by animal hormones that activate plasma membrane enzymes. (One widely quoted report of growth induction by auxin without any latent period [10] was probably an artefactual result of low wall pH, rather than a true auxin effect). It thus seems likely that auxin induces H^+ extrusion indirectly (as a result of some other primary action within the cell). Consistent with this, the principal membrane-localized auxin-binding sites (putative receptors), at least in corn coleoptiles, occur on ER membranes, although a minor proportion of possibly qualitatively different sites may occur on other membranes, including the plasma membrane (1).

It was recently proposed (6) that auxin acts by causing cytoplasmic acidification (e.g., by stimulating acid-producing metabolism, or ion exchange between intracellular compartments). This fall in cytoplasmic pH would then stimulate the H^+ extrusion pump that serves to keep cytoplasmic pH within its normal range. The hypothesis thus regards the auxin effect on H^+ extrusion as indirect. It is founded on microelectrode data suggesting that cytoplasmic pH falls by about 0.1 unit, in 5-20 min (in different experiments), after giving auxin to corn coleoptile epidermal cells. ^{31}P-NMR measurements of cytoplasmic pH, capable of detecting the inferred acidification, have thus far failed to confirm it. Cytoplasmic pH seems too general a signal for highly specific effects such as the auxin induction of specific gene expression (next section), and one can doubt whether an 0.1 unit change in pH could really have any biologically important consequence. However, the proponents of this hypothesis (6) advocate involvement also of Ca^{2+} and/or other ions, which they feel might have more specific, and pleiotropic, effects. It remains to be seen whether this hypothesis for primary auxin action has more merit than others that preceded it.

Auxin stimulation of H^+ extrusion is blocked by RNA and protein synthesis inhibitors, which might suggest that rather than resulting from an effector action, it is due to auxin-induced gene expression. We shall return to this point in the next section.

Action on gene expression. Theologis, in this volume, reviews the evidence for rapid, selective induction of gene activity by auxin in pea stem tissue. About half a dozen mRNAs increase substantially within 2-3 hr after exposure to auxin. A couple of these begin to increase within 10 min, a couple more by about 30 min, and others not until more than 1 hr. The induction is clearly non-coordinate, suggesting either multiple mechanisms, or a control cascade leading from the earlier to the later responses. Similar results have been obtained with soybean hypocotyl and oat coleoptile tissue. There seems little doubt that the earliest mRNA responses are due either directly to a primary auxin action, or to some process very closely coupled to it, and that they are *not* caused by auxin-stimulated H^+ extrusion (see Theologis' article).

Since the earliest-detected increases in mRNAs induced by auxin apparently begin within the latent period of about 10 min for the cell enlargement response, one can see in them a possible explanation of growth induction by auxin. For this to be a realistic possibility, several criteria must be met. First, the translation products of at least one of the early-induced mRNAs must have some biological action that directly causes, or indirectly stimulates, a primary process in wall loosening. The translation product might, for example, be a wall-loosening enzyme, an H^+-ATPase or positive effector thereof, a wall polymer synthase, or a protein that influences how new wall polymers are transported into, or interact with, the wall. Second, a physiologically significant amount of this mRNA must appear early enough that there is still time within the latent period of auxin action on growth for the following to occur: (a) mRNA transport into the cytoplasm; (b) initiation and completion of translation product polypeptide chains (which evidence from other organisms indicates would probably require at least a couple of minutes); (c) transport of translation product chains to their site of action, e.g., via the Golgi system, to the plasma membrane or cell wall (which evidence from plants indicates would require 5 minutes or more); and (d) sufficient action by the translation product (e.g., accumulation of enough H^+ or new wall polymers in the cell wall) to significantly increase the rate of wall loosening. Third, after auxin withdrawal the auxin-induced mRNA must decline to the uninduced level earlier than the growth rate does, with enough time to spare for its translation product also to decline to a level that no longer stimulates wall loosening.

As yet the function of none of the early auxin-induced mRNAs' gene products has been identified. Nor is it yet known whether any of them even begins to appear within the latent period, let alone appears in physiologically significant amounts. Considering how late within the latent period the first slight increases in these mRNAs occurs, it can be doubted whether any of them could satisfy the second set of criteria. What has been reported about the kinetics of auxin-inducible mRNAs actually conflicts with the third type of criterion. The levels of the earliest auxin-induced mRNAs rise steadily for at least 2 hours, whereas within a few *minutes* after the end of the 10 min latent period (at which time only a small mRNA increase has yet occurred) the growth rate typically has risen to its maximum. Most of the auxin-induced rise in mRNA is, therefore, evidently superfluous to growth. The halflife of the early auxin-inducible mRNAs in the absence of auxin, or after inhibition of RNA synthesis, is at least 1 hr (see Theologis' article). At 20 or 30 min after auxin withdrawal, when the elongation rate has declined to the control value, the cells still contain an auxin-inducible mRNA level high in the "superfluous" range, and much greater than exists at the time the growth rate becomes maximally stimulated during growth induction.

There may of course be other, as yet undetected, early auxin-inducible mRNAs the timing of whose induction and turnover better matches that of the auxin effect on growth. However, the previously noted kinetic demands on their action, to account for growth induction by auxin, are such that this role can by no means be taken for granted. As noted for possible mechanisms of wall loosening, these kinetic demands are considerably relaxed if the induced mRNAs, via their gene products, simply *contribute* to growth, rather than being wholly responsible for the auxin stimulation.

Accessory roles of gene expression. Some auxin-inducible mRNAs probably code for polypeptides with developmentally significant functions other than stimulating the growth rate, such as enzymes for ethylene synthesis. Some, however, may well code for gene products that participate in previously defined accessory processes in wall loosening, which can influence growth over the longer term but are not responsible for its rapid initiation by auxin.

Vanderhoef (15) advanced the concept that auxin action on growth consists of two phases. The first phase, comprising the latent period and early stimulation of growth rate, he regarded as due to an effector action, possibly stimulation of H^+ extrusion. The second phase, which begins 0.5-1 hr after exposure to auxin, he considered to depend on auxin-regulated gene activity involved in accessory processes needed to maintain continued growth over a time scale of hours or days. The timing of the early auxin mRNA inductions thus far detected conforms well with this concept of the second phase. Vanderhoef suggested auxin-stimulated cell wall synthesis as an example of a gene activity-regulated accessory process. As noted already, polysaccharide synthases might well be coded for by some of the auxin-induced mRNAs (although this remains to be demonstrated). The timing of auxin enhancement of polysaccharide synthase activities in pea conforms rather well with that of the early mRNA inductions, but at least part of the auxin effect on synthase activity seems to be an activation tied to H^+ extrusion (12).

As noted above, because of inhibitor effects, auxin-induced H^+ extrusion might be attributed to induction of gene transcription. Since according to the best measurements the auxin induction of H^+ extrusion occurs with kinetics similar to growth induction, the currently detected auxin induction of mRNAs is unlikely to account for it, for the reasons given above. However, maintenance of auxin-stimulatable H^+ extrusion capacity by control of gene expression seems a likely accessory process.

Remarkably, auxin-induced H^+ extrusion can very rapidly (within about 5 min) be suppressed, post induction, by protein synthesis inhibitors such as cycloheximide (this is also true of auxin-stimulated growth, another piece of evidence linking the two). This may mean either that the mechanism involves a protein with a very unusually short halflife (efforts to detect such a protein have been unsuccessful), or that auxin-stimulated H^+ extrusion is actually tied to the simple occurrence of protein (and perhaps also RNA) synthesis, rather than being due to a conventional induction. No satisfactory mechanistic explanation of this intriguing feature has yet emerged. A possible explanation, at the regulatory level, arises from questions about how cell growth is related to a plant's overall metabolic performance, briefly discussed next.

Regulation of cell expansion relative to assimilation. Although, as explained at the outset, plant cell expansion cannot be attributed to general synthesis of cytoplasmic protein, in order for growth to yield physiologically normal cells, wall extension must be regulated to keep it in proper pace with the synthesis of cell components. In a meristem, the rate of wall extension is such that each cell on the average just doubles in size in the time it takes to replicate all the cell's components, after the preceding division. In a non-dividing growth zone ("cell enlargement zone") biosynthesis and wall extension are not coupled so strictly, but even here the plant needs regulatory mechanisms that can keep the rate of wall extension in physiologically appropriate balance with the rate of protein (and other biomolecule) synthesis. These mechanisms might involve substrate levels, hormones or other effectors, or gene activity, but whatever they are they must act on some of the processes discussed in previous sections of this article. They comprise the link between growth at the cellular level, considered in this volume, and growth at the whole plant level, usually viewed as an accumulation of plant biomass resulting from photosynthesis plus mineral nutrient uptake and assimilation.

When protein synthesis cannot occur, such a control mechanism should shut down wall extension and cell growth. If wall loosening depends on a low wall pH, extension could be suppressed by stopping auxin-stimulated H^+ extrusion. This stoppage would reflect a regulatory response of the cell to its lack of protein synthesis, rather than a mechanistic dependence of auxin-stimulated H^+ extrusion on transcription and translation of a specific mRNA. Possible complications like this really ought to be considered in interpreting any RNA or protein synthesis inhibitor effect.

Possibility of growth mutants. To definitively establish mechanisms of wall loosening and reveal their genetic foundations, cell expansion-deficient mutants, analogous to the hormone and photoreceptor mutants discussed by MacMillan and Phinney and by Koornneef et al. in this volume, would be invaluable. Ideally these would be temperature-sensitive mutants blocked non-nutritionally in wall extension and thus in growth when the gene is expressed, but which can grow and be propagated at a permissive temperature. But if, as suggested above, the cell has at its disposal more than one biochemical means of accomplishing wall loosening, wall-extension mutants might be difficult or impossible to detect, because blockage of one mechanism might be made up by operation of others, so the cell (and plant) could continue to grow despite the defect. Mutants might more readily be obtained if one could identify a simpler system that has but a single wall-loosening mechanism.

To use such mutants to further our understanding of cell growth we shall need to apply the physiological principles outlined above, and discussed further in other articles in this volume. Most of the topics covered in this overview are also considered in more depth, with extensive citations of the pertinent literature, in the review by Taiz (14).

LITERATURE CITED

1. DOHRMANN UC, R HERTEL, H KOWALIK 1978 Properties of auxin binding sites in different subcellular fractions from maize coleoptiles. Planta 140: 97-106
2. EISINGER W 1983 Regulation of pea internode expansion by ethylene. Annu Rev Plant Physiol 34: 225-240
3. GREEN PB, RO ERICKSON, J BUGGY 1971 Metabolic and physical control of cell elongation rate. *In vivo* studies in *Nitella*. Plant Physiol 47: 423-430
4. GREEN PB, WR CUMMINS 1974 Growth rate and turgor pressure. Auxin effect studied with an automated apparatus for single coleoptiles. Plant Physiol 54: 863-869
5. GREEN PB, K BAUER, WR CUMMINS 1977 Biophysical model for plant cell growth: auxin effects. *In* AM Jungreis, ed, Water relations in membrane transport in plants and animals. Academic Press, NY, pp 30-45

6. FELLE H, B BRUMMER, A BERTL, RW PARISH 1986 Indole-3-acetic acid and fusicoccin cause cytosolic acidification of corn coleoptile cells. Proc Natl Acad Sci USA 83: 8992-8995
7. INOUHE M, R YAMAMOTO, Y MASUDA 1987 UDP-glucose level as a limiting factor for IAA-induced elongation in Avena coleoptile segments. Physiol Plant 63: 406-412
8. KEEGSTRA K, K TALMADGE, WD BAUER, P ALBERSHEIM 1973 The structure of plant cell walls III. A model of the walls of suspension-cultured sycamore cells based on the interconnections of the macromolecular components. Plant Physiol 51: 188-196
9. LOCKHART JA 1965 An analysis of irreversible plant cell elongation. J Theor Biol 8: 264-275
10. NISSL D, MH ZENK 1969 Evidence against induction of protein synthesis during auxin-induced initial elongation of Avena coleoptiles. Planta 89: 323-341
11. RAY PM 1967 Radioautographic study of cell wall deposition in growing plant cells. J Cell Biol 35: 659-674
12. RAY PM 1985 Auxin and fusicoccin enhancement of β-glucan synthase in peas. An intracellular activity apparently modulated by proton extrusion. Plant Physiol 78: 466-472
13. RAY PM, AW RUESINK 1963 Kinetic experiments on the nature of the growth mechanism in oat coleoptile cells. Devel Biol 4: 377-397
14. TAIZ L 1984 Plant cell expansion: regulation of cell wall mechanical properties. Annu Rev Plant Physiol 35: 585-657
15. VANDERHOEF LN, RR DUTE 1981 Auxin-regulated wall loosening and sustained growth in elongation. Plant Physiol 67: 146-149

Physiology of Cell Expansion During Plant Growth, *D.J. Cosgrove and D.P. Knievel* Eds.,
Copyright ©1987, The American Society of Plant Physiologists

THE MECHANISM OF WALL LOOSENING AND WALL EXTENSION

ROBERT E. CLELAND

*Department of Botany, University of Washington,
Seattle, WA 98195*

INTRODUCTION

The ability of a plant cell to enlarge depends, in part, on the extensibility of the cell wall. Lockhart (18) first recognized that cell enlargement can be described by the equation:

$$dV/dt = \mathbf{m} \, (P-Y) \qquad\qquad \text{Eq. 1}$$

where the rate of cell elongation is the product of the wall extensibility (**m**) and the difference between turgor pressure (P) and the wall yield pressure (Y). Any change in **m** will result in a change in the growth rate, as long as P and Y remain constant. In order to understand how cell expansion is regulated, it is necessary to understand what **m** represents, and how it can be controlled. To do this, four questions must be asked. First, what are the intrinsic extensibility properties of plant cell walls? Second, are they identical to or different from the characteristics of **m** as deduced from the properties of *in vivo* cell extension? Third, is it possible to obtain *in vitro* wall extension with characteristics identical to those of *in vivo* extension? And, finally, how does wall loosening (i.e., an increase in **m**) occur *in vivo*?

THE INTRINSIC EXTENSIBILITY PROPERTIES OF PLANT CELL WALLS

Any polymeric substance, such as the cell wall, will have a set of intrinsic extensibility properties (4, 11, 25). Are these properties identical to **m**? To answer this, one must first determine the mechanical properties of the walls, and compare them with the properties of **m** as infered from growth data. This is simple in theory, but difficult in practice. There are four problems which can complicate any measurement of the intrinsic extensibility properties of the walls. First, the mechanical properties of walls are always different quantitatively, and possibly qualitatively, when subjected to unidirectional stress (applied force) as compared with a multidirectional stress (turgor). With applied force the extension can be up to 3-times greater than with the same longitudinal stress produced by turgor. It has been possible to determine this relationship experimentally only with giant algal cells (14). As a result, it is not possible to compare quantitatively *in vivo* (turgor-driven) with *in vitro* (applied force-driven) extension rates.

A second problem concerns the conditioning of the walls, and the direction of the force vectors. When a material is extended twice to the same length with the same force vectors, the extension characteristics of the second extension (conditioned) are different from those of the first extension (unconditioned). Subsequent extensions will show characteristics of conditioned walls, as long as the original length is not exceeded, but whenever the length is exceeded, the extension will revert to unconditioned characteristics (7). A cell wall *in vivo* may act as a conditioned or an unconditioned material, depending upon its past history of turgor. The walls of a cell which has experienced no history of reduced turgor will probably appear to be unconditioned, but the walls of a cell which has shrunk slightly due to a loss of turgor would be expected to act as a conditioned material. Walls extended *in vitro* will always act as an unconditioned material during the first extension because the force vectors are different from turgor. Most studies on the extensibility of polymeric substances have been restricted to conditioned material, while almost all studies on cell walls have been confined to the unconditioned state.

A third problem is that cell walls in different tissues in an organ may have different extensibility properties, and different biochemical capacities. Thus when a dicot stem section is partly bisected and placed in water, the halves bend outwards due to the greater mechanical rigidity of the walls in the outer layers which restricts their ability to expand (see also chapter by Kutschera). However, in those same cells, auxin causes a greater increase in extensibility, with the result that the halves curve inwards due to excess growth of the outer sets of cells (the basis for the "slit-pea" assay for auxin). Any measurement of the average mechanical properties of an organ may obscure differences in extensibility between cell layers.

Finally, there is the question as to the ideal technique for measuring the intrinsic extensibility of the cell wall. In general, three main techniques have been employed; creep, stress relaxation and the Instron technique (8). In a creep test (5), the walls are subjected to a constant applied force, and the extension is measured as a function of time. In stress relaxation (19), the walls are rapidly extended, then held at a fixed length, and the decay in stress across the walls is measured as a function of time. In the Instron assay (3), a tissue is extended at a constant rate of extension, and the stress along the walls is measured as a function of the amount of extension. Which test is preferable? In theory, all three give the same qualitative information (12), and to date, experimental data support this conclusion (7, 25), despite some claims to the contrary (e.g., 16).

What are the intrinsic extensibility properties of cell walls, then? When a polymeric substance is subjected to a constant stress, it can in theory undergo one or more of three types of extension, which can be distinguished on the basis of reversibility and relationship to time (Fig. 1). Elastic extension involves an instantaneous, time-independent extension, which is completely reversed upon removal of the stress. With viscous flow, on the other hand, the amount of extension is proportional to time, and is completely non-reversible. A viscoelastic extension starts rapidly upon imposition of stress, but then

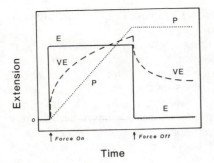

FIG. 1. The three types of extension of a polymeric material under constant stress. In elastic extension (E), the extension is instantaneous, but remains constant with time until the stress is removed, whereupon it is completely reversed. In plastic flow (P) extension is directly proportional to time. When the stress is removed the extension undergoes no reversal. With viscoelastic extension (VE) the extension rate decreases with time, but never falls to zero. Upon removal of stress there is a time-dependent contraction.

decreases in rate; in cross-linked polymeric substances such as cell walls the extension is more nearly proportional to the log of time than to time (4, 5, 11, 25).

Despite the difficulties mentioned above, it is encouraging to note that basically identical information has been obtained from all higher plant walls tested to date (4, 8, 25). When cell walls are subjected to a constant applied force, the extension shows both elastic and viscoelastic characteristics (5). There is a large, instantaneous and reversible elastic extension, followed by an extension whose rate decreases with time. If the walls are unconditioned, this extension is largely irreversible, but if conditioned, it is reversible. The extension, after completion of the elastic extension, is nearly proportional to the log of time, and the Q_{10} is less than 1.1 over the temperature range of 5-60 °C. This extension is unaffected by metabolic inhibitors such as KCN or N_2. The wall behaves, then, like a typical cross-linked viscoelastic substance.

WALL EXTENSIBILITY *IN VIVO*

An indication of the extensibility characteristics of cell walls *in vivo* can be obtained by consideration of the growth kinetics, over short time periods, when changes in P and Y would be relatively minimal. For a rapidly elongating tissue, such an auxin-treated *Avena* coleoptile section, the rate of extension is constant with time (25). There is no indication of the decline in extension rate with time that is found with *in vitro* extension. However, if oxidative respiration is blocked with inhibitors such as KCN or N_2, the growth rate rapidly declines, as if the extension changes from a metabolically-controlled viscous flow to a non-metabolic viscoelastic extension. With *in vivo* extension, the Q_{10} is somewhat unusual: 2 to 3 in the 15 to 25 °C range, but about 1.1 between 25 and 35 °C (23).

Clearly the extensibility of walls *in vivo* (**m**) is different from that of walls *in vitro* (Table I). There are two ways to reconcile these differences. The first is to assume that cell elongation *in vivo* does not involve any extension of the wall already present, but only the intussusception of new wall, in a manner similar to the growth of bacteria (see chapter by Koch). The strong evidence for cellulose microfibril reorientation during cell elongation speaks against such an

Table I. *Comparison of the characteristics of in vivo cell wall extension an in vitro wall extensibility of Avena coleoptiles.*

Characteristic		*In vivo*	*In vitro*
Extension proportional to:		Time	Log time
Q_{10},	15-25 °C	2-3	1.1
	25-35 °C	1.1	1.1
1 mM KCN		Inhibits, 3-5 min	No effect

idea (21). The second possibility is that cell elongation involves a continual series of extension events. Each event would be initiated by the breakage of some load-bearing bond (wall loosening event), followed by a small increment of viscoelastic extension (4). The extensibility, **m**, would be the product of the rate of wall loosening events (dB/dt) times the amount of extension which can occur as the consequence of that one event (VE) (Fig. 2). The rate of extension would be altered by any change in either dB/dt or in VE. The Q_{10} data suggest that for *Avena* coleoptiles, the rate of auxin-induced extension is determined by dB/dt in the 15-25 °C range, but by VE at higher temperatures.

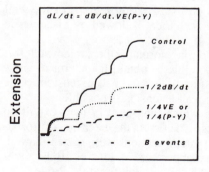

FIG 2. The role of extension events in controlling the rate of cell expansion. Each event is initiated by a breakage of a load-bearing bond (B event), followed by an increment of extension determined by the potential for viscoelastic extension (VE) times the effective turgor (P-Y). The rate of extension is a function of three factors; dB/dt, VE and P-Y. The two lower curves show the effect on the rate of a change in dB/dt or in VE or P-Y.

If this analysis is correct, it follows that measurement of the mechanical properties of walls by the creep, stress relaxation or Instron techniques cannot actually measure **m** directly. However, there is clearly a close relationship between the measured mechanical properties and **m**. Cleland has argued (7) that the mechanical tests, because they employ different force vectors than turgor, actually measure an immediate past average of **m** rather than either VE or the instantaneous value of **m**. The length of time over which **m** is averaged would presumedly differ from tissue to tissue. While it is apparently about 60-90 minutes in *Avena* coleoptiles, it appears to be a shorter period for maize coleoptiles (16).

LONG-TERM WALL EXTENSION *IN VITRO*

Cell wall extension is dependent upon the wall loosening events. What are the possible mechanisms by which these events might occur? Basically, there are three mechanisms (Fig. 3). In the first, **bond breakage**, load bearing

Possible Mechanisms of Wall Loosening

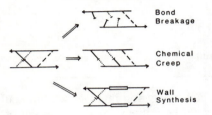

Bond Breakage

Chemical Creep

Wall Synthesis

FIG 3. Possible mechanisms of wall loosening. In **bond breakage**, cleavage of load-bearing bonds leads to an increment of viscoelastic extension. In **chemical creep**, load-bearing bonds are cleaved, but after extension has occurred they reform with new partners in an unstressed configuration. In **wall synthesis**, stress is released by intussusception of new wall polymers, permitting viscoelastic extension to occur.

bonds in the wall are broken either enzymatically or non-enzymatically. Four types of bonds are possibilities: glycosidic bonds, bonds within or between peptide chains (including isodityrosine bonds), hydrogen bonds and calcium bridges between pectic chains. For the first two, the breakage would require participation of enzymes, while for the latter two it would be non-enzymatic. A second possibility, **chemical creep**, would be an enzymatic breakage of load-bearing glycosidic bonds, which is followed by subsequent reformation of those bonds with new partners in a non-stressed configuration. Finally, **wall synthesis** might simply release the stress on load-bearing regions of the wall by intussusception of new pieces of polysaccharide, followed by viscoelastic extension of that area of wall.

Which of these mechanisms occurs during normal cell enlargement? If it is wall synthesis, then it should not be possible to obtain long-term *in vitro* wall extension which has the characteristics of *in vivo* cell extension. One appoach is to see if such long-term *in vitro* wall extension is possible.

Two systems have been studied in which long-term *in vitro* extension can occurs, but with different characteristics. The first is the extension of frozen-thawed *Avena* coleoptile sections. In this system, sections were frozen-thawed, and then clamped in a constant-stress creep apparatus. A load of 10-30 g was imposed at pH 7, whereupon the walls extended viscoelastically. Rayle et al. (24) first showed that when the pH was reduced below 6 for sections without the barrier of the cuticle, or pH 4 for ones with an intact cuticle, rapid extension (acid-growth) took place. Because the acid-extension was initially rapid, but then steadily declined in rate over the next hour, it appeared that this extension was different from *in vivo* growth. However, when the extension was followed over longer time periods (usually 6 hours), it became apparent that the acid extension did continue at a constant rate over long times, and that the extension consisted of two components; a "burst" of extension (exponential phase) with a constantly decreasing rate of extension plus a linear phase with a constant extension rate (Fig. 4). In fact, the extension could be fitted closely to the equation:

$$l = l_0 + a - ae^{-kt} + ct \qquad \text{Eq. 2}$$

where l_0 is the initial length, a is the total extension that occurs during the exponential phase, k is the rate constant for this phase, and c is the rate constant of extension during the linear phase (10).

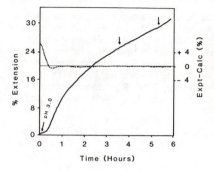

FIG. 4. Long-term in vitro extension of Avena coleoptile cell walls. Frozen-thawed intact sections were incubated for 30 min under constant load of 20 g at pH 7, then the pH was changed to 3.0 and the extension over the next 6 hrs was recorded. The data were fitted to Equation 2. The closeness of fit can be seen from the fact that a curve generated from the calculated constants deviates from the experimental curve (Exptl-Calc) only in the first minutes where acid penetration is a problem.

The characteristics of this long-term in vitro acid extension were remarkably similar to those of in vivo cell elongation. After the initial burst, which may be a consequence of the fact that applied force has different force vectors than turgor, the extension rate was constant for up to 10 hours, and led to extensions of greater than 50% of the initial length. When the pH was raised to 7 (as will occur in the apoplastic solution upon addition of respiratory inhibitors), the elongation rate rapidly declined in a manner similar to in vivo extension after addition of inhibitors. The Q_{10} was 2-3 in the 15-25 °C range, but about 1.1 at higher temperatures.

The nature of these acid-induced wall loosening events is not known. However, it is almost certainly the hydrolysis of some glycosidic bond, since treatment of the tissues with proteolytic enzymes, or conditions that denature proteins led to a loss in the ability of the wall to undergo acid-extension (23). Proteolytic enzymes caused no extension themselves, suggesting that peptide-associated bonds are not involved. Removal of calcium from the wall (9) or treatment with agents that break hydrogen bonds (23) were without effect as well. The effect of acid seems to be the activation of some polysaccharide hydrolase which has an acidic pH optimum.

The second system was either frozen-thawed (FT) or methanol-boiled (MB) sections of red light grown soybean hypocotyls (semi-etiolated). Sections were incubated in solutions for varying times, bisected (to overcome the permeability barrier of the cuticle), and then boiled or frozen. The sections were then incubated in various solutions in vitro. In some experiments, sections were subjected to analysis with the Instron technique, and their plastic extensibility (PEx) was determined. Others were placed in the creep apparatus under 10-40 g load, at pH 7 for 30 minutes to undergo viscoelastic extension, and then the pH was reduced to 3-6 and the acid-extension was recorded. After either Instron or creep tests, the wall calcium was extracted from the sections and determined by atomic absorption spectroscopy.

Unlike *Avena* coleoptiles, the mechanical properties of soybean hypocotyl walls were strongly influenced by the presence of wall calcium (26). Removal of calcium with EGTA increased PEx; addition of calcium decreased it (Fig. 5). Acidic buffers also increased PEx, but the main effect of the acid was

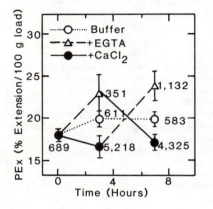

FIG. 5. Reversible effect of addition or removal of calcium on Instron plastic extensibility (PEx) of soybean hypocotyl cell walls. Bisected, methanol-boiled sections were incubated in 5 mM Na-acetate buffer, pH 6, + 0.I mM EGTA or 50 mM CaCl$_2$ as indicated. After 0, 3 or 7 hrs PEx was determined and wall calcium was determined (values in μg Ca/g DW).

removal of wall calcium, as seen by the strong correlation between wall calcium and PEx after treatment with different acidic buffers which vary in effectiveness in removing calcium.

Soybean hypocotyl walls, like *Avena* coleoptile walls, underwent long-term acid extension (Fig. 6). Again, the extension could be fitted closely to equation 2, indicating that both the exponential and linear phases of acid extension were present. However, there was a major difference between soybean and *Avena* cell walls. After boiling in methanol, soybean walls were still capable of undergoing acid extension, while those of *Avena* were not. This suggests that the acid was not activating polysaccharide hydrolases in this case. The Q_{10} for the acid extension was about 2 in both FT and MB walls. Acid must be causing a chemical breakage of bonds, but whether the bonds in question are hydrogen bonds or calcium crosslinks is still unknown.

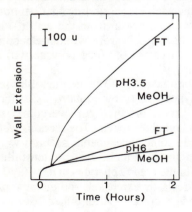

FIG. 6. Long-term acid *in vitro* extension of soybean hypcotyl cell walls. Bisected sections were either frozen-thawed or methanol-boiled, then placed in the creep apparatus under 20 g load at pH 6. After 10 min, the pH was changed to 3.5 or 6.0 and extension was recorded for the next 2 hours.

THE MECHANISM OF *IN VIVO* WALL LOOSENING

The studies with *in vitro* extension have indicated that it is possible for cell walls to undergo expansion with characteristics nearly identical to those of *in vivo* cell elongation, without any involvement of wall synthesis. This in no way eliminates wall synthesis as the mechanism of *in vivo* wall loosening, however. Synthesis of cellulose does not seem to be necessary, since the cellulose synthesis inhibititor 2,6-dichlorobenzonitrile failed to stop auxin-induced expansion of pea epicotyl sections (2). But synthesis of one or more component of the hemicelluloses could be involved. Increases in the rate of incorporation of exogenous sugars into wall hemicelluloses have been found to occur coincident with the onset of rapid cell elongation (1); whether this is the cause or the consequence of wall extension is difficult to determine. The correlation between the rate of wall synthesis and the growth rate is not good; *Avena* coleoptile sections grown in the presence or absence of exogenous sugars had the same initial rate of auxin-induced growth, but widely different rates of wall synthesis (22). But one must look on these data with caution, since it may be only a small component of wall synthesis that is required for wall loosening, and it may occur in only a minority of the cells; namely, in the cells responsible for the control of growth, such as the outer cell layers in dicot stems.

Wall synthesis may have a second, but equally important role. This is to maintain the structure of the wall so that bond breakage will lead to viscoelastic extension. Without wall synthesis, as bond breakage and extension occur, the wall polymers may become so entangled that further bond breakage is impossible. Wall synthesis, by intercalating new wall material, may maintain the ability of the enzymes to act upon the wall.

Bond breakage in response to acid has been shown to occur *in vitro* in both coleoptile and dicot stem cell walls. Does a similar mechanism exist *in vivo*? Whenever auxin induces rapid cell enlargement, there is an increased acidity of the apoplastic solution (6). But whether this acidity is sufficient to account for the *in vivo* wall loosening is a matter of considerable controversy (see 6, 15, 25); there is insufficient space here to consider the pros and cons of this argument.

The bonds that are cleaved in response to acid may vary, depending upon the tissue involved. In pea stems there is a release of xyloglucans from the wall in response to either auxin or exogenous acid (17); this suggests that some bonds linking xyloglucans to other wall polysaccharides are cleaved under these conditions. A decrease in molecular weight of xyloglucans of *Avena* coleoptiles has also been reported (13). But in other tissues, such as the *Vigna angularis* epicotyl, the bond cleaved may be one involving galactose (20). As yet, it is premature to specify which glycosidic bond may be cleaved under acidic conditions.

A second possibility, at least with dicot walls, is that the load-bearing bonds may include calcium bridges between pectic carboxyl groups. Removal of calcium *in vitro* leads to some wall loosening, but whether this also occurs *in vivo* is not yet certain. The role of acid in these walls might be, in part, to facilitate the solubilization of calcium involved in crossbridges. Alternatively, removal of calcium bridges may be required before a different acid-mediated wall loosening event can occur.

CONCLUSIONS

1. The plant cell wall has intrinsic extensibility properties; when under stress, they undergo a combination of elastic and viscoelastic extension. The extension is characterized by a Q_{10} close to 1 and an extension proportional to the log of time.
2. The ability of cell walls to extend *in vivo* depends, to a large extent, upon the parameter **m**, the wall extensibility. This extensibility is not identical to the intrinsic wall extensibility, as it has a Q_{10} of greater than 2 between 15 and 25 °C, and an extension proportional to time.
3. It is concluded that *in vivo* extension occurs via a series of extension events, each of which is initiated by the cleavage of some load-bearing wall bond, followed by a small increment of viscoelastic extension.
4. Long-term *in vitro* wall extension has been obtained with both *Avena* coleoptile and soybean hypocotyl cell walls, but the mechanism is different. In *Avena* coleoptiles the extension is in response to acidic pH and appears to involve the enzymatic cleavage of some glycosidic bond. In soybean walls the extension occurs in response to either the removal of wall calcium crosslinks or to low pH, and has a non-enzymatic component.
5. The mechanism of wall loosening *in vivo* is uncertain. It could involve acid-induced cleavage of hemicellulose bonds, since both the acidic conditions and the ability of walls to respond to these conditions are known to occur. However, other possible mechanisms, such as loosening by intussusception of wall polysaccharides cannot be ruled out at present. The mechanisms of wall loosening may differ from tissue to tissue, just as in vitro long-term extension mechanisms differ between tissues.

Acknowledgments--The research described here was supported by the U.S. Department of Energy, contract DE-AM06-76ER73019, and by the National Science Foundation, grant DMB 8502021. The assistance and advice of S. Virk, D. Cosgrove, M. Tepfer and E. Van Volkenburgh in this research is gratefully acknowledged.

REFERENCES

1. ABDUL-BAKI AA, PM RAY 1971 Regulation by auxin of carbohydrate metabolism involved in cell wall synthesis by pea stem tissue. Plant Physiol 47: 537-544
2. BRUMMELL DA, JL HALL 1985 The role of cell wall synthesis in sustained auxin-induced growth. Physiol Plant 63: 406-412
3. CLELAND RE 1967 Extensibility of isolated cell walls: measurement and changes during cell elongation. Planta 74: 197-209
4. CLELAND RE 1971 Cell wall extension. Annu Rev Plant Physiol 22: 197-222
5. CLELAND RE 1971 Mechanical behavior of isolated *Avena* coleoptile walls subjected to constant stress. Plant Physiol 47: 805-811
6. CLELAND RE 1980 Auxin and H^+-excretion: the state of our knowledge. *In* F Skoog, ed, Plant Growth Substances 1979. Springer-Verlag, New York, pp 71-78

7. CLELAND RE 1984 The Instron technique as a measure of immediate-past wall extensibility. Planta 160: 514-520
8. CLELAND RE 1986 The role of hormones in wall loosening and plant growth. Aust J Plant Physiol 13: 93-103
9. CLELAND RE, DL RAYLE 1977 Reevaluation of the effect of calcium ions on auxin-induced elongation. Plant Physiol 60: 709-712
10. CLELAND RE, D COSGROVE, M TEPFER 1987 Long-term acid-induced wall extension in an *in-vitro* system. Planta 170: 379-385
11. FERRY JD 1970 Viscoelastic Properties of Polymers. Wiley, New York, 2nd ed, 671 pp
12. FUJIHARA S, R YAMAMOTO, Y MASUDA 1978 Viscoelastic properties of plant cell walls. III. Hysteresis loop in the stress-strain curve at constant strain rate. Biorheology 15: 87-97
13. INOUHE M, R YAMAMOTO, Y MASUDA 1984 Auxin-induced changes in the molecular weight distribution of cell wall xyloglucans in *Avena* coleoptiles. Plant Cell Physiol 25: 1341-1351
14. KAMIYA N, M TAZAWA, T TAKATA 1963 The relation of turgor pressure to cell volume in *Nitella* with special reference to mechanical properties of the cell wall. Protoplasma 57: 501-521
15. KUTSCHERA U, P SCHOPFER 1985 Evidence against the acid growth theory of auxin action. Planta 163: 483-493
16. KUTSCHERA U, P SCHOPFER 1986 *In-vivo* measurement of cell-wall extensibility in maize coleoptiles. Effect of auxin and abscisic acid. Planta 169: 437-447
17. LABOVITCH JM, PM RAY 1974 Relationship between promotion of xyloglucan metabolism and induction of elongation by IAA. Plant Physiol 54: 499-502
18. LOCKHART JA 1965 An analysis of irreversible plant cell elongation. J Theor Biol 8: 264-275
19. MASUDA Y 1978 Auxin-induced cell wall loosening. Bot Mag Tokyo, Spec. Issue 1: 103-123
20. NISHITANI K, Y MASUDA 1980 Modifications of cell wall polysaccharides during auxin-induced growth in azuki bean epicotyl segments. Plant Cell Phyiol 21: 169-181
21. PRESTON RD 1974 The Physical Biology of Plant Cell Walls. Chapman & Hall, London.
22. RAY PM 1962 Cell wall synthesis and cell elongation in oat coleoptile tissue. Amer J Bot 49: 928-939
23. RAYLE DL, RE CLELAND 1972 The *in-vitro* acid-growth response: relation to *in-vivo* growth responses and auxin action. Planta 104: 282-296
24. RAYLE DL, PM HAUGHTON, RE CLELAND 1970 An *in-vitro* system that simulates plant cell extension growth. Proc Nat Acad Sci US 67: 1814-1817
25. TAIZ, L 1984 Plant cell expansion; regulation of cell wall mechanical properties. Annu Rev Plant Physiol 35: 585-657
26. VIRK S, RE CLELAND 1986 The role of calcium in the mechanical strength of soybean hypocotyl cell walls. Plant Physiol 80: S76

Physiology of Cell Expansion During Plant Growth, *D.J. Cosgrove and D.P. Knievel* Eds.,

THE BIOCHEMISTRY OF "GROWING" CELL WALLS

NICHOLAS C. CARPITA

Department of Botany and Plant Pathology, Purdue University West Lafayette, IN 47907

INTRODUCTION

It is very difficult to discuss plant cell enlargement and development and only speak about the cell wall, but it is equally difficult to discuss such development and not focus on the physical and biochemical dynamics responsible for this development. Much of this development is made possible as the phragmoplast of a dividing cell is transformed into a dynamic, expanding matrix as enlargement or elongation occurs. Fine-tuning of cell expansion is controlled by orientation of the deposition of cellulose microfibrils and by directed "wall loosening" in which interactions between cellulose and non-cellulosic polymers are modified, perhaps by polysaccharide hydrolases that cleave specific load-bearing bonds.

Control of cell expansion and elongation by phytohormones has been a principal subject of research from the very early investigations in plant physiology. And today, studies focus on two important aspects. First, it is difficult to imagine that such morphogenesis cannot involve changes in gene expression to induce or enhance expression of regulatory or structural proteins that, among many possible functions, must include the synthesis, assembly, and even turnover of cell wall components. In fact, investigations have documented that natural and synthetic auxins induce or enhance, and even repress, the synthesis of several soluble and membrane-associated proteins, and this alteration is a result of increased rates of translation from mRNA (75, 84). Using cDNA probes from RNAs that must encode some of these proteins, increases in total amount of some mRNAs for auxin-specific proteins were documented (27, 78; also see chapter by Theologis). Induction by gibberellic acid of α-amylase and other hydrolytic enzymes involved in mobilization of reserve materials in graminaceous seeds is also well-documented (3, 12, 34), but only recently has the involvement of gibberellic acid (GA) in the induction of synthesis of novel proteins related to growth been investigated (13).

The second researchable aspect is the physical and biochemical factors that directly participate in the enlargement, and this aspect has received most of the attention throughout the years. The role of auxin and gibberellin in causing increased "extensibility" (15), and the relationship between this physical property and the loss of sugar or sugar fragments from polysaccharides comprising the wall has been established (70). A true appreciation for the dynamic nature of the cell wall has emerged only recently. These studies have been aided immensely by revelations of the chemical structures of specific polymers comprising the primary cell walls. While accurate three-dimensional views of the wall, necessary to understand fully the physical constraints and the actual mechanisms of microfibril separation and wall expansion, are a bit far from reach, considerable information is available on the chemical structure of polysaccharides and proteins that function in formation of this rigid but dynamic matrix. From such studies, we have learned two important general facts. One, the cell walls of monocots of the Family Gramineae (grasses and cereals) and some related species are radically different from those of dicots and even other monocots. Two, the composition of the primary walls of these graminaceous species changes markedly as the developmental program changes from cell division to cell expansion. The purpose of this review is to summarize the relationship between the chemical composition of cell walls of dicots and graminaceous monocots and the hydrolytic events correlated with cell expansion and to discuss the relevance of the synthesis, and possible turnover, of stage-specific polysaccharides in the graminaceous species.

CELL WALL COMPOSITION AND TURNOVER IN DICOTYLEDONOUS SPECIES

From many studies using cells in suspension culture or tissues in which primary cell walls predominate, generalized wall structures have emerged (Fig. 1 a and b). Cellulose microfibrils are covered by hemicelluloses tightly held through hydrogen bonds, and this framework is embedded in a gel matrix of pectic polysaccharides (18,44). Besides polygalacturonic acid, a simple polymer composed of (1-4)-α-D-galactosyl uronic acid units, there are more complex rhamnogalacturonans (RG I) containing regions of repeating (1-2)-α-L-rhamnosyl-(1-4)-α-D-galactosyluronic acid units producing a contorted rod-like structure (50). Upon the O-4 of about half of the 2-linked rhamnosyl units are side branches of either helical (1-5)-α-L-arabinofuranosyl chains or more highly branched structures with both 2,5- and 3,5-linked arabinosyl units. Galactans with 4-linkages and arabinogalactans containing 3- and 6-linked galactosyl backbone structures highly substituted with terminal arabinofuranosyl units also constitute common polymeric side-groups upon the rhamnogalacturonans (18). Rhamnogalacturonan II, a remarkably complex pectic polysaccharide, also comprises a small proportion of the gel matrix of cells of sycamore maple and other species (50). Amounts of this polymer are probably too low to play a major structural role, and the physiological function of this molecule is still a mystery.

The pectin gel is formed primarily through cross-bridging with Ca^{2+} and contortion of the (1-4)-α-D-galactosyluronic acids, linking once soluble polymers into an "egg-box" rigid structure (62). These immobilized gels are dissolved by Ca^{2+}-chelating agents, such as EDTA, EGTA and ammonium oxalate. "Pectic substances" are actually defined as material dissolved by such

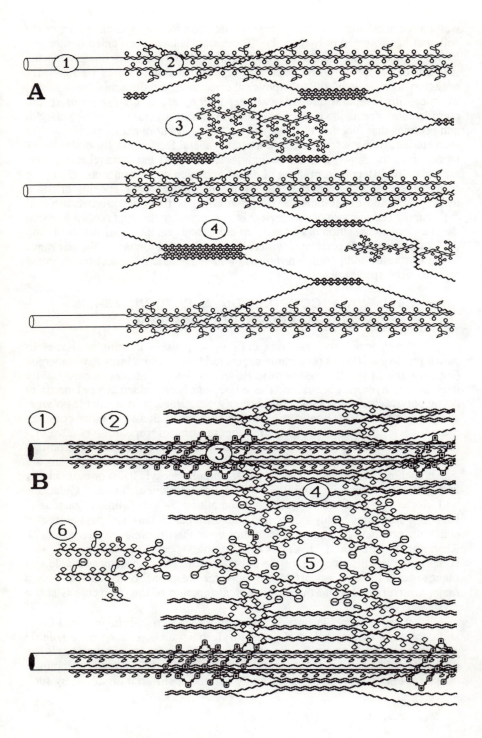

treatment regardless of chemical structure. Once this material is extracted exhaustively, additional uronic acid-rich material is still retained by the matrix and is removed by dilute alkali, such as Na_2CO_3, $NaBH_4$-Na_2BO_4, or 0.1 M KOH solutions. It is likely that this second group of pectic polymers is immobilized by alkali-labile ester linkages, perhaps through cross-bridging via hydroxycinnamic acids (40). Fry (22) has identified such hydroxycinnamic acids, such as ferulic acid, that may participate in cross-bridging through formation of ester bonds with neutral sugars of the side branches of rhamnogalacturonans and subsequent biphenyl or ether bond formation of the these aromatics. The extent of Ca^{2+} cross-bridging, of esterification through aromatic linkages, and even of branching and size of neutral sugar side-chains can each influence gel flexibility, porosity, interaction with hemicellulosic polymers, and perhaps mobility of protein components (see also chapter by Biggs and Fry).

The principal hemicellulose of dicotyledonous species is xyloglucan, a linear (1-4)-β-D-glucosyl chain substituted regularly at the O-6 with xylopyranosyl units, some of which are further substituted with (1-2)-β-D-galactosyl and 2-O-α-L-fucosyl-(1-2)-β-D-galactosyl units (18). A repeating unit structure of hepta- and nona saccharides has been proposed from chemical linkage analysis and the use of "restriction endoglycosidases", such as the (1-4)-β-D-glucanohydrolase from *Aspergillus* whose activity is restricted to unbranched 4-glucosyl units and a similar glucanohydrolase from *Trichoderma* in which hydrolysis of the 4-linkage of O-6 substituted glucosyl units also occurs (41, 42) (Fig. 2). The substituted xylosyl units constitute a "molecular hook" of yet undetermined significance (Fig. 2a), although there is a suggestion that this nonasaccharide may suppress growth upon cleavage from xyloglucan polysaccharide through an unidentified feed-back system (23, 83). Although enzymic digestions of carefully purified xyloglucan preparations suggest that much of the xyloglucan may exist in this form, deviation from this structure to a less substituted one are evident (18); this may reflect limited hydrolysis of the xyloglucan during growth, although very little information is now available on the structure of nascent xyloglucan in secretory vesicles of the Golgi apparatus. An acidic arabinoxylan also constitutes a portion of the hemicellulosic material. It consists of a (1-4)-β-D-xylosyl backbone substituted mostly at the O-2 with

FIG. 1 (facing page). Comparison of models of the primary cell walls of dicots and graminaceous monocots. A. The dicot cell wall: Cellulose microfibrils (1) are coated with a monolayer of xyloglucan (2). Additional xyloglucan and arabinoxylan may span the microfibrils (not shown). The cellulose-xyloglucan framework is embedded in a gel matrix of polygalacturonic acids, cross-linked in part by Ca^{2+} (4). Additional polymers containing mostly neutral sugars constitute the major side-groups and are attached to the rhamnosyl units of rhamnogalacturonan (3). Not shown is the hydroxyproline-rich extensin which can cross-stich the cellulose fibrillar network. B. The graminaceous monocot wall: The cellulose microfibrils (1) are coated primarily with arabinoxylans and some xyloglucan (2), a portion of which are immobilized by phenolic ether cross-stiches (3). Other acidic arabinoxylans are hydrogen bonded to each other and may span the matrix (4); additional glucuronoarabinoxylan (GAX) may define the pores of the matrix, replacing the function of pectic substances in these species (5). The highly substituted GAX (6) may be the nascent xylan and is cross-linked loosely through diferulic acid. Not shown is the developmental stage-specific β–glucan which is synthesized primarily when cell expansion begins.

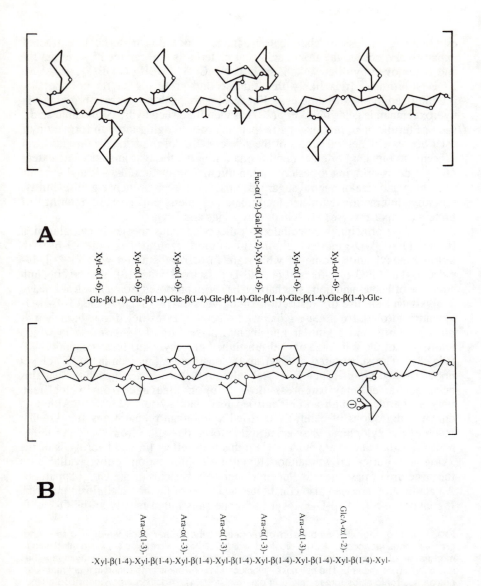

FIG. 2. A. Molecular model of the hepta- and nonasaccharide repeating structure of dicot xyloglucan. B. Molecular model of a highly substituted GAX from graminaceous monocots.

arabinofuranosyl linkages (19). Xyloglucans hydrogen bond tightly to the cellulose microfibrils and are probably organized onto the microfibrils shortly after cellulose synthesis (31). Amounts of xyloglucan are nearly equal to those of cellulose (18, 31). Considering that the cellulose microfibril consists of a crystal of at least 36 glucan chains, xyloglucans not only comprise a "monolayer" on these crystals, but also may span neighboring microfibrils to constitute a true rigid matrix.

Extensin, a structural hydroxyproline-rich glycoprotein, also comprises a significant amount of the cell wall of dicots (48). The "polyproline-like" tight helix is reinforced by intramolecular isodityrosine residues, presumably formed through an extracellular peroxidase, producing a rod-like protein long enough to span the wall anticlinally (68). Intermolecular isodityrosine linkages or covalent linkages of protein-protein or protein-polysaccharide are suspected of producing a rigid, inextensible wall (49). Although extensin accumulation more likely signals irreversible cessation of growth (63), Fry (21) suggests that hormonal suppression of peroxidase secretion might delay cross-linking during cell expansion.

The dynamic nature of the synthesis of cell wall polymers, their assembly into the cell wall, and their eventual hydrolysis and turnover is subtle. The physical and biochemical bases for microfibril separation resulting in wall expansion have been open to many suggestions, mostly in connection with studies concerning auxin- and acid-induced enlargement. An early attractive hypothesis that acid alters hydrogen bonding of xyloglucan to cellulose, inducing the chemical "creep" of the wall (44), has been laid to rest. Absolutely no effect of acid on either acetone-stabilized binding of xyloglucan fragments (76) or fluorescence- and ^{125}I-labelled purified xyloglucan (31) to cellulose microfibrils could be demonstrated *in vitro*. The fact that Ca^{2+} strongly inhibits growth and wall extensibility (17, 56) coupled with the fact that Ca^{2+} forms strong cross-bridges of polygalacturonic acid (62) led to an equally attractive suggestion that strength of the wall is controlled by the stiffness of the pectin gel imparted by this Ca^{2+} cross-bridging function. Auxin, through acid, would then increase extensibility by displacement of the Ca^{2+} with protons. In earlier studies, Clelend (14) was unable to detect release of $^{45}Ca^{2+}$ from coleoptile sections, but as will be described later, the walls of these graminaceous species are very different from those of dicots. Because of the high cation exchange capacity of the polygalacturonic acid, displacement of Ca^{2+} from the inner lamellae of the wall matrix could essentially go undetected, anyway. Data showing strict correlation between gel rigidity or viscosity and wall stiffness are conflicting (16, 55, 72). How the gel matrix participates in cell expansion is unknown. A physical alteration in the rigidity of the gel is a likely result of reduced Ca^{2+}, but it is unlikely that the rigidity of the gel matrix alone is sufficient to control expansion considering that tension of several thousand bar are generated by turgor pressures of only 10 bar in an "average" cell 100 μm in diameter (9). Terry and Bonner (73), using a low-speed centrifugation technique designed to release polymers in the cell wall solution of tissue sections without cytoplasmic contamination, found that auxin induced remarkable increases in soluble uronic acid-rich material and arabinogalactan(protein)s. Even though pectins themselves may not participate directly in the growth-sensitive load-bearing bonds, auxin clearly has an impact on the matrix structure.

While the interactions of Ca^{2+} and pectate gelation with wall stiffness and growth are a bit muddled, more convincing evidence has accumulated indicating that glycosidase-catalyzed hydrolysis of hemicellulosic polymers is directly involved in expansion. Auxin or acid induces release of xyloglucans from cell wall preparations (24, 46, 47), and this effect is negated by incubation in neutral buffers (39). Terry and colleagues (73, 74) confirmed these data in living tissues by use of the centrifugation technique. Wall extensibility has always been suspected of being under enzymic control (15, 70), and a considerable body of data directly implicates endo-glycosidases that participate in the digestion of xyloglucans comprising a load-bearing function. A reduction in molecular size of xyloglucans is induced *in vivo* by auxin or acid (59, 60), and the effects are also blocked by incubation of sections in a neutral buffer (58). "Cellulases", or (1-4)-β-D-glucanohydrolases, have been purified from walls of growing tissues (32, 80), and Goldberg and her colleagues (25, 26) find fair correlation of their activities with cell extension. The endo-glucanases from walls of growing tissues display good activity toward purified xyloglucan *in vitro*, and an endo-glucanase from pea cleaves the xyloglucan into hepta- and nonasaccharide fragments (32). These fragments also appear *in vivo* in growing cells in suspension culture (23).

Although the turnover of xyloglucan seems to be under auxin control, how that control is exerted is yet unknown. Levels of the mRNA for the "cellulases" that hydrolyze xyloglucan is increased by auxin (77), and activities of the endo-glucanase increase markedly in pea (20, 32). The activity is not under strict control by pH, and coupled with observations that auxin also induces alterations in pectin structure, the interaction of pectin and xyloglucan could constitute another control. The Ca^{2+}/H^+ exchange may induce subtle changes in porosity or solubility that permits access of the glucanohydrolases to the substrate. These potential interactions are supported by some data (2, 71), and these kinds of studies certainly deserve future attention.

CELL WALL COMPOSITION AND TURNOVER IN GRAMINACEOUS MONOCOTS

From the studies of Hirst and Haworth and their colleagues in the 1920's and 30's, we should appreciate that the walls of graminaceous monocots are quite different from those of dicots (28, 30). More recently, methylation analyses have produced a generalized picture of the primary walls of these species (Fig. 1b). Pectic polysaccharides constitute only about 10% of the wall, half of this material comprising polygalacturonic acid and rhamnogalacturonan (67; Carpita, unpublished results) and the other half an acidic arabinoxylan similar to that extracted by alkali. At least 50% of the wall is hemicellulose, and a majority of these polymers are acidic arabinoxylans consisting of (1-4)-β-D-xylosyl backbones substituted primarily at the O-3 with terminal-α–L-arabinofuranosyl units and at the O-2 with a smaller number of terminal glucosyluronic and 4-O-methyl glucosyluronic acid units (4, 10, 30, 79) (Fig. 2b). Xyloglucan similar to that of dicots is also found, but total amounts are much smaller than those comprising walls of dicots; the α-D-xylosyl

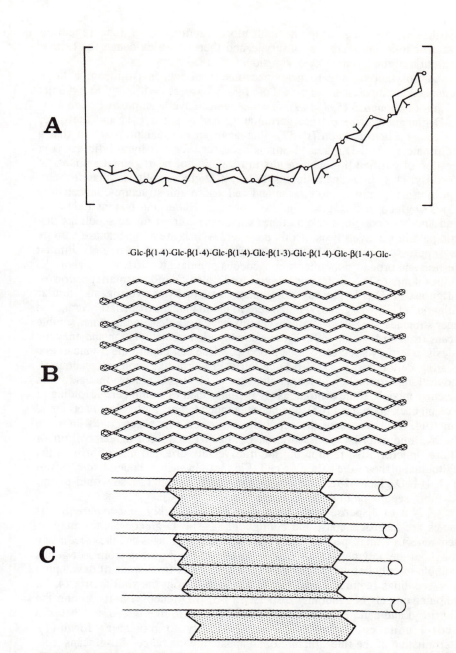

-Glc-β(1-4)-Glc-β(1-4)-Glc-β(1-4)-Glc-β(1-3)-Glc-β(1-4)-Glc-β(1-4)-Glc-

FIG. 3. A. Molecular model of the repeating units of β–glucan from graminaceous monocots. B. Proposed molecular "sheet" produced by regularly-spaced contiguous (1-3)-β-D-glucosyl units (43). C. Proposed integration of the β–glucan sheet into the cellulosic fibrillar matrix.

substitutions are less frequent and more random such that repeating heptasaccharide units are not observed and there is no evidence for further substitution of the terminal xylosyl residues (4, 18, 79).

A polymer unique to graminaceous monocots is β-glucan, a large unbranched homopolymer composed of (1-3)-β-D- and (1-4)-β-D-glucosyl units in a ratio of about 1:3 (5, 18, 79). Two enzymes have been purified, one a (1-4)-β-D-glucanohydrolase from germinating barley seeds (81) and the other secreted by *Bacillus subtilis* (57), that have an unusual property in that a penultimate (1-3)-β-D-glucosyl unit is necessary for full hydrolytic activity. Digestion of purified β-glucan results in formation of nearly equal amounts of cellotriosyl-(1-3)-β-D- and cellobiosyl-(1-3)-β-D-glucose oligosaccharides (69), suggesting a repeating cellotetraose and cellotriose unit structure connected by (1-3)-β-D-glucosyl linkages (Fig. 3a). Further, Huber and Nevins (37) have found another endoglucanase and an exoglucanase from maize seedlings that participate in the hydrolysis of this polymer. When the exoglucanase activity was suppressed with Hg^{2+}, the endoglucanase was capable of only limited digestion and produced a rather homogeneous population of fragments about 50 residues long (37). Kato and Nevins (43) also found a small portion of contiguous (1-3)-β-D-glucosyl oligomers in more extensive digests; similar contiguous (1-3)-β-D-glucosyl units were found in glucans from barley bran, and longer stretches of (1-4)-β-D-glucosyl units have been detected in water soluble glucans from barley bran (69). From a combination of methylation and enzymic analysis, a tertiary structure can be proposed. One can picture the β-glucan as a repeating cellotetraosyl-(1-3)-β-D-cellotriosyl worm-like molecules about 50 residues long terminating in longer stretches of (1-4)-β-D-glucosyl units. From molecular models, three contiguous (1-3)-β-D- linkages would permit folding of the chain back upon itself (43), and hence, a much larger "molecular sheet" could form (Fig. 3b). To carry the supposition further, this sheet could interlace cellulose microfibrils upon secretion of the β-glucan during deposition of cellulose microfibrils, reinforcing the transversely oriented microfibrils in the longitudinal plane of the enlarging cell (Fig. 3c). In such a model, hydrolysis of the (1-4)-β-D-glucosyl-rich regions by the maize endoglucanase would permit microfibril separation in the appropriate plane.

Not to dispense with the hypothesis so quickly, but apparently β-glucans are heterogeneous, and contiguous (1-3)-β-D-glucosyl units may be heterogeneous with respect to number of consecutive linkages and their frequency throughout the polymer (1). A sizable portion of the β-glucan from soft grains is soluble in only hot water (18, 69, 79), whereas that from walls of developing maize seedlings requires 2 M KOH to dislodge it from the wall matrix (4, 5). Computer models of randomly dispersed (1-3)-β-D-glucosyl units among the cellotriosyl and cellotetraosyl units generated a worm-like thread, the 3-linked glucosyl units conferring increased flexibility to molecular form (1). Determination of the fine structure of an apparent wide range of β-glucans from graminaceous monocot species is truly necessary to obtain a clear picture of the heterogeneity of molecular form and how this structure interacts directly with cellulose microfibrils.

Unlike in dicots, the hydroxyproline-rich extensin is virtually absent from the walls of graminaceous monocots, where the cross-linking function has been replaced by esterified and etherified aromatic compounds (6, 66). The diphenyl diferulic acid may cross-link a small proportion of glucurono-arabinoxylan (GAX) (54), but most of the GAX is anchored by stronger covalent linkages than provided by these esters (6, 7, 66). Loss of auxin-induced expansion is correlated with a large increase in incorporation of these phenolic compounds into the cell wall (7), hence, like with extensin, this aromatic cross-linking likely reinforces the primary wall irreversibly.

The specific involvement of β-glucans in cell enlargement in graminaceous species is supported by several lines of evidence. First, studies of the dynamics of wall formation during development of the maize coleoptile revealed that β-glucan was virtually absent from walls of the embryonal coleoptile (5). Similar studies of walls of an embryonic line of proso millet cells in suspension culture, in which little cell expansion occurs except for that following cell division, showed that these walls are also nearly devoid of β-glucan (8). β-Glucan synthesis begins as elongation commences, and amounts determined by linkage analysis reach nearly 30% of the alkali-extractable hemicellulosic fraction (Fig. 4). Similar results were obtained when the β-glucan was quantified using the *B. subtilis* glucanohydrolase (53). The glucan

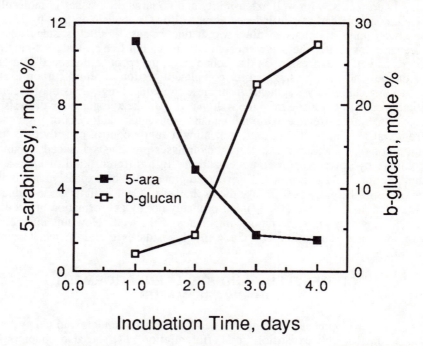

FIG 4. Comparison of the mole fractions of 5-arabinosyl units and β–glucan during coleoptile development.

disappeared from the walls of senescent coleoptiles specifically (53). Second, autolysis of the wall was demonstrated in excised tissues from graminaceous species, glucose being the principal product (35, 51), and auxin accelerated the disappearance of the glucans in excised sections of coleoptiles deprived of sugar (52, 64, 65). Third, the detection and purification of the endo- and exoglucanases that participate in this hydrolysis has been demonstrated (36, 37). Wall autolysis and purified exoglucanase exhibit relatively broad, weakly acidic pH optima, whereas the purified endoglucanohydrolase from maize seedlings has a narrower pH optimum of about 4.5-5, making it a more likely candidate for participating in pH-dependant selective cleavage (29). The exoglucanase may be involved only in recycling of sugar for the expanding cell.

Additional data indicate that cell enlargement is more complicated than activation of an endo-β-glucanase that cleaves load-bearing bonds of the β-glucan. Although an extracellular fungal (1-3)-β-D-glucanase can produce limited growth (82), the maize endoglucanase cannot when added back to abraded coleoptile sections (37). The endoglucanase requires 3 M LiCl for extraction, indicating that it adheres tightly either to the β-glucan substrate or to the acidic GAX (37). Considering that growth is controlled by activities at the innermost lamellae, endoglucanase may adsorb to the outer lamellae and fail to reach the critical substrate. There is additional data that suggest that glucan hydrolysis is not closely correlated with wall extensibility and no large differences in molecular size accompany early auxin-induced growth (70). Catabolism of these β-glucans and other polymers may constitute two distinct stages. In the first stage, subtle cleavages may leave the polymer essentially intact but free critical load-bearing bonds and allow expansion. In the second stage, the spent or "loosened" regions of the polymer may be targeted for complete digestion to return carbon to the expanding cell. Only fragments from these subsequent digestions may become soluble enough to escape from the wall matrix. Unlike xyloglucan hydrolysis in dicots, auxin pre-treatments do not enhance subsequent autolysis of β-glucans from walls of macerated tissues, indicating a subtle control of activity is lost upon homogenization or that glucanohydrolases previously induced by auxin continue hydrolyses upon homogenization. In a different kind of approach, antibodies raised against the LiCl-extracted proteins inhibit expansion and block autolysis of the glucan (38). Like dicots, however, Ca^{2+} inhibits auxin-induced extension and autolysis (14, 17), but probably not by gel formation of pectins, considering they are such minor components of the wall. How this interaction can influence the activity of the wall glycosidases certainly merits more attention.

GIBBERELLIN AND CELL ENLARGEMENT IN DWARF MAIZE

Developmental mutants are perhaps the most underexploited systems for studies of cell expansion and differentiation (11; see also chapters by MacMillan and Phinney and by Koornneef et al.). Studies of gene activation-repression by auxin have relied on reintroduction of auxin to excised tissues depleted of endogenous hormone (27, 75, 78). The dwarf mutants of dicots (*Pisum* and *Arabidopsis*) and graminaceous monocots (*Zea*) have been established (45, 61), and the mutation traced to lesions in the synthesis of gibberellins (33). In intact seedlings, exogenous gibberellin (A3) induces remarkable increases in growth rates of primary leaves and a change in the pattern of enlargement in cells of mesocotyl and leaves from isodiametric expansion to elongation. The chemical composition of cell walls of leaf, mesocotyl, and coleoptile of the dwarf maize is nevertheless quite similar to that of a normal hybrid. The walls comprise substantial amounts of β-glucan albeit at lower proportions (20% of hemicellulose compared to about 30% for normal maize), so synthesis of this stage-specific polymer is not under strict gibberellin control. Application of GA3 into the coleoptile cavity induces a 5- to 7-fold increase in elongation of the primary leaves, and dry weight accumulation keeps pace with this elongation. Methylation analyses demonstrated that the proportion of β-glucan in the leaves actually decreases to about 75% of the original level, while remaining nearly constant in the slowly expanding leaves of the controls without GA (Fig. 5). Slight decreases in the proportion of xyloglucan were observed in walls of both GA-treated and control seedlings, whereas compensation in percentage was maintained by increases in the levels of acidic arabinoxylans. The decrease in proportion of β-glucan induced by GA was also observed by enzymic determination with the *B. subtilis* glucanohydrolase (Fig. 5).

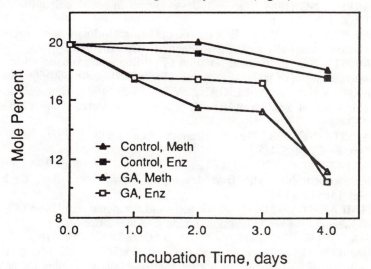

FIG. 5. Influence of gibberellin on the mole fraction of β–glucan in primary walls of dwarf maize during cell expansion.

Because there was marked accumulation of polysaccharide during the growth period, one cannot deduce from these studies whether the synthesis of β-glucan is suppressed relative to synthesis of xylans, whether synthesis of xylans is enhanced relative to β-glucans, or whether β-glucan undergoes specific turnover. Considering the dynamic nature of the synthesis of β-glucan and its specific hydrolysis, it is tempting to suggest the later possibility. We now are examining autolysis in dwarf maize to determine if activities of hydrolases responsible for turnover of the β-glucan are under gibberellin control.

Acknowledgment--I thank the National Science Foundation for their continued support of the work on cell wall structure in graminaceous monocots.

LITERATURE CITED

1. BULIGA GS, DA BRANDT, GB FINCHER 1986 Sequence statistics and solution conformation of a barley (1-3,1-4)-β-D-glucan. Carbohydr Res 157: 139-159
2. BATES GW, PM RAY 1981 pH dependent interactions between pea cell wall polymers possibly involved in wall deposition and growth. Plant Physiol 68: 158-164
3. BAULCOMBE DC, D BUFFARD 1983 Gibberellic-acid-regulated expression of α-amylase and six other genes in wheat aleurone layers. Planta 157: 493-501
4. CARPITA NC 1983 Hemicellulosic polymers of cell walls of *Zea* coleoptiles. Plant Physiol 72: 515-521
5. CARPITA NC 1984 Cell wall development in maize coleoptiles. Plant Physiol 76: 205-212
6. CARPITA NC 1984 Extraction of hemicelluloses with increasing concentrations of alkali. Phytochemistry 23: 1089-1093
7. CARPITA NC 1986 Incorporation of proline and aromatic amino acids into cell walls of maize coleoptiles. Plant Physiol 80: 660-666
8. CARPITA NC, JA MULLIGAN, JW HEYSER 1985 Hemicelluloses of cell walls of a proso millet cell suspension culture. Plant Physiol 79: 480-484
9. CARPITA NC 1985 Tensile strength of cell walls of living cells. Plant Physiol 79: 485-488
10. CARPITA NC, D WHITTERN 1986 A highly-substituted glucuronoarabinoxylan from developing maize coleoptiles. Carbohydr Res 146: 129-140
11. CHAILAKHYAN MK 1979 Genetic and hormonal regulation of growth, flowering, and sex expression in plants. Amer J Bot 66: 717-736
12. CHANDLER PM, JA ZWAR, JV JACOBSEN, TJV HIGGINS, AS INGLIS 1984 The effects of gibberellic acid and abscisic acid on α-amylase mRNA levels in barley aleurone layers: studies using an α-amylase cDNA clone. Plant Molec Biol 3: 407-418
13. CHORY J, DF VOYTAS, NE OLSZEWSKI, FM AUSUBEL 1987 Gibberellin-induced changes in the populations of translatable mRNAs and accumulated polypeptides in dwarfs of maize and pea. Plant Physiol 83: 15-23

14. CLELAND RE 1960 Effect of auxin on loss of calcium from cell walls. Nature (London) 185: 44
15. CLELAND RE 1981 Wall extensibility: hormones and wall extension. *In*: W Tanner, FA Loewus, eds., Encyclopedia of Plant Physiology, New Series, Plant Carbohydrates II, Vol. 13B. Springer-Verlag, Berlin, pp. 255-273
16. CLELAND RE, DL RAYLE 1977 Reevaluation of the effect of calcium ions on auxin induced elongation. Plant Physiol 60: 709-712
17. COOIL B, J BONNER 1957 Effects of calcium and potassium ions on the auxin-induced growth of *Avena* coleoptile section. Planta 48: 696-723
18. DARVILL AG, M McNEIL, P ALBERSHEIM, DP DELMER 1980 The primary cell walls of flowering plants. *In*: PK Stumpf, EE Conn, eds., The Biochemistry of Plants, Vol 1. Academic, New York, pp. 91-161
19. DARVILL JE, M McNEIL, AG DARVILL, P ALBERSHEIM 1980 The structure of plant cell walls XI. Glucuronoarabinoxylan, a second hemicellulose in the primary cell walls of suspension-cultured sycamore cells. Plant Physiol 66: 1135-1139
20. FAN DF, GA MACLACHLAN 1967 Massive synthesis of RNA and cellulase in the pea epicotyl in response to IAA with or without concurrent cell division. Plant Physiol 42: 1114-1122
21. FRY SC 1980 Gibberellin-controlled pectinic acid and protein secretion in growing cells. Phytochemistry 19: 735-740
22. FRY SC 1983 Feruloylated pectins from the primary cell walls: their structure and possible functions. Planta 157: 111-123
23. FRY SC 1986 *In-vivo* formation of xyloglucan nonasaccharide: A possible biologically active cell-wall fragment. Planta 169: 443-453
24. GILKES NR, MA HALL 1977 The hormonal control of wall turnover in *Pisum sativum*. New Phytol 78: 1-15.
25. GOLDBERG R 1977 On possible connections between auxin induced growth and cell wall glucanase activities. Plant Sci Lett 8: 233-242
26. GOLDBERG R, R PRAT 1982 Involvement of cell wall characteristics in growth processes along the mung bean hypocotyl. Plant Cell Physiol 23: 1145-1154
27. HAGEN G, A KLEINSCHMIDT, T GUILFOYLE 1984 Auxin-regulated gene expression in intact soybean hypocotyl and excised hypocotyl sections. Planta 162: 147-153
28. HAMPTON HA, WN HAWORTH, EL HIRST 1929 Polysaccharides. Part IV. The constitution of xylan. J Chem Soc, 1739-1753
29. HATFIELD R, DJ NEVINS 1986 Purification and properties of an endoglucanase isolated from the cell walls of *Zea mays* seedlings. Carbohydr Res 148: 265-278
30. HAWORTH WN, EL HIRST, E OLIVER 1934 Polysaccharides Part XVIII. The constitution of xylan. J Chem Soc, 1917-1923
31. HAYASHI T, MPF MARSDEN, DP DELMER 1987 Pea xyloglucan and cellulose V. Xyloglucan-cellulose interactions *in vitro* and *in vivo*. Plant Physiol 83: 384-389

32. HAYASHI T, Y-S WONG, G MACLACHLAN 1984 Pea xyloglucan and cellulose II. Hydrolysis by pea endo-1,4-β-glucanases. Plant Physiol 75: 605-610

33. HEDDEN P, BO PHINNEY 1979 Comparison of *ent*-kaurene and *ent*-isokaurene synthesis in cell-free systems from etiolated shoots of normal and *dwarf*-5 maize seedlings. Phytochemistry 18: 1475-1479

34. HIGGINS TJV, JA ZWAR, JV JACOBSEN 1976 Gibberellic acid enhances the level of translatable mRNA for α-amylase in barley aleurone layers. Nature (London) 260: 166-169

35. HUBER DJ, DJ NEVINS 1979 Autolysis of cell wall β-D-glucan in corn coleoptile. Plant Cell Physiol 20: 201-212

36. HUBER DJ, DJ NEVINS 1980 ß-D-glucan hydrolase activity in *Zea* coleoptile walls. Plant Physiol 65: 768-773

37. HUBER DJ, DJ NEVINS 1981 Partial purification of endo- and exo-β-D-glucanase enzymes from *Zea mays* L. seedlings and their involvement in cell wall autohydrolysis. Plant 151: 206-214

38. HUBER DJ, DJ NEVINS 1981 Wall protein antibodies as inhibitors of growth and of autolytic reactions of isolated cell wall. Physiol Plant 53: 533-539

39. JACOBS M, PM RAY 1975 Promotion of xyloglucan metabolism by acid pH. Plant Physiol 56: 373-376

40. JARVIS MC 1982 The proportion of calcium-bound pectin in plant cell walls. Planta 154: 344-346

41. KATO Y, K MATSUDA 1980 Structure of oligosaccharides obtained by controlled degradation of mung bean xyloglucan with acid and *Aspergillus oryzae* enzyme preparation. Agric Biol Chem 44: 1751-1758

42. KATO Y, K MATSUDA 1980 Structure of oligosaccharides obtained by hydrolysis of mung bean xyloglucan with *Trichoderma viride* cellulase. Agric Biol Chem 44: 1759-1766

43. KATO Y, DJ NEVINS 1984 Enzymic dissociation of *Zea* shoot cell wall polysaccharides. II. Dissociation of (1-3),(1-4)-β-D-glucan by purified (1-3),(1-4)-β-D-glucan 4-glucanohydrolase from *Bacillus subtilis*. Plant Physiol 75: 745-750

44. KEEGSTRA K, KW TALMADGE, WD BAUER, P ALBERSHEIM 1973 The structure of plant cell walls III. A model of the walls of suspension-cultured sycamore cells based on the interactions of the macromolecular components. Plant Physiol 51: 188-197

45. KORNNEEF M, JH VAN DER VEEN 1980 Induction and analysis of gibberellin-sensitive mutants of *Arabidopsis thaliana* L. Heynh. Theor Appl Genet 58: 257-263

46. LABAVITCH JM, PM RAY 1974 Turnover of cell wall polysaccharides in elongating pea stem segments. Plant Physiol 53: 669-673

47. LABAVITCH JM, PM RAY 1974 Relationship between promotion of xyloglucan metabolism and induction of elongation by IAA. Plant Physiol 54: 499-502

48. LAMPORT DTA 1980 Structure and function of plant glycoproteins. *In*: J Preiss, ed., The Biochemistry of Plants, Vol 3. Academic, New York, pp. 501-541
49. LAMPORT DTA, L EPSTEIN 1983 A new model for the primary cell wall: A concatenated extensin-cellulose network. *In*: DD Randell, et al. eds., Current Topics in Plant Biochemistry and Physiology, Vol 2. Univ Missouri, Columbia, pp. 73-83
50. LAU JM, M McNEIL, AG DARVILL, P ALBERSHEIM 1985 Structure of the backbone of rhamnogalacturonan I, a pectic polysaccharide in the primary-cell walls of plants. Carbohydr Res 137: 111-125
51. LEE S, A KIVILAAN, RS BANDURSKI 1967 *In vitro* autolysis of plant cell walls. Plant Physiol 42: 968-972
52. LOESCHER W, DJ NEVINS 1972 Auxin induced changes in *Avena* cell wall composition. Plant Physiol 50: 556-563
53. LUTTENEGGER DG, DJ NEVINS 1985 Transient nature of a (1-3),(1-4)-β-D-glucan in *Zea mays* coleoptile walls. Plant Physiol 77: 175-178
54. MARKWALDER H-V, H NEUKOM 1976 Diferulic acid as a possible crosslink in hemicelluloses of wheat germ. Phytochemistry 15: 836-837
55. METRAUX J-P, L TAIZ 1977 Cell wall extension in *Nitella* as influenced by acids and ions. Proc Natl Acad Sci USA 74: 1565-1569
56. MOLL C, RL JONES 1981 Calcium and gibberellin-induced elongation of lettuce hypocotyl sections. Plant 152: 450-456
57. MOSCATELLI EA, EA HAM, EL RICKES 1961 Enzymic properties of a β-glucanase from *Bacillus subtilis*. J Biol Chem 236: 2858-2862
58. NISHITANI K, Y MASUDA 1982 Auxin-induced structural changes in cell wall xyloglucans in *Vigna angularis* epicotyl segments. Plant Sci Lett 28: 87-94
59. NISHITANI K, Y MASUDA 1983 Auxin-induced changes in cell wall xyloglucans. Effects of auxin on the two different subfractions of xyloglucans in the epicotyl cell wall of *Vigna angularis*. Plant Cell Physiol 24: 345-355
60. NISHITANI K, Y MASUDA 1981 Auxin-induced changes in the cell wall structure: changes in the sugar compositions, intrinsic viscosity, and molecular weight distribution of matrix polysaccharides of the epicotyl cell wall of *Vigna angularis*. Physiol Plant 52: 482-494
61. PHINNEY BO 1956 The growth response of a single gene dwarf mutants of *Zea mays* to gibberellic acid. Proc Natl Acad Sci USA 42: 185-189
62. REES DA 1977 Polysaccharide shapes. Outline Studies in Botany Series. Chapman and Hall, London.
63. SADAVA D, MJ CHRISPEELS 1973 Hydroxyproline-rich cell wall protein extensin role in the cessation of elongation in excised pea epicotyls. Devel Biol 30: 49-55
64. SAKURAI N, Y MASUDA 1977 Effect of IAA of cell wall loosening: changes in mechanical properties and noncellulosic glucose content of *Avena sativa* coleoptile wall. Plant Cell Physiol 18: 587-594

65. SAKURAI N, Y MASUDA 1978 Auxin induced extension, cell wall loosening and changes in the wall polysaccharide content of barley coleoptile segments. Plant Cell Physiol 19: 1225-1234

66. SCALBERT A, B MONTIES, J-Y LALLEMAND, E GUITTET, C ROLANDO 1985 Ether linkage between phenolic acids and lignin fractions from wheat straw. Phytochemistry 24: 1359-1362

67. SHIBUYA N, R NAKANE 1984 Pectic polysaccharides of rice endosperm walls. Phytochemistry 23: 1425-1429

68. STAFSTROM JP, LA STAEHELIN 1986 Cross-linking patterns in salt-extractable extensin from carrot cell walls. Plant Physiol 81: 234-241

69. STAUDTE RG, JR WOODWARD, GB FINCHER, BA STONE 1983 Water-soluble (1-3),(1-4)-β-D-glucans from barley (*Hordeum vulgare*) endosperm. III. Distribution of cellotriosyl and cellotetraosyl residues. Carbohydr Polym 3: 299-312

70. TAIZ L 1984 Plant cell expansion: regulation of cell wall mechanical properties. Ann Rev Plant Physiol 35: 585-657

71. TAYLOR IEP, M TEPFER, PT CALLAGHAN, AL MACKAY, M BLOOM 1983 Use of [1]H-n.m.r. to study molecular motion in cellulose, pectin, and bean cell walls. J Appl Polym Sci, Appl Polym Symp 37: 377-384

72. TEPFER M, IEP TAYLOR 1981 The interaction of divalent cations , the pectic substances and their influence on acid-induced cell wall loosening. Can J Bot 59: 1522-1525

73. TERRY ME, BA BONNER 1980 An examination of centrifugation as a method of extracting an extracellular solution from peas, and its use for the study of indoleacetic acid-induced growth. Plant Physiol 66: 321-325

74. TERRY ME, RL JONES, B BONNER 1981 Soluble cell wall polysaccharides released by pea stems by centrifugation. I. Effect of auxin. Plant Physiol 68: 531-537

75. THEOLOGIS A, PM RAY 1982 Early auxin-regulated polyadenylated mRNA sequences in pea stem tissue. Proc Natl Acad Sci USA 79: 418-421

76. VALENT BS, P ALBERSHEIM 1974 The structure of plant cell walls V. On the binding of xyloglucan to cellulose fibers. Plant Physiol 54: 105-108

77. VERMA DPS, GA MACLACHLAN, H BYRNE, D EWINGS 1975 Regulation and *in vitro* translation of mRNA for cellulase from auxin-treated pea epicotyls. J Biol Chem 250: 1019-1026

78. WALKER JC, JL KEY 1982 Isolation of cloned cDNAs to auxin-responsive poly(A)[+] RNAs of elongating soybean hypocotyl. Proc Natl Acad Sci USA 79: 7185-7189

79. WILKIE KCB 1979 The hemicelluloses of grasses and cereals. Adv Carbohydr Chem Biochem 36: 215-264

80. WONG YS, GB FINCHER, GA MACLACHLAN 1977 Kinetic properties and substrate specificities of two cellulases from auxin-treated pea epicotyls. J. Biol Chem 252: 1402-1407

81. WOODWARD JR, GB FINCHER 1982 Substrate specificities and kinetic properties of two (1-3),(1-4)-β-D-glucan endo-hydrolases from germinating barley (*Hordeum vulgare*). Carbohydr Res 106: 111-122

82. YAMAMOTO R, DJ NEVINS 1981 Coleoptile growth-inducing capacities of exo-β-(1-3)-glucanases from fungi. Physiol Plant 51: 118-122

83. YORK WS, AG DARVILL, P ALBERSHEIM 1984 Inhibition of 2,4-dichlorophenoxyacetic acid-stimulated elongation of pea stem segments by a xyloglucan oligosaccharide. Plant Physiol 75: 295-297

84. ZURFLUH LL, TJ GUILFOYLE 1982 Auxin-induced changes in the population of translatable messenger RNA in elongating sections of soybean hypocotyl. Plant Physiol 69: 332-337

Physiology of Cell Expansion During Plant Growth, *D.J. Cosgrove and D.P. Knievel* Eds.,

PHENOLIC CROSS-LINKING IN THE CELL WALL

KAREN J. BIGGS AND STEPHEN C. FRY

Department of Botany, University of Edinburgh
The King's Buildings, Mayfield Road
Edinburgh EH9 3JH, U.K.

INTRODUCTION

The primary cell walls of various plants have been shown to contain at least two types of polymer-bound phenolic dimers (14). The first, isodityrosine, is formed from the phenolic amino acid tyrosine, and is found in the cell wall glycoprotein, extensin. The second involves phenylpropanoid compounds derived from *p*-coumarate, especially diferulate, which is associated with wall matrix polysaccharides. It has been suggested that both tyrosine- and *p*-coumarate-derived dimers may cross-link wall polymers and help form the cell wall matrix (14). A brief summary of experimental evidence for this will be given.

Formation of phenolic cross-links is believed to be under the control of peroxidases located in the cell wall and so a discussion of the enzyme's probable role will be included. This will then be extended into a consideration of the importance of phenolic cross-linking to our current understanding of how cell walls can regulate cell expansion, and of the possible involvement of plant hormones in such regulation. Attention will be focused on the importance of the epidermis in the control of organ growth.

TYPES OF PHENOLIC CROSS-LINKS

Isodityrosine. The structure and properties of this dimeric phenolic amino acid were first described in 1982 (10) when a tyrosine derivative was isolated from cell wall hydrolysates of various plant calli. Isodityrosine was shown to be an oxidatively coupled dimer of tyrosine linked through a diphenyl ether bridge and it was suggested that isodityrosine could be responsible for cross-linking the cell wall glycoprotein extensin (Fig. 1a).

Cooper and Varner (5) established that extensin, which is secreted as a salt-soluble glycoprotein 'monomer' (M_r *ca.* 10^5) is gradually insolubilised in the cell wall and that this insolubilisation is accompanied by an increase in isodityrosine content. Their work did not, however, demonstrate that the isodityrosine formed was acting as an interpolypeptide cross-link between pairs of extensin molecules. Within extensin both intrapolypeptide and interpolypeptide isodityrosine bridges could exist but at present we only have

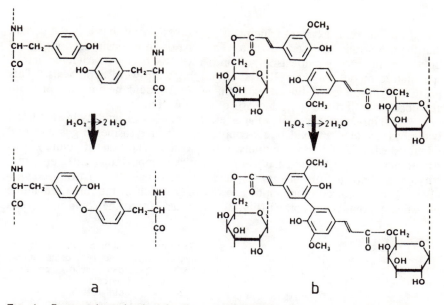

FIG. 1. Proposed mechanism for the cross-linking of wall polymers by peroxidase-catalysed coupling of phenolic side-chains.

    a.   Cross-linking of two glycoprotein-bound tyrosine residues to form an isodityrosine bridge.

    b.   Cross-linking of two polysaccharide-bound ferulate groups to form a diferulate bridge [each ferulate is shown attached to the 6-position of a non-reducing terminal ß-galactose residue of the polysaccharide chain, a structure known from spinach pectins].

    --------- = Rest of glycoprotein or polysaccharide

definite evidence for intrapolypeptide loops (7). Looped peptides were isolated from among extensin tryptides (peptide sequences generated by trypsin cleavage of deglycosylated walls). Analysis showed that intramolecular isodityrosine (Idt) was present in the sequence:

$$\overbrace{\text{-Tyr-Lys-Tyr-Lys-}}^{O} \quad \text{or} \quad \text{-1/2Idt-Lys-1/2Idt-Lys-}$$

(7). Recent peptide mapping of extensin precursors from tomato suspension cultures revealed a hexapeptide sequence,

-Val-Lys-Pro-Tyr-His-Pro-

(37), which, it has been speculated, contains the tyrosine residue most likely to participate in formation of an intermolecular isodityrosine cross-link. Such cross-links, although plausible, remain conjectural.

    Salt-extractable extensin molecules have been viewed under the electron microscope (38) and monomer extensin is visualised as a kinked rod. The position of the kinks was suggested to correlate with the -Tyr-Lys-Tyr-Lys- regions for intramolecular linkages. Interestingly, the EM preparations also reveal oligomers of extensin which could be cross-linked, but it is not certain that the cross-links are isodityrosine. The idea of an extensin network cross-linked by interpolypeptide isodityrosines (10) has been embroidered into a speculative model for the cell wall (27): the 'warp-weft' hypothesis. This postulates that a cross-linked extensin network could be interwoven with cellulose microfibrils to form the basis of a cell wall matrix.

Trimers of tyrosine. Paper electrophoresis of acid-hydrolysed cell walls has revealed an oxidatively-coupled trimer of tyrosine (Biggs and Fry, unpublished) in addition to the dimer, isodityrosine. The exact structure of this trimer has yet to be elucidated, although it is demonstrably different from trityrosine [known from insects (1)] and isotrityrosine [known from *Ascaris* (17)]. We are currently attempting to discover which of the three remaining plausible alternatives is correct (Fig. 2). A trimer of tyrosine could not form a

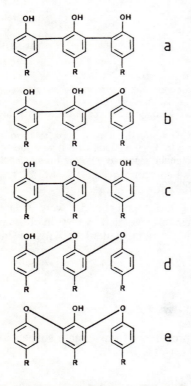

FIG. 2. Five plausible structures for an oxidatively coupled trimer of tyrosine found in hydrolysates of potato cell wall glycoprotein.

a.	Trityrosine
b.	Isotrityrosine
c-e	Un-named

$-R = -CH_2-CH(NH_2)-COOH$

tight intrapolypeptide loop like that described for isodityrosine (7); a trimer would have to be either interpolypeptide or a *loose* intrapolypeptide loop, and either of these dispositions would be of great interest in building a model of the cell wall.

Diferulate. Wall polysaccharides have been shown to contain derivatives of *p*-coumarate and ferulate. A phenolic coupling product, diferulate, has been detected by hydrolysis of polysaccharides from various monocotyledonous plants and from members of the dicotyledonous Centrospermae (Caryophyllales) (8, 9, 11, 13, 18, 22, 23, 31, 34). Diferulate is difficult to localise directly, but is presumably formed from polymer-bound ferulate, which is brightly fluorescent and can therefore be localised in the cell wall by simple u.v. fluorescence microscopy, as shown for grasses (21) and spinach (13). Analysis of sugar beet pectins by Rombouts and Thibault has also revealed feruloyl ester groups (31). Pectins isolated by these authors from non-members of the Centrospermae including potato, citrus and cherries did not contain ferulate. However, cell walls of potato appeared to contain a feruloylated galactan (35) and the walls of growing cell cultures of *Capsicum* are a major sink for phenylpropanoid metabolism (20).

Work on spinach cell suspension cultures demonstrated that, in these Dicotyledons, the phenolic groups were associated with pectins, both neutral (arabinose and galactose-rich) and acidic (galacturonate, rhamnose, arabinose and galactose-rich) pectins being feruloylated (11). Enzymic digestion of cell walls with 'Driselase' revealed that the feruloyl groups were attached to the non-reducing termini of pectin molecules. The presence of ferulate on these non-reducing termini of both neutral arabinose and galactose domains, which are exposed regions of the pectin molecules, means that the phenolic groups are well positioned to undergo oxidative coupling because they are accessible to peroxidase action. It was therefore suggested that the ferulate groups could cross-link spinach pectins by the mechanism in Fig. 1b and an estimation of *ca.* 1 feruloyl group per 50 pectin sugar residues would provide ample scope for such coupling. The cross-linking product, diferulate, has been isolated from spinach cell walls in small amounts (12, 13, 14).

In members of the Gramineae enzymic digestion of cell wall polysaccharides showed that the feruloyl groups were linked to arabinoxylans (19). Shibuya (34) found high levels of diferulic acid attached to arabinoxylans in rice endosperm cell walls. He noted that high-M_r fractions obtained by enzymic digestion of cell walls contained correspondingly greater levels of phenolic compounds, particularly diferulate. This is consistent with the idea of diferuloyl bridges cross-linking polysaccharides.

Carpita (2) has analysed the phenylpropanoids of maize coleoptile cell walls and, on the basis of a relatively low diferulate content, has concluded that only a small proportion of the hemicellulose can possibly be immobilised by diferulate bridges. However, his labelling data for day-3 coleoptiles indicate a (1/2-diferulate):ferulate ratio of 1:15 [Fig. 4B of (2)]; the same walls contain ferulate at 25 nmol/mg and arabinoxylan-bound sugar residues [solubilised by 0.01-1.0 M KOH] at 1000 nmol/mg [Fig. 2B of (2)]. It can therefore be estimated that there was one 1/2-diferulate group per 600 arabinoxylan sugar residues. The degree of polymerisation of the arabinoxylan is not known, but, if it is in the order of 1500 [reported for oat coleoptile arabinoxylan (40)], then each average arabinoxylan molecule was associated with 2.5 differulate bridges. This would be quite sufficient to make a major contribution to wall architecture.

Nevertheless, as with isodityrosine in extensin cross-linking, conclusive evidence for a role of diferulate in polysaccharide cross-linking awaits the isolation and identification of a phenolic cross-link with its backbone molecules still attached.

Quinone-methide reaction-products. Although diferulate has been isolated from cell walls in amounts large enough to play a significant architectural role (2, 14, 22, 23), chromatography of alkali hydrolysates of cell walls (2, 12, 13) yielded larger amounts of lower-R_F phenylpropanoid substances, suggesting that coupling-products larger or more complex than dimers predominated. These putative cross-links were especially conspicuous in the most Driselase-resistant components of the wall (12) and substantial amounts were not released from the polysaccharides by treatment with alkali (2, 12). The nature of the bond(s) between these phenylpropanoid derivatives and the polysaccharides is unknown, but resistance to both alkali and Driselase

49

(which attacks all known glycosidic linkages of the spinach cell wall except α–xylosyl) strongly suggests *ether* bonds rather than ester or glycosidic bonds. Similar bonds were formed when free ferulic acid was incubated with isolated cell walls that contained active peroxidases (41).

How could ether bonds be formed? One attractive possibility is that some of the initially-formed dimers of the feruloyl esters are semi-stable quinone-methides (Fig. 3). These are highly reactive molecules which can react with water to form benzyl alcohols, with carboxylic acids to form benzyl esters and with alcohols to form benzyl *ethers* (Fig. 4). In the cell wall, which is about 40% carbohydrate and 60% water, the high concentration of polysaccharide-bound -OH groups relative to H_2O might favour the formation of benzyl-sugar ether cross-links. It is not inconceivable that a novel enzyme would further channel quinone-methides into ether formation, and the need to achieve controlled formation of ether bonds could provide a clue as to why the feruloyl residues are linked to polysaccharides at such specific positions (9). *By quinone-methide/polysaccharide bonding, a single dimer of ferulate could potentially cross-link up to four different polysaccharide chains*: the two already attached via ester bonds to the participating ferulate groups, and one or two more subsequently grafted via ether bonds to the C_3-side-chains of the quinone-methide dimer.

By analogy, grass lignins form ether linkages with the -OH group of ferulate itself, possibly via quinone-methide intermediates (33).

ROLE FOR PEROXIDASES IN PHENOLIC CROSS-LINKING

Peroxidases are known to be located within the cell wall in primary plant tissues (13) and are also known to catalyse reactions involving phenolic groups. Peroxidases catalyse the generation of free radicals from phenols and the highly reactive radicals formed will easily undergo coupling to form dimers and larger oligomers. Peroxidases are therefore likely candidates to catalyse the coupling of tyrosine groups to form isodityrosine in extensin or coupling of ferulate groups to form diferulate in pectins and arabinoxylans.

Cooper and Varner's work (5) showed that insolubilisation of extensin was under enzymic control and that insolubilisation *in vivo* could be inhibited by various antioxidants, especially L-ascorbate, a well-known peroxidase blocker. Lower concentrations of ascorbate appeared to inhibit extensin cross-linking for shorter times and it was suggested that ascorbate, which is normally synthesised by cells, had been oxidised within the wall. This led to the speculation of a wall bound peroxidase/ascorbate oxidase system that could control the rate of phenolic free radical generation and, consequently, of phenolic cross-linking of cell wall glycoprotein. Another reducing agent, dithiothreitol, has been shown to inhibit isodityrosine formation in potato cells (10). Lamport (26) has observed in tomato cell suspension cultures an inverse relationship between peroxidase activity and the size of the pool of extensin monomer molecules. This is further evidence that peroxidase activity could be responsible for recruiting extensin monomers into an insoluble glycoprotein network.

Evidence for peroxidase-catalysed coupling of feruloyl residues includes the finding that sugar beet pectins will gel *in vitro* upon treatment with peroxidase and H_2O_2 (31). Following on from the observations of Geissmann

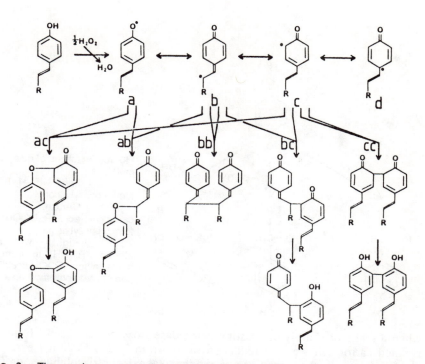

FIG. 3. The random nature of oxidative coupling, illustrated by a *p*-coumaroyl-polysaccharide ester such as is found in many primary cell walls.

Top row: The *p*-coumaroyl groups (top left) are oxidised by H_2O_2 in the presence of peroxidase, to yield free radicals (position indicated by the heavy dot). These are non-enzymically interconverted to produce an equilibrium mixture of four isomers (mesomers), a-d.

Middle row: Three of the mesomers (a, b and c) take part in coupling, again non-enzymically, to form five different dimers (ac, ab, bb, bc and cc) [aa dimers are unknown]. Three of the dimers (ab, bb and bc) are *quinone-methides,* possessing the structure

O $=$ $=$ CH---, and one (bb) has 2 quinone-methide moieties.

Bottom row: All the C=O groups *other* than those of quinone-methides quickly revert to phenolic -OH groups.
-R = -CO -O -polysaccharide

and Neukom (18) it was proposed that gel formation results from enzymic cross-linking of polysaccharide-bound feruloyl ester groups.

POSSIBLE CONTROL POINTS IN PHENOLIC CROSS-LINKING

In vitro cross-linking of tyrosine with horseradish peroxidase and H_2O_2 yields different products depending upon the pH of the reaction. At the pH optimum of about 9.5 the main reaction product is dityrosine, which has not been found in cell walls, but at pH 3-5, where the rate of oxidation is slower, some isodityrosine is produced (15). The products obtained are very similar

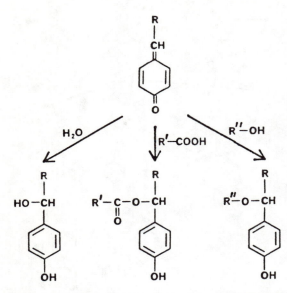

FIG. 4. Reactions of quinone-methides –

Left: with H_2O to form a benzyl alcohol

Centre: with a carboxylic acid (R'-COOH) to form a benzyl ester

Right: with an alcohol (R"-OH) to form a benzyl ether. [R"-OH could be part of a wall polysaccharide.]

when a range of sharply contrasting peroxidase isozymes are used. In intact isolated carrot cell walls, a reduction in pH from 7 to 4 inhibited extensin insolubilisation, both in terms of rate and the total extent of insolubilisation (5). These observations taken together lead to the idea that *in vivo* the cell wall environment is important to phenolic cross-linking; for example, the interaction of basic extensin with acidic pectins may be involved in the control of coupling.

The biological control of phenolic cross-linking could be imposed at any of a range of points. These include

(a) the synthesis and secretion of peroxidase,
(b) the supply of a reductant to convert O_2 into H_2O_2,
(c) the supply of the enzyme that converts O_2 into H_2O_2 [though this might be a direct trans-plasmalemmar redox reaction (29)],
(d) the number of phenolic groups per wall polymer molecule [a constant in extensin, but variable in feruloyl-polysaccharides, the feruloylation reaction occurring within the endomembrane traffic system (16)],
(e) cell wall pH,
(f) the presence of a template for cross-linking [pectin? -- see (15)],
(g) the levels of antioxidant peroxidase blockers [e.g., ascorbate (5)].

PHENOLIC CROSS-LINKING AND CELL EXPANSION

Coupling of phenolic groups such as tyrosine and ferulate in wall polymers suggests an important structural role in building cross-linked networks that can interact to form the intact cell wall matrix. That such cross-linking appears to be under enzymic control means that cross-linking of wall polymers can be regulated and this has significant consequences for cell properties such as wall expansion during growth. We envisage that a low number of isodityrosine cross-links in young cell walls would give a relatively open extensin network which would allow for wall extension, while in older cell walls we would expect a higher incidence of cross-links to produce a tighter network, more rigid walls and, consequently, less scope for wall expansion. "A reduction in wall extension is more likely to be related to the amount of hydroxyproline rigidly bound in the wall than to the total secreted to the wall" (42).

The question of whether phenolic coupling in the wall is a reversible process has also to be addressed. Once phenolic cross-links are made they are unlikely to be broken *in vivo*. Plants have no known enzymes to split either the diphenyl ether bonds of isodityrosine or the biphenyl bond of diferulate. Nevertheless, reductions of growth rate in plant tissues are rarely irreversible, and there are a number of ways in which a wall moderately tightened by phenolic cross-linking could subsequently be loosened, e.g., (a) proteolysis of the backbone of extensin, (b) hydrolysis of the backbone of pectin or arabinoxylan, (c) hydrolysis of the ester bridge between diferulate and one or both polysaccharide chains, or (d) simply maintaining a slow rate of growth until eventually the tightly cross-linked wall layer has become displaced to the outer part of the cell wall which may be architecturally irrelevant since it is believed that (in giant internodal cells of certain algae and perhaps also in the epidermis of land plants) only the inner region of the wall may be load-bearing (30).

A tightening of the cell wall through increased phenolic coupling could be triggered by increased peroxidase activity, the enzyme that catalyses the coupling and possibly is also involved in the generation of H_2O_2. This idea is supported by a number of cases in which there is a negative correlation between peroxidase levels and rate of cell expansion (Table I). Work on spinach cell cultures (8) led to the suggestion that the observed growth promotion stimulated by gibberellic acid treatments could be due to a suppression of peroxidase secretion into the cell walls. This, in turn, could lower the rate of the peroxidase-catalysed formation of bridges through dimerisation of feruloyl-polysaccharides (and any associated grafting of polysaccharides to quinone-methide intermediates). A decrease in the number of cross-links formed would allow a rapid rate of cell expansion to be maintained.

A recent investigation into cell wall peroxidase activity during elongation that suggests a role for peroxidase in growth control involved

Table I. *Conditions affecting* growth rate and peroxidase levels.*

Condition	Peroxidase levels	Growth	Reference
Ca^{2+}	+	-	(39)
Chlorocholine chloride	+	-	see (8)
Genetic dwarfism	+	-	see (8)
Methotrexate resistance	+	-	(4)
Blue light	+	-	(36)
Maturation (ageing)	+	-	(3)
Gibberellic acid	-	+	(8)

*Key: + = increase, - = decrease.

53

monitoring the effects of blue light on cucumber hypocotyls (36). Irradiation with blue light had been previously shown to inhibit stem growth (6) and this inhibition was found to be correlated with an increase in apoplastic peroxidase activity (36). The effect on peroxidase was reversible so that when the blue light was turned off peroxidase levels decreased and growth resumed. Although this effect is an interesting one, the technical problems encountered in monitoring peroxidase levels [especially in the outer epidermal wall--see below] prevented a direct comparison with the rapid effects (1-2 minutes) that blue light had upon stem elongation and therefore do not allow us, at present, to conclude that peroxidase action can control cell expansion. Nevertheless, it was of interest that the inhibitory effect of blue light on growth could be relieved by treatment of the tissue with the peroxidase blocker, ascorbate (36).

Investigations into wall-bound peroxidase activity in pearl millet internodes with relation to the cessation of elongation growth (3) led Chanda et al. to conclude that wall peroxidases are not a control site for halting cell growth during maturation. They observed an increase in the specific activity [total activity was not stated] of wall-bound peroxidase during elongation with a sharp peak in activity at 20 mm back from the internodal plate. Chanda et al. point out that, at this level in the internode, elongation was still occurring; it did not cease until about 40 mm back. They therefore discount a role for peroxidase in stopping growth. However, the peak in wall bound peroxidase does coincide well with the onset of a *deceleration* in cell growth. We therefore do not find the authors' objection convincing.

ROLE OF SPECIFIC CELL LAYERS IN ORGAN GROWTH

In growing stems, hypocotyls, coleoptiles, some petioles and possibly the lamina, the epidermis appears to be rate-limiting for the elongation of the whole organ [see (25)]. This is the basis of the split pea bioassay for auxin. It can also be seen very convincingly by cutting a 25-cm length of a young rhubarb stalk, peeling it, and measuring it again: it will have rapidly elongated to 26 or 27 cm (32). Epidermal walls are very different from the others in a stem: they are about 10 times thicker, and they have waxes, cutin, and a thick layer of cutin-polysaccharide complex. Cuticles are autofluorescent, even in those dicotyledonous plants that do not possess ferulate groups in the parenchymatous cell walls (23). Chemical analysis of cuticles demonstrates the presence of esterified phenylpropanoids, e.g., ferulate and *p*- and *m*-coumarate (24), although it has not been established whether these are esterified to cutin or polysaccharides. Cuticles are not restricted to mature tissue, being present even on the summit of the apical meristem, and synthesis of the aliphatic components of cutin in *Clivia* leaves peaked in the zone of most rapid growth (28). Cutin clearly does not preclude growth, but cross-linking of cutin-associated phenolics might limit growth. The outer wall of the epidermis is thus of great interest in the control of plant growth, but has been sadly neglected. Its high content of phenolic material, in all land plants, lends weight to the thesis of this paper.

CONCLUSIONS

Whilst the peroxidase-catalysed coupling of cell wall phenolics to form structural polymer networks is undoubtedly significant to our understanding of the mechanisms behind cell expansion, it is apparent that several areas require further investigation. We still await conclusive identification of an inter-polypeptide or inter-polysaccharide phenolic coupling product. We need to delineate more precisely the coupling reactions involved, and to verify the participation of quinone-methide intermediates. The contribution of the cell wall environment (e.g., polymer molecules neighbouring those being cross-linked) to the occurrence and steering of peroxidase catalysis of the formation of cross-links *in vivo* must be established. We can then begin to consider the range of factors and their interactions that might influence the cross-linking of wall polymers during cell growth and in other processes such as cell defence. Detailed investigations into the chemistry and metabolism of epidermal cell walls are called for.

Acknowledgments--We are grateful to Mrs. Janice Miller for excellent technical assistance. KJB thanks the Leatherhead Food RA and the Agricultural and Food Research Council (AFRC) for a studentship, and SCF thanks the AFRC for a grant in support of this research.

LITERATURE CITED

1. ANDERSEN SO 1964 The cross-links in resilin identified as dityrosine and trityrosine. Biochim Biophys Acta 93: 213-215
2. CARPITA NC 1986 Incorporation of proline and aromatic amino acids into cell walls of maize coleoptiles. Plant Physiol 80: 660-666
3. CHANDA SV, AK JOSHI, PN KRISHNAN, PP VAISHNAV, YD SINGH 1986 Distribution of indoleacetic acid oxidase, peroxidase and esterase activities in relation to elongation growth in pearl millet internode. J Plant Physiol 122: 373-383
4. CHIBBAR RN, R CELLA, D ALBANI, RB VAN HUYSTEE 1984 The growth of and peroxidase synthesis by two carrot lines. J Exp Bot 35: 1846-1852
5. COOPER JB, JE VARNER 1984 Cross-linking of soluble extensin in isolated cell walls. Plant Physiol 76: 414-417
6. COSGROVE DJ 1981 Rapid suppression of growth by blue light. Occurrence, time course, and general characteristics. Plant Physiol 67: 584-590
7. EPSTEIN L, DTA LAMPORT 1984 An intramolecular linkage involving isodityrosine in extensin. Phytochemistry 23: 1241-1246
8. FRY SC 1979 Phenolic components of the primary cell wall and their possible rôle in the hormonal control of growth. Planta 146: 343-351
9. FRY SC 1982 Phenolic components of the primary cell wall. Feruloylated disaccharides of D-galactose and L-arabinose from spinach polysaccharide. Biochem J 203: 493-504
10. FRY SC 1982 Isodityrosine, a new cross-linking amino acid from plant cell wall glycoprotein. Biochem J 204: 449-455

11. FRY SC 1983 Feruloylated pectins from the primary cell wall: their structure and possible functions. Planta 157: 111-123

12. FRY SC 1984 Incorporation of [^{14}C]cinnamate into hydrolase-resistant components of the primary cell wall. Phytochemistry 23: 59-64

13. FRY SC 1986 Polymer-bound phenols as natural substrates of peroxidases. *In* H GREPPIN, C PENEL, Th GASPAR, eds, Molecular and Physiological Aspects of Plant Peroxidases. Université de Genève, pp 169-182

14. FRY SC 1986 Cross-linking of matrix polymers in the growing cell walls of angiosperms. Annu Rev Plant Physiol 37: 165-186

15. FRY SC 1987 Formation of isodityrosine by peroxidase isozymes. J. Exp Bot 38: 853-862

16. FRY SC 1987 Intracellular feruloylation of pectic polysaccharides. Planta 171: 205-211

17. FUJIMOTO D, K HORIUCHI, M HIRAMA 1981 Isotrityrosine, a new cross-linking amino acid isolated from *Ascaris* cuticle collagen. Biochem Biophys Res Commun 99: 637-643

18. GEISSMANN T, H NEUKOM 1973 On the composition of the water-soluble wheat flour pentosans and their oxidative gelation. Lebensm-Wiss u -Technol 6: 59-62

19. GUBLER F, AE ASHFORD 1985 Release of ferulic acid esters from barley aleurone. Aust J Plant Physiol 12:297-305

20. HALL RD, MA HOLDEN, MM YEOMAN 1987 The accumulation of phenylpropanoid and capsaicinoid compounds in cell cultures and whole fruit of the chilli pepper, *Capsicum frutescens* Mill. Plant Cell Tissue & Organ Culture 8: 163-176

21. HARRIS PJ, RD HARTLEY 1976 Detection of bound ferulic acid in cell walls of the Gramineae by ultraviolet fluorescence microscopy. Nature 259: 508-510

22. HARRIS PJ, RD HARTLEY 1980 Phenolic constituents of the cell walls of Monocotyledons. Biochem Syst Ecol 8: 153-160

23. HARTLEY RD, PJ HARRIS 1981 Phenolic constituents of the cell walls of Dicotyledons. Biochem Syst Ecol 9: 189-203

24. HUNT GM, EA BAKER 1980 Phenolic constituents of tomato fruit cuticles. Phytochemistry 19: 1415-1419

25. KUTSCHERA U, R BERGFELD, P SCHOPFER 1987 Cooperation of epidermis and inner tissues in auxin-mediated growth of maize coleoptiles. Planta 170: 168-180

26. LAMPORT DTA 1986 Roles for peroxidases in cell wall genesis. *In* H GREPPIN, C PENEL, Th GASPAR, eds, Molecular and Physiological Aspects of Plant Peroxidases. Université de Genève, pp 169-182

27. LAMPORT DTA 1986 The primary cell wall: a new model. *In* RA YOUNG, RM ROWELL, eds, Cellulose: Structure, Modification and Hydrolysis, pp 77-87

28. LENDZIAN KJ, J SCHÖNHERR 1983 In-vivo study of cutin synthesis in leaves of *Clivia miniata* Reg. Planta 158: 70-75

29. MARIGO G, M BELKOURA 1985 Cation stimulation of the proton-translocating redox activity at the plasmalemma of *Catharanthus roseus* cells. Plant Cell Reports 4: 311-314

30. RICHMOND PA, J-P MÉTRAUX, L TAIZ 1980 Cell expansion patterns and directionality of wall mechanical properties in *Nitella*. Plant Physiol 65: 211-217
31. ROMBOUTS FM, JF THIBAULT 1986 Feruloylated pectin substances from sugar beet pulp. Carbohydr Res 154: 177-187
32. SACHS J 1865 Handbuch der Experimentalphysiologie der Pflanzen. Engelmann, Leipzig
33. SCALBERT A, B MONTIES, J-Y LALLEMAND, E GUITTET, C ROLANDO 1985 Ether linkage between phenolic acids and lignin fractions from wheat straw. Phytochemistry 24: 1359-1362
34. SHIBUYA N 1984 Phenolic acids and their carbohydrate esters in rice endosperm cell walls. Phytochemistry 23: 2233-2237
35. SHIMONY C, J FRIEND 1976 Ultrastructure of the interaction between *Phytophthora infestans* and tuber slices of resistant and susceptible cultivars of potato (*Solanum tuberosum* L.) Orion and Majestic. Isr J Bot 25: 174-183
36. SHINKLE JR, RL JONES 1986 Cell wall peroxidases and stem elongation in *Cucumis* seedling hypocotyl: a rapid effect of blue light. *In* B VIAN, D REIS, R GOLDBERG, eds, Cell walls '86 [Proceedings of the fourth cell wall meeting, Paris, Sept 10-12, 1986]. Université M. & P. Curie, École Normale Supérieure, Paris, pp 190-193
37. SMITH JJ, EP MULDOON, JJ WILLARD, DTA LAMPORT 1986 Tomato extensin precursors P1 and P2 are highly periodic structures. Phytochemistry 25: 1021-1030
38. STAFSTROM JP, LA STAEHELIN 1986 Cross-linking patterns in salt-extractable extensin from carrot cell walls. Plant Physiol 81: 234-241
39. STICHER L, C PENEL, H GREPPIN 1981 Calcium requirement for the secretion of peroxidases by plant cell suspensions. J Cell Sci 48: 345-353
40. WADA S, PM RAY 1978 Matrix polysaccharides of oat coleoptile cell walls. Phytochemistry 17: 923-931
41. WHITMORE FW 1976 Binding of ferulic acid to cell walls by peroxidases of *Pinus elliottii*. Phytochemistry 15: 375-378
42. WILSON LG, JC FRY 1986 Extensin--a major cell wall glycoprotein. Plant Cell Env 9: 239-260

Physiology of Cell Expansion During Plant Growth, *D.J. Cosgrove and D.P. Knievel* Eds.,
Copyright ©1987, The American Society of Plant Physiologists

CHANGES IN CELLULOSE MICROFIBRIL ORIENTATION DURING DIFFERENTIAL GROWTH IN OAT COLEOPTILES

DEBRA B. FOLSOM AND R. MALCOLM BROWN, JR.

*Botany Department, The University of Texas at Austin,
Austin, Texas 78713-7640*

INTRODUCTION

Synthesis of new wall elements and their concomitant organization into a functional cell wall are fundamental to plant morphogenesis (33). Especially critical for the wall assembly process is the synthesis of cellulose. Although the "typical" vascular plant cell wall contains many other important biomolecules (e.g., pectins, hemicelluloses, proteins, etc.), it is the cellulose microfibrillar network which constitutes the load-bearing portion of the wall (19, 35, 36). We have examined this microfibrillar network in *Avena sativa* coleoptiles to determine whether microfibril orientation may perform a role in mediating differential growth of gravistimulated plants.

ORIENTATION AND SYNTHESIS OF CELLULOSE MICROFIBRILS

In 1958, Roelofsen, drawing upon his own research as well as data compiled from other investigators, was the first plant biologist to speculate about the critical role of the innermost (most recently synthesized) cellulose microfibril layers in providing physical controls of plant cell expansion (36,37). Since then, Richmond (34,35), applying biophysical methods to analyze light microscopical evidence, and Srivastava et al. (39), Vian et al. (42), and Eisinger et al. (15), interpreting chemical disruption of wall order, have made clear demonstrations of the significance of the innermost wall.

Since organization of the innermost cellulose microfibril layers is important in cell expansion, it is useful to review briefly what is known about cellulose microfibril assembly and its spatial control in vascular plants. In 1976, Mueller et al. (30) published the first account of a putative cellulose-synthesizing complex in a vascular plant. Using freeze fracture coupled with electron microscopy, they discovered a transmembrane structure in the plasma membrane of *Allium cepa* which consists of two distinct elements: on the exoplasmic fracture face (EF; for nomenclature, see 4), a 23-25 nm structure termed the globule; on the protoplasmic face (PF), a hexagonal array of particles (22-24 nm), called the rosette. The two structures together constitute a *terminal, cellulose-synthesizing complex*, or TC. There is good correlation between the number and distribution of globules and rosettes (24), supporting the hypothesis that these two structures are indeed complementary. Single TCs can be seen to terminate individual cellulose microfibril impressions; thus it is believed that one TC is responsible for the generation of a single microfibril. Terminal complexes are thought to be unidirectionally mobile in the plane of the fluid membrane, driven by the energy of crystallization of the polymerized cellulose (21, 23, 25). Since the original description of vascular plant TCs in onion, similar structures have been observed by various investigators in many other plants (summarized in 9).

How is the direction of TC movement in the plasma membrane mediated? Clearly, spatial control of microfibril deposition is crucial to the assembly of a properly expandable wall. The primary candidate for membrane control of microfibril deposition is the microtubule (MT). Since 1963, when Ledbetter and Porter observed cytoplasmic MTs parallel to the innermost layer of cellulose microfibrils, investigators have been seeking to detect and/or define the role of MTs in controlling cell wall formation (15, 22). Staehelin and Giddings (40) considered previous studies showing that inhibition of microtubule formation resulted in disruption of microfibril patterns in some green algae and wrote "...we may deduce that microtubules exert their control on microfibril orientation by influencing the path taken by the synthesizing complexes..." Although disruption of MTs by colchicine and other agents can cause major alterations in microfibril orientation and cell shape, there is no evidence to indicate any absolute requirement for MTs in cell growth or fibrillogenesis (22, 39). Herth (23, 24), Mueller and Brown (29), and Staehelin and Giddings (40) have proposed that membrane-associated MTs define pathways in the membrane in which TCs are confined to move. This model would allow a relatively small number of MTs to control the direction of movement of large numbers of TCs. To date, there is no cytological evidence that MTs form any direct contact with terminal complexes or cellulose microfibrils.

MICROFIBRIL ORIENTATION DURING GRAVITROPISM

In light of the convincing arguments demonstrating the importance of cellulose microfibril orientation in "normal" plant growth (35, 36), we became interested in investigating wall organization during gravitropic curvature in oat coleoptiles (*Avena sativa* cv. Garry). Gravicurvature in coleoptiles results from the cessation of growth in the upper hemicylinder of the gravistimulated plant, while the lower hemicylinder continues elongation at a rate equal to or slightly in excess of the prestimulation rate (13). Thus, in a single gravistimulated plant, one can examine the walls of elongating and non-elongating cells which were morphologically and functionally identical in the upright plant.

Most previous investigations of the gravitropic phenomenon were directed at elucidating the earliest steps in the transduction pathway immediately following graviperception. Iwami and Masuda (26) and Bandurski and Schulze (2) have considered auxin redistribution; Goswami and Audus (18) and Slocum and Roux (38) have examined establishment of ion gradients. Possible changes in wall polysaccharide composition have been investigated by Carrington and Firn (11); Tanada and Vinten-Johansen (41) and Behrens et al. (3) have measured electrical potential differences. Mulkey and Evans (31) have demonstrated that tropic curvature is preceded by proton efflux. These names represent only a few of the many scientists who have probed intermediary steps in gravitropism. To complement these investigations, we undertook to approach the transduction pathway from the opposite direction. Clearly, such a profound alteration in relative growth rates, as seen when comparing upper and lower hemicylinders of gravistimulated coleoptiles, may be expected to correlate with changes in the load-bearing, cellulosic portion of the cell wall (8). We tested this hypothesis using the freeze fracture technique coupled with transmission electron microscopy to detect changes in the wall. It has already been shown (above) that the innermost wall layers are those which determine the properties of the total wall composite (35). Freeze fracture is an ideal tool for visualizing not only the innermost cellulose microfibril impressions but also the putative synthesizing complexes in place in the plasma membrane.

Freeze fracture of coleoptiles from upright-grown, actively elongating oat seedlings (control plants) shows cellulose microfibrils to be arranged in parallel, transverse to the longitudinal axis of elongation (Fig. 1) (6). Terminal

0.5 µm

FIG. 1. This EF plasma membrane view demonstrates the highly parallel, ordered nature of cellulose microfibril impressions typically seen in a cell from a rapidly elongating region in a control (upright-grown) *Avena* coleoptile.

complexes were seen in clusters or rows (Fig. 2), with 81.8% of the total number of TCs occurring within 50 nm of their nearest neighbors (5, 17). When a large membrane face is viewed, it is apparent that there are large areas of membrane that have no clusters or even single TCs. Thus, the distribution of TCs and subsequent assembly of microfibrils are not random processes in the

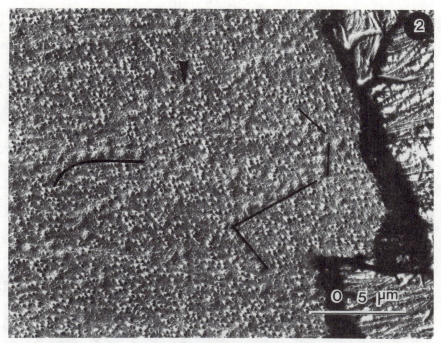

FIG. 2. Freeze fracture of an upright-grown coleoptile reveals terminal complexes occurring in rows or files, as designated by the black bars in this micrograph of the EF plasma membrane. Not every TC in the picture is incorporated into a row or cluster. Some, such as the one designated by the arrowhead, are isolated by at least 50 nm from their nearest neighbors; however, the majority of TCs in the picture are located within 50 nm of the nearest TC. At the right of the picture, some older wall layers may be observed.

coleoptile (5). An extreme example of localized distribution of TCs has been shown in the development of wall thickenings in xylem vessel elements of *Lepidium* (24) and *Zinnia* (20). In these cases, the occurrence of terminal complex globules and rosettes is restricted only to the regions of secondary wall thickening.

Cellulose microfibrils generated by TCs in close proximity to one another have greater probabilities of forming intermicrofibrillar hydrogen bonds (8). Such a network of closely packed, parallel microfibrils, tightly associated by hydrogen bonds, would form a relatively constrictive innermost wall layer (8). This microfibrillar band would be resistant to lateral expansion, permitting extension only along the longitudinal axis (19). Such a situation is certainly compatible with the known normal growth pattern of coleoptiles: they elongate rapidly (about 1 mm/hr) but show little change in diameter.

In the case of the differential growth observed in gravistimulated coleoptiles, one might expect differences in microfibril orientation when upper hemicylinder walls are compared to lower hemicylinder walls. Such is indeed the case, but with a surprising twist. Despite the fact that elongation ceases immediately following gravistimulation, coleoptile upper hemicylinders show normal microfibril orientation and TC distribution; freeze fracture views of epidermal and subepidermal parenchyma cells from tissue frozen 5-35 minutes after gravistimulation (Figs. 3, 4) are indistinguishable from replicas of control (upright-grown) material.

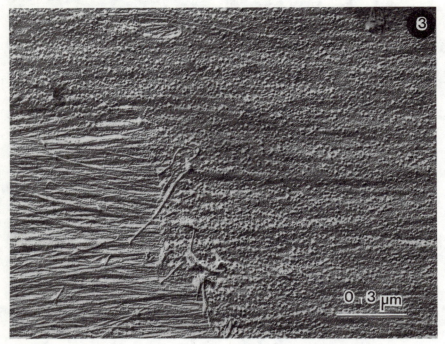

FIG. 3. In a gravistimulated coleoptile upper hemicylinder, 25 minutes post-stimulation, an EF plasma membrane view reveals ordered, parallel cellulose microfibril impressions. In the lower left corner of the micrograph, the plasma membrane has been torn away, revealing the ordered, innermost wall microfibril layer itself.

Cells from the actively elongating regions of lower hemicylinders show alteration of terminal complex arrangement (Fig. 5), as well as changes in the orientation of the innermost layer of cellulose microfibrils (Fig. 6). Terminal complexes no longer occur in clusters, but instead are seen, individually, randomly distributed across the entire cell plasma membrane surface (Fig. 5). Presumably as a result of dissociation of TCs, the cellulose microfibrils are more randomly oriented. These changes are evident within five minutes, the earliest time period yet tested. When the innermost wall layer is examined thirty minutes after gravistimulation, it is difficult to discern any single, predominant direction of microfibril orientation.

In addition to revealing differences in cellulose microfibril orientation and TC distribution in opposite gravistimulated coleoptile hemicylinders, freeze fracture shows a distinct asymmetry in the number of intramembranous particles (IMPs) seen in EF plasma membrane views. Cells from coleoptile upper hemicylinders of plants gravistimulated for 25 minutes had 987.0 ±43.8 IMPs/μm^2, which is not significantly different from control (upright-grown) coleoptile plasma membrane cells which had 987.1 ±64.7 IMPs/μm^2. In sharp

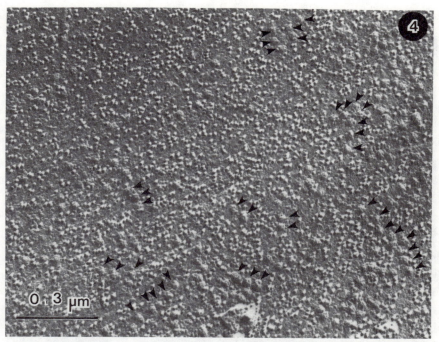

FIG. 4. Terminal complexes (arrowheads) are seen to be aggregated in this plasma membrane view from a coleoptile upper hemicylinder gravistimulated for 25 minutes.

contrast, lower hemicylinders had only 344.7 ±26.6 IMPs/μm^2 (5, 7). The reduced density of IMPs in the lower hemicylinder may be a result of the more rapid elongation of that region. We suspected, at first, that the change in orientation (due to gravistimulation) might perhaps disturb the normal, Golgi-derived vesicle exocytosis required to provide assorted membrane components to the elongating cell. Results of preliminary experiments with the calcium chelator, ethyleneglycol-bis-(ß-aminoethyl ether)-N,N'-tetraacetic acid (EGTA) suggest otherwise.

We applied EGTA exogenously to gravistimulated *Avena* seedlings by reorienting the plants onto agar plates containing a bathing solution of EGTA (10^{-3} M) Gravitropism was effectively inhibited at 10^{-3}M EGTA, as we predicted based upon Daye et al. (12). Tissue from the upper and the lower hemicylinders was prepared for freeze fracture from 5 min to 2 h after the onset of EGTA treatment. After short EGTA exposure (less that 30 minutes), we saw no disruption of cell wall order in either the upper or the lower hemicylinder. (Plants were pretreated with 0.1% Calcofluor to allow visualization of the microfibrils synthesized during the EGTA treatment. See below.) However, we did observe that EGTA treatment appeared to curtail or entirely retard the fusion of vesicles with the plasma membrane (Figs. 7,8). Longer exposure to EGTA brings about changes in microfibril pattern, but we suspect that these changes may be only a secondary effect of the membrane disruption brought about by the inhibition of vesicle fusion. After 50 minutes of EGTA exposure, the number of intramembranous particles decreases drastically, with only 674.2 ±112.4

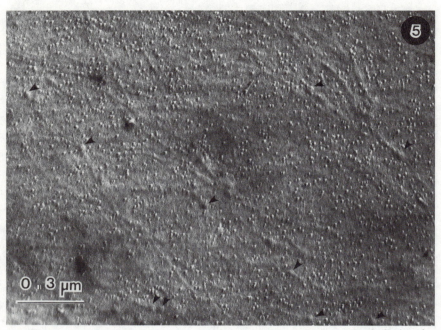

FIG. 5. This is an EF plasma membrane view of a cell from the lower hemicylinder of a coleoptile gravistimulated for 25 minutes. Cellulose microfibril impressions are disordered, and terminal complexes (arrows) are dissociated from their normal clustered distribution (most are separated by more that 50 nm from their nearest neighbor.) A relatively low density of intramembranous particles is observed.

IMPs/μm^2 (control = 995.7 $\pm$13.5 IMPs/μm^2); in addition, there are actual holes in the membrane, suggesting that not only wall but also membrane synthetic processes have been adversely affected. Effects of prolonged EGTA treatment are not reversible once membrane integrity is destroyed. The primary and specific means by which EGTA exposure brings about the inhibition of gravitropism in oat coleoptiles is probably not through the mediation of cell wall synthesis. However, in using EGTA and impeding vesicle fusion, it has been possible to discover more about the relative numbers of Golgi-derived vesicles destined for the plasma membranes of upper vs. lower hemicylinders. As would be predicted by our knowledge of the relative growth rates of the opposite hemicylinders, the lower hemicylinder (Fig. 7) is the recipient of a far greater number of vesicles than is the upper hemicylinder (Fig. 8). This supports Digby and Firn (13) and others who have clearly demonstrated that the lower hemicylinder grows at a greater rate than the upper half and is thus certain to require more building materials.

INHIBITION OF MICROFIBRIL CRYSTALLIZATION

After observing differences in the patterns of cellulose microfibril deposition following gravistimulation, we became interested in the possibility

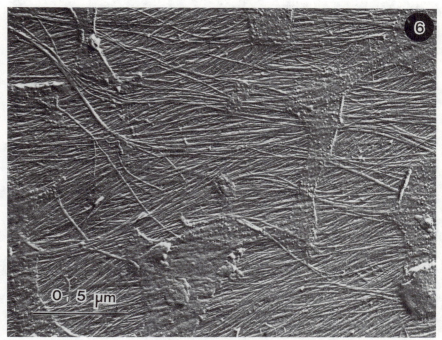

FIG. 6. In this cell from the lower hemicylinder of a coleoptile gravistimulated for 25 minutes, some patches of EF plasma membrane may be seen. In many regions, the plasma membrane has been torn away, and the innermost cellulose microfibril layers are clearly visible. Orientation of cellulose microfibrils is much altered; deposition is clearly not parallel and highly ordered as demonstrated in control plants.

that the TC association/dissociation phenomenon might provide some level of physical mediation of gravitropic curvature. Inhibitor studies have shown that normal plant elongation, *per se*, does not require cellulose synthesis (10). But is this also the case in *differential* growth? To test this question, we reoriented seedlings on agar plates in a bathing solution of the fluorescent brightening agent, Calcofluor, at a concentration of 0.01% (w/v). Calcofluor is known to interfere with the synthesis of microfibrils by preventing inter-glucan chain hydrogen bonding, thus disrupting the process of microfibril crystallization (21, 23). We found that Calcofluor inhibits the gravitropic response (6). Horizontally oriented coleoptiles and roots elongate in the presence of Calcofluor, but neither organ shows its characteristic tropic curvature. Growth rates of plants bathed in Calcofluor are comparable to those seen in control, upright-grown plants, supporting the idea that Calcofluor has no toxic action when added exogenously to intact cells, and also indicating that microfibril assembly is not required for cell elongation. Freeze fracture of Calcofluor-exposed cells reveals no microfibril impressions on the plasma membrane (Fig. 9). Instead, an amorphous, apparently non-crystalline material (Fig. 9) is observed between the older wall layers and the plasma membrane (where cellulose microfibrils would have been expected under normal conditions) in cells of both the upper and the lower coleoptile hemicylinders. We believe this

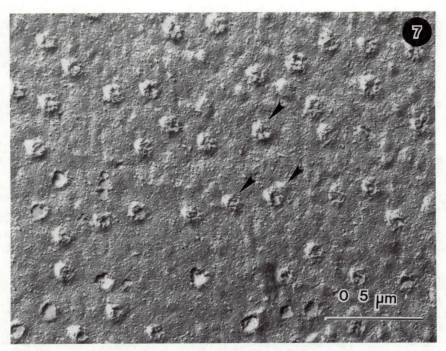

FIG. 7. This micrograph is a freeze fracture view from the lower hemicylinder of a coleoptile treated for 30 minutes with 1% Calcofluor (to eliminate old wall layer microfibril impressions), rinsed, then gravistimulated in the presence of 10^{-3}M EGTA for 50 minutes. Note the large number of vesicles (a few are designated with arrows) which have accumulated at the plasma membrane. No microfibril impressions are evident.

amorphous material to be polymerized, but uncrystallized, cellulose (7). The presence of this amorphous material between the plasma membrane and older cell wall layers in cells of the upper hemicylinder suggests that cellulose synthesis does continue in that region of the gravistimulated coleoptile, even though the cells are not actually elongating. If the Calcofluor is rinsed from the coleoptile (using buffer or distilled water), normal gravitropism will occur with curvature visible to the unaided eye within 25 minutes. Reversibility of the inhibitory effect indicates that the *older* wall layers, already synthesized before the application of Calcofluor, have no critical role in the control of gravitropic curvature. Thus, by its disruption of cellulose microfibril crystallization *at the time of synthesis*, Calcofluor inhibits gravitropism. These results strongly imply that crystalline cellulose microfibril assembly is required for differential growth in *Avena* coleoptiles.

ACID pH ALTERS WALL MICROFIBRIL PATTERN

After observing that disruption of microfibril assembly with Calcofluor inhibits gravicurvature, we became interested in what effects other agents known to affect gravitropism might have on microfibril orientation. Thus we considered the effects of exogenously added protons, in an effort to mimic the

FIG. 8. The upper hemicylinder of an *Avena* coleoptile gravistimulated in the presence of 10^{-3}M EGTA for 50 minutes following a pretreatment with 1% Calcofluor for 30 minutes shows no microfibril impressions. A few vesicles (V) are observed in the vicinity of the plasma membrane in this EF view. In the lower right corner of the micrograph, older wall layers may be seen.

naturally occurring efflux of protons which is seen to follow gravistimulation in corn roots and coleoptiles. It has been shown by Evans and Vesper (16), Mulkey et al. (32), and others that, during differential growth, the region of the plant organ undergoing elongation (the upper hemicylinder in roots, the lower hemicylinder in stems and coleoptiles) undergoes a localized excretion of protons. Mulkey and Evans (31) demonstrated that differential growth could be induced in corn roots by the unilateral application of agar blocks of low pH. The acidified side of the root grew faster than the side opposite the agar block. In our laboratory, we applied an agar paste at pH 4 to one side of an actively growing, upright coleoptile. Visible curvature of the coleoptile away from the side of acidification occurred within 25 minutes. Tissue samples for freeze fracture were prepared from the acid-treated side as well as the opposite hemicylinder after 5-35 minutes of acid exposure. Freeze fracture of the acidified, more rapidly elongating side revealed disorganized cellulose microfibril orientation (Figs. 10, 11) and dissociated terminal synthesizing complexes (Fig. 10). In appearance, coleoptile cell plasma membranes from the artificially acidified portion of the upright plant were strongly reminiscent of plasma membrane views of cells from the lower hemicylinder of gravistimulated coleoptiles (Fig. 7). Another point of similarity between the acid-treated plant

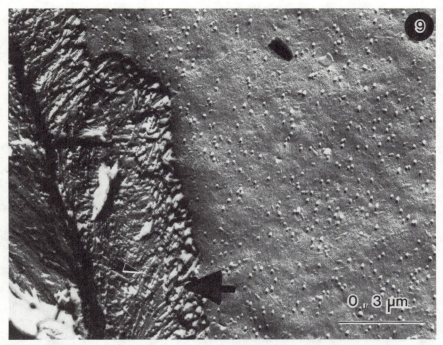

FIG. 9. Freeze fracture of cells from the lower hemicylinder of an *Avena* coleoptile placed in a horizontal position on Calcofluor-containing agar for 20-30 minutes reveals the loss of microfibril impressions (EF). A large arrow designates a region of non-fibrillar material (believed to be uncrystallized cellulose) located between the plasma membrane and the outer, fibrillar wall layers (arrowhead.)

and the gravistimulated plant was the lower density of intramembranous particles observed on the plasma membrane of acidified cells.

Since naturally occurring proton effluxes have been observed following gravistimulation (16, 31), it may be hypothesized that such fluxes may play a key physiological role in directly modulating the establishment of differential growth by regulating aggregation/disaggregation of terminal complexes and thus controlling the character of the newly forming cell wall. Of course, disruption of cellulose synthesis may be only a secondary manifestation of another process which is modulated by protons. It should be stressed, however, that only a change in the load-bearing portion of the wall composite, namely cellulose microfibrils, can permit the curvature that is observed after localized acid application.

SUMMARY

Freeze fracture of non-cryoprotected *Avena* coleoptiles reveals differences in the distribution of terminal, cellulose-synthesizing complexes and in the pattern of cellulose microfibril deposition when epidermal and subepidermal (parenchyma) cells from gravistimulated upper and lower hemicylinders are compared. Cells from the upper hemicylinder are

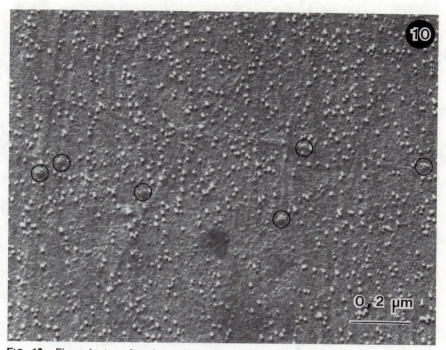

FIG. 10. Five minutes after the application of pH 4 agar paste, this cell of an upright coleoptile shows disruption of microfibril impressions and dissociation of TCs. (More than 50 nm separate TCs from their nearest neighbors.)

indistinguishable from cells of control (upright) plants, having clustered terminal complexes and microfibrils closely appressed to one another and transverse to the longitudinal axis of cell expansion. In contrast, cells from the lower hemicylinder have terminal complexes dissociated from the clustered configuration and cellulose microfibril orientation is more random. We suggest that the close proximity of TCs in normal, upright coleoptiles and gravistimulated upper hemicylinders allows intermicrofibrillar hydrogen bonding and results in the synthesis of a highly ordered, constrictive wall. A greater distance between TCs in the lower hemicylinder precludes hydrogen bond formation at the time of microfibril synthesis, and the resultant wall is less constrictive.

Studies with the cellulose microfibril crystallization inhibitor, Calcofluor, suggest that crystalline cellulose microfibril assembly is required for differential growth in *Avena* coleoptiles. The localized unilateral application of acid (pH 4) induces differential growth and provokes changes in terminal complex distribution and cellulose microfibril orientation similar to those seen in gravistimulated lower hemicylinders. We hypothesize that the naturally occurring efflux of protons following gravistimulation may play a role in modulating aggregation/disaggregation of terminal complexes, thus determining

FIG. 11. Freeze fracture clearly reveals disruption of cellulose microfibril impressions in this upright grown cell which was exposed to unilateral application of pH 4 agar paste for 12 minutes, as compared to the highly ordered, parallel condition common in control plant microfibril impressions. The arrow in the center region of the micrograph indicates the direction of cellulose synthesis prior to the low pH treatment.

the orientation of the nascent cellulose microfibrils and ultimately determining the load-bearing character of the newly formed wall.

Future directions. Differential growth in plants is prompted not only by gravity, but by other stimuli as well. Phototropism is the macroscopic plant growth response whose transduction pathway has been described most extensively in the literature. Some steps in its pathway are common to both gravitropism and phototropism, including the pattern of differential growth (14) and the efflux of protons on the faster growing side of the responding tissue (32). Other differential growth responses await further documentation (magnetotropism, 1; hydrotropism, 27). In no case has the organization of the innermost cellulose microfibril layer been examined. It would be worthwhile to investigate this critical portion of the wall in order to ascertain whether this phenomenon of terminal complex association/dissociation and corresponding microfibril pattern is a common prerequisite for differential growth induced by *any* stimulus.

LITERATURE CITED

1. AUDUS LJ 1960 Magnetotropism: A new plant-growth response. Nature 185: 132-134

2. BANDURSKI RS, A SCHULZE 1985 A working theory for the mechanism of the gravity-induced asymmetric distribution of IAA in the *Zea mays* hypocotyl. Plant Physiol 77: S307

3. BEHRENS HM, D GRADMAN, A SIEVERS 1985 Membrane-potential responses following gravistimulation in roots of *Lepidium sativum* L. Planta 163: 463-472

4. BRANTON D, S BULLIVANT, NB GILULA, MJ KARNOVSKY, H MOOR, K MÜHLETHALER, DH NORTHCOTE, L PARKER, B SATIR, P SATIR, V SPETH, LA STAEHELIN, RL STEERE, RS WEINSTEIN 1975 Freeze etching nomenclature. Science 190: 54-56

5. BROWN DS 1986 Patterns of cellulose microfibril deposition during the gravitropic response in *Avena* coleoptiles. Ph.D. Dissertation, The University of Texas at Austin, Austin, Texas

6. BROWN DS, RM BROWN, JR 1983 The role of cellulose microfibril synthesis in the gravitropic response. J Cell Biol 97: S415

7. BROWN DS, RM BROWN, JR 1985 Cellulose microfibril synthesis: Its role in the gravitropic response. Plant Physiol 77: S58

8. BROWN, JR RM 1981 Integration of biochemical and visual approaches to the study of cellulose biosynthesis and degradation. *In* Chemistry and Biochemistry of Wood-based Processes and Products, Vol. 3, pp 3-15. Proc. Intl. Symp. of Wood and Pulping Chem., Stockholm, 1981

9. BROWN, JR RM 1985 Cellulose microfibril assembly and orientation: Recent developments. J Cell Sci Suppl 2: 13-32

10. BRUMMELL DA, JL HALL 1985 The role of cell wall synthesis in sustained auxin-induced growth. Physiol Plant 63: 406-412

11. CARRINGTON CMS, RD FIRN 1985 Polysaccharide synthesis and turnover in the cell walls of growing and non-growing cells of gravistimulated sunflower hypocotyls. J Plant Physiol 118: 49-59

12. DAYE S, RL BIRO, SJ ROUX 1984 Inhibition of gravitropism in oat coleoptiles by the calcium chelator ethylene glycol-bis-ß-aminoethyl ether)-N,N'-tetraacetic acid. Physiol Plant 61: 449-454

13. DIGBY J, RD FIRN 1979 An analysis of the changes in growth rate occurring during the initial stages of geocurvature in shoots. Plant, Cell Environ 2: 145-148

14. DIGBY J, RD FIRN, S CARRINGTON 1982 Studies on differential growth causing tropic curvature in shoots. *In* Plant Growth Substances 1982, PF Wareing (ed.), pp 519-528. New York: Academic Press

15. EISINGER W, L CRONER, L TAIZ 1983 Ethylene-induced lateral expansion in etiolated pea stems. Kinetics, cell wall synthesis, and osmotic potential. Plant Physiol 73: 407-412

16. EVANS ML, MJ VESPER 1980 An improved method for detecting auxin-induced hydrogen-ion efflux from corn coleoptile segments. Plant Physiol 66: 561-565

17. FOLSOM DB, RM BROWN, JR 1987 Changes in the distribution of terminal cellulose-synthesizing complexes and resultant microfibril patterns observed following gravistimulation of *Avena* coleoptiles. Protoplasma (In Press)

18. GOSWAMI KKA, LJ AUDUS 1976 Distribution of calcium, potassium, and phosphorous in *Helianthus anuus* hypocotyls and *Zea mays* coleoptiles in relation to tropic stimuli and curvature. Ann Bot 40: 49-64

19. GREEN PB 1969 Cell morphogenesis. Ann Rev Plant Physiol 20: 365-394

20. HAIGLER CH, RM BROWN, JR 1986 Transport of rosettes from the Golgi apparatus to the plasma membrane in isolated mesophyll cells of *Zinnia elegans* during differentiation to tracheary elements in suspension culture. Protoplasma 134: 111-120

21. HAIGLER CH, RM BROWN, JR, M BENZIMAN 1980 Calcofluor white alters the *in vivo* assembly of cellulose microfibrils. Science 210: 903-906

22. HEPLER PK, BA PALEVITZ 1974 Microtubules and microfilaments. Ann Rev Plant Physiol 25: 309-362

23. HERTH W 1980 Calcofluor White and Congo Red inhibit chitin microfibril assembly of *Poterioochromonas*: evidence for a gap between polymerization and microfibril formation. J Cell Biol 87: 442-450

24. HERTH W 1985 Plasma membrane "rosettes" involved in localized wall thickening during xylem vessel formation of *Lepidium*. Planta 164: 12-21

25. HERTH W, I HAUSSER 1984 Chitin and cellulose fibrillogenesis in vivo and their experimental alteration. *In* Structure, Function, and Biosynthesis of Plant Cell Walls. WM Dugger S Bartnicki-Garcia (eds.), pp. 89-119. Proc. VII Annual Symposium in Botany

26. IWAMI S, Y MASUDA 1976 Distribution of labelled auxin in geotropically stimulated stems of cucumber and pea. Plant Cell Physiol 17: 227-237

27. JAFFE MJ, H TAKAHASHI, RL BIRO 1985 A pea mutant for the study of hydrotropism in roots. Science 230: 445-447

28. LEDBETTER MC, KR PORTER 1963 A "microtubule" in plant cell fine structure. J Cell Biol 19: 239-250

29. MUELLER SC, RM BROWN JR 1982 The control of cellulose microfibril deposition in the cell wall of higher plants. I. and II. Planta 154: 489-515

30. MUELLER SC, RM BROWN JR, TK SCOTT 1976 Cellulose microfibrils: nascent stages of synthesis in a higher plant cell. Science 194: 949-950

31. MULKEY TJ, ML EVANS 1981 Geotropism in corn roots: evidence for its mediation by differential acid efflux. Science 212: 70-71

32. MULKEY TJ, KM KUZMANOFF, ML EVANS 1981 Correlations between proton-efflux patterns and growth patterns during geotropism and phototropism in maize and sunflower. Planta 152: 239-241

33. PRESTON RD 1974 The Physical Biology of Plant Cell Walls. London: Chapman and Hall

34. RICHMOND PS 1983 Patterns of cellulose microfibril deposition and rearrangement in *Nitella*. In vivo analysis by a birefringence index. J Appl Polym Sci: Appl Polym Symp 37: 107-122

35. RICHMOND PS, JP MÉTRAUX, L TAIZ 1980 Cell expansion patterns and directionality of wall mechanical properties in *Nitella*. Plant Physiol 65: 211-217
36. ROELOFSEN PA 1958 Cell-wall structure as related to surface growth. Some supplementary remarks on multi-net growth. Acta Bot Neerl 7: 77-91
37. ROELOFSEN PA 1965 Ultrastructure of the wall in growing cells. Adv Botan Res 2: 69-149
38. SLOCUM RD, SJ ROUX 1983 Cellular and subcellular localization of calcium in gravistimulated oat coleoptiles and its possible significance in the establishment of tropic curvature. Planta 157: 481-492
39. SRIVASTAVA LM, VK SAWHNEY, M BONNETTEMAKER 1982 Cell growth, wall deposition, and correlated fine structure of colchicine-treated lettuce hypocotyl cells. Can J Bot 55: 902-917
40. STAEHELIN LA, TH GIDDINGS 1982 Membrane-mediated control of cell wall microfibrillar order. *In* Developmental Order: Its Origin and Regulation, pp 133-147. New York: Alan R. Liss, Inc.
41. TANADA T, C VINTEN-JOHANSEN 1980 Gravity induces fast electrical field change in soybean hypocotyls. Plant Cell Environ 3: 127-130
42. VIAN B, M MOSINIAK, J-C ROLAND 1982 Dissipative process and experimental retardation of the twisting in the growing plant cell wall. Effect of ethylene-generating agents and colchicine: a morphogenetic evaluation. Biol Cell 46: 310-310

Physiology of Cell Expansion During Plant Growth, *D.J. Cosgrove and D.P. Knievel* Eds.,
Copyright ©1987, The American Society of Plant Physiologists

THE BACTERIA'S SOLUTION TO THE PROBLEM OF GROWING WITHIN ITS OWN PRESSURIZED SPACE

ARTHUR L. KOCH

Dept. Biology
Indiana University
Bloomington, IN 47405

"... I am, in point of fact, a particularly haughty and exclusive person, of pre-Adamite ancestral descent. You will understand this when I tell you that I can trace my ancestry back to a protoplasmal primordial atomic globule. Consequently, my family pride is something inconceivable..."

Pooh-Bah
in *The Mikado*
by W. S. Gilbert & Sir Arthur Sullivan
1885

INTRODUCTION

Poo-Bah in the Mikado of Gilbert and Sullivan states that he is descended from the first globule of atomic protoplasm. I, too, am proud to make the same claim—but so can your plants and my bacteria. Poo-Bah had the key physical-chemical fact correctly stated—protoplasm not otherwise constrained would form itself into a more or less spherical globule. This can be seen in the coascervates of De Jong and Oparin (36) and in the proteinoids of Fox (4). It is simply a matter of surface tension.

Surface tension forces are seemingly weak. For example, very large soap bubbles are rarely spherical and become distorted by the slightest breeze. For the size typical of bacterial cells these forces can become important, but this works against the most important biological precept: divide and multiply. A soap bubble only grows bigger as more air is introduced and shows no tendency to divide unless an additional mechanical force is applied from the outside. But bacterial cells grow and then split in two. I would like to describe the problem in more detail and outline the solutions that various cell types have adopted as evolution has taken place.

For the more evolutionarily advanced animal cells, the solution depends on contractile proteins. Actin, tubulins, dyneins, etc., are highly evolved substances required for the mitotic apparatus, cytoskeleton formation and muscle contraction. Amoeboid cells (without rigid walls and with contractile proteins) allow mechanical movement and pinocytosis and phagocytosis.

But most plant cells like most bacteria have a strong wall outside of the cytoplasmic membrane. A part of the reason for this structure is that it allows the cells to resist the tendency to swell indefinitely in an aqueous environment in response to osmotic pressure. [Metazoans largely solve this same problem biochemically, using metabolic energy to excrete certain products.] The turgor pressures in plant and bacterial cells can be quite large; this means that the cell walls have to be strong. Cell walls have rigidity as well as strength. The rigidity of the wall allows the maintenance of a non-spherical shape. But this rigidity in turn creates the critical problem of how the cells can grow and divide in spite of the pressure. In order to enlarge, the stress-resisting wall must be split at least locally, but the cell must not rupture. What are the ways that this is done? I will listen carefully to hear details of the higher plant solution presented at this symposium. For my part, I am going to focus on what we can surmise about the first successful living cell and on the strategies of several major types of procaryotes. These archetypal organisms are interesting because they grow and divide without the aid of mechanoproteins and without external scaffolding such as some viruses use in their self-assembly process (9).

THE BACTERIAL STRATEGIES

At present, a large body of microbiological data is available that is consistent with an hypothesis called the *surface stress theory* (SST) (1, 2, 10, 12-29, 33, 43, 44). Its primary premise is that the bacterial cell must lay down new wall material and crosslink it with adjacent older material before the wall is weakened by autolytic cleavage. This requirement should not be surprising--it is just good engineering practice. The bridge isn't laden with traffic until after the suspension is complete. So it is in building bacterial cell wall. But before we go into detail, a word about the stress bearing component of the bacterial wall is necessary.

The peptidoglycan (or murein) is composed of carbohydrate chains of two alternating components, N-acetyl glucosamine and N-acetyl muramic acid (See Ref 41). Attached to the carboxyl group of the side chain of the muramic acid component is a peptide chain. Two peptide chains coming from different oligosaccharide chains are covalently linked together in an unconventional linkage between a carboxyl group of an amino acid and the second amino group of diaminopimelic acid, lysine, or ornithine. This important feature means that the entire wall fabric is held together by covalent bonds and does not depend on hydrogen bonding. This is quite different from the walls of plant cells which depend on H-bonds to maintain the microfibrilar arrangement of the carbohydrate chains which provides the tensile strength of the wall necessary to contain the cell's internal pressure. This means that peptidoglycan is much stronger than chitin or cellulose per unit thickness of fungal or plant cell wall and has little tendency to creep. Aggregates of bacteria do creep like fungal and plant cell walls (5, also see chapter by Cleland) and other materials (34). Of course, strength to resist stretch in each of two dimensions is a highly important advantage, if a high turgor pressure is to be contained by a cell with only a thin tension-resisting layer.

The comparison between a gram-negative heterotroph and a bicycle tire is useful. The turgor pressure inside a gram-negative heterotroph is several atmospheres, say 2 atm (200 kPa) and the stress-bearing wall is probably one molecular layer thick. A bicycle tire, however, holds the same pressure with vulcanized rubber that is 4 or 5 mm thick. In both cases the walls are sufficiently strong without being wasteful. One would surmise from these facts alone that the bonds in the bacterial wall are more efficiently arranged for resisting two-dimensional stress than the random three-dimensional vulcanized elastomer that Harvey Firestone invented.

With such a 'corset' or 'suit of chain-mail' [technically called a sacculus] around the cell, the problem of explaining the 'loosening of the wall' required for cell wall growth is a particularly bothersome one. It means that the cell must have autolytic [hydrolytic] enzymes in order to grow; but it must control them very critically in relation to cell wall building enzymes or the cell will commit suicide.

An understanding about such biologically inappropriate behavior requires a detour to discuss a major class of antibiotics. Many antibacterial compounds act upon the cell wall. Among these are the penicillins which act by blocking the transpeptidases that form the peptide cross-links (See 41). This means that in the presence of penicillins, the cells can not make stress-bearing wall. Of course, the cells do not need to form stress bearing wall if they are not growing. This is the oft stated explanation of failure of penicillins to work when cell growth is also blocked in any other way (say by a different type of antibiotic or inhibitor). This also means that the autolysins are usually so well controlled that no harm is done. But sometimes cell self-destruction does occur. We will return to this problem below.

What kind of enzymes can attack the cell wall? I would insist that the exact nature of the residues of the bacterial murein have changed considerably since procaryotes were invented. Of course, the concept and production of the two-dimensional sacculus was a prerequisite of a successful procaryote with an internal osmotic pressure higher than that of its environment. Today the typical bacterial cell wall is composed of unusual carbohydrates and amino acids in unusual optical configurations. These unique chemical properties are the result of evolutionary history and make them resistant to most of the proteases and glycosidases from eucaryotic sources. Some bacteria are spectacularly sensitive to the lysozymes of avian eggs, tears of mammals, and placental tissue—but this suggests that, though the bacteria adapted to their animal hosts' defenses, lysozyme is a secondary defense.

Bacteria produce their own enzymes which hydrolyze peptidoglycan. At first sight it appears strange that the bacteria have created their own "autolytic" enzymes. But hydrolases are needed to cleave the peptidoglycan to permit wall enlargement. [Although some enzymes, perceived as autolysins, may indeed serve a role directly in this manner in the enlargement of wall, yet others may serve in the degradation of penicillins, and yet others in hydrolyzing the walls of other bacteria to be used as nutrients.] The major types that are assumed to have a role in wall enlargement are amidases, glucosaminidases, and specific endopeptidases.

BACTERIAL PARADIGMS

We can learn about the process of growth and division in bacteria in several ways. One is by studying the biochemistry and identifying the structure of the wall, its biosynthetic and autolytic fate, and by studying the properties of the enzymes involved. Another way is to use the electron microscope (EM) and well chosen fixation and staining procedures. Yet a third general approach is to carry out physiological experiments possibly involving the loss of radioactive label. Finally the biophysical predictions of proposed models of bacterial growth can be explored and compared with the results of the previously listed approaches. We now seem to have a reasonable outline of the growth strategies of four disparate microorganisms.

The first organism that was analyzed in all these ways was *Streptococcus faecium* (25). This is a gram-positive coccus which has a thick wall (the equivalent of 30 layers of murein). The key point of the biochemical studies was that wall, once made, is stable. The key point from the EM studies was that the wall, once made, retained its shape. And the key point from the physiological studies was that new wall was laid down only as invaginating septa developed (and thus wall growth took place in zones). The biophysical analysis assumed that the septum was still plastic and capable of responding to the physical forces as a soap bubble would, while the septum was splitting and septal wall was becoming externalized. The mathematical analysis led to the prediction of the exact shape of the poles observed with the EM. Thus the strategy of this organism is to develop a thick septum by ingrowth, split it, let physical forces bulge it out when it is newly exteriorized, after which its shape freezes. These are the essentials of the process. Of course, there are many details and the following references should be consulted (15, 16, 24).

The second organism examined in this way is *Bacillus subtilis*. It is a gram-positive rod. Its pole is flatter than that of *S. faecsium*, but the mode of pole formation is the same. The more pointed pole of *S. faecium* arises because as the pole septum splits, the murein stretches to enlarge the annuli or surface where new murein is being deposited. The slope of the pole of *B. subtilis* results from the way that a disk can stretch when pressure is applied to one side, given that its perimeter must remain fixed (1, 7, 21, 27, 28, 33, 43, 44). The side wall growth is quite different. Perhaps it should be contrasted with the 'multinet' theory of Roelofsen (37-40). What happens is that a new layer of linked, but unstretched murein is laid down immediately outside the cytoplasmic membrane (CM). As more growth ensues, another new layer is formed between it and the CM, the cell elongates, and the bonds in the existing murein fabric stretch. Later, a given layer becomes stretched to its elastic limit and bears a good deal of the stress. Still later the autolysins cleave the now outermost murein that remains attached with few stress-bearing bonds. Eventually, in most cells, the hydrolysed fragments are discarded into the medium.

The biophysics leading to cylindrical growth requires: i) that the pole perimeter remain constant; ii) that the cell lay down new crosslinked, but relaxed, peptidoglycan all over the cylinder region; and iii) that autolysis of the outermost layer is permitted all over the cylinder region. In this circumstance turgor pressure causes cell elongation and does not lead to bulging or to constriction (1, 15, 16, 18, 22, 23, 27, 28).

For *Escherichia coli*, a gram-negative rod, the analysis is much more difficult, because the wall is so thin (12, 14, 15, 17, 19, 29). There is active discussion in the literature about how thick the murein layer is (8). If it is more than one layer, perhaps the model for the gram-positive rod presented above is followed. But I believe that it is substantially one layer thick and some quite different process is involved. I also believe that this more elaborate strategy was the precursor of apical growth in fungi and plants.

Hyphal fungi and actinomycetes, like plant roots, explore and exploit the soil by growing *through* it. In both cases, the plan is the same, elongate the tip by expanding the previous apical material and adding new at the tip of the apex. This requires a gradient of wall growth decreasing behind the current tip. The organism must orchestrate the loosening, mobilization, and synthetic steps very exactly. This is because the turgor pressure that applies throughout the cytoplasm would cause inappropriate bulges or ruptures if the rates of these processes were not properly balanced.

The gram-negative rod faces a similar problem with its very thin wall. It must enlarge by inserting new oligosaccharide chains all over the cylindrical region through the membrane and then link them via both peptide and glycosyl bonds to the preexisting stress-bearing wall. Then and only then, can previously stress-bearing wall be judiciously cleaved. Under such conditions the biophysics described for the side wall of the gram-positive rod will apply and the process of elongation without (much) change in diameter can be accomplished. However gram-negative cell division cannot follow the 'split and stretch' process of the gram-positive organisms because the wall is so thin and nothing like the septum of gram-positive organisms has ever been found in gram-negative bacteria. I proposed that what the cell does to achieve division is the same thing that organisms with hyphal morphology do to extend the axial tip; i.e., change the rules locally. The biophysical analysis shows that the cell need only continue in an exacerbated way the insertion/cleavage process in the region where the cell division is to occur. Wall enlargement leads to constriction and to cell division if there is an alteration in the biochemical energetics of external wall formation. Only a two-fold decrease in the quantity that corresponds to surface tension is required to cause division. I do not know how this is achieved, but we have listed a number of possibilities (19).

In summary of this section: bacteria come in different shapes and wall thicknesses and have accordingly different growth strategies. All the strategies keep the wall strong while it is being enlarged. The simplest type of cell considered above (the gram-positive coccus, *S. faecium*) requires only that the cell is able to cause a septum to develop at the right place and split it. The next simplest (the gram-positive rod, *B. subtilis*) requires that in addition, a cylindrical region is made by laying down new complete layers and "molting" outer layers analogously to a tree shedding bark. Still more complex (the gram-negative rod, *E. coli*) is the gram-negative strategy of inserting new wall and cleaving locally stressed walls, but with different energetics at the point where constriction is to take place. These strategies depend on both extant poles being sufficiently rigid so that new growth maintains the diameter of the organism in the space between. Finally, the actinomycetes (and certain other procaryotes) grow apically. They must do this by balancing the rates of several processes in order to maintain a constant radius. We have no clue as to how spirilla shapes

78

arise. However, I doubt that this morphology results from the chemical structure of the peptidoglycan building blocks, because the murein is too amorphous and irregular.

NEW MEASUREMENTS OF PRESSURE IN BACTERIA

The problem to be discussed in this section has been long studied in plant physiology, but not in bacteriology (6, 7, 35, 47). This is in large part because of a difference in scale. With advances in miniaturization, it has been possible to measure the turgor in giant algal cells and higher plant cells (see chapters by Cosgrove and by Hsiao), but it is still not possible to cannulate a bacterium. It is also not easy to measure the size of a living bacterium. Fixation for electron microscopy causes size changes and phase microscopy has large errors due to inherent light halos and the Heisenberg uncertainty principle. But now it is possible to make these important measurements. I will present our current studies of the pressure of *Ancylobacter aquaticus* and of strain in *E. coli* (J. Bacteriol. 166: 1979-1984; 169: 3654-3663; 169: 4737-4742).

The first to estimate the osmotic pressure of bacteria were Mitchell and Moyle in 1956 (32). They took bacterial pellets, dried them as films and then exposed them to different humidities to measure how much water was taken up. The humidity was provided by solutions of known concentrations and osmotic pressure. Then it was a simple matter of applying van't Hoff's law to calculate the osmotic pressure of the cells. But the leap to asserting that this had been the osmotic pressure when the bacteria had been growing is a large one. Later studies were not much more reliable (see review in Ref. 17). Basically, they depended on measuring the volume of cells in a suspension. These measurements could be done by dilution experiments with probes that cannot penetrate certain cell boundaries, such as dextrans labeled with a blue dye. More sophisticated studies used radioactive probes and several probes such as heavy H_2O, sucrose, and inulin (i.e., the poly fructose from the Jerusalem Artichoke) simultaneously. These probes each penetrated the cell to different degrees. Inulin cannot pass the outer membrane or through the murein; sucrose can enter the wall and periplasmic space; and water enters all compartments. Osmotic pressure was measured by the effect of high external sucrose concentrations in causing plasmolysis. But all of these kinds of studies suffered from the fact that they could only be applied to dense suspensions of bacteria that certainly could not have a normal physiology.

Our approach to measuring the osmotic pressure of bacteria was to apply the method devised by Walsby (46) to a gram-negative heterotroph. This method depends on studying organisms with gas vesicles. These are hollow capsules with conical ends made by certain aquatic microorganisms, ostensibly to cause them to rise in the water column to regions where growth conditions are optimal. These vesicles are composed of a special protein with an amino acid arrangement such that when the subunits assemble, the inner surface is so hydrophobic that water leaves and gases at the partial pressure of the ambient fluid replace them. These vesicles can be collapsed by applying a critical external pressure, the collapse pressure of a vesicle, C, which varies from vesicle to vesicle. Since there are very many vesicles per cell the average C in each cell is nearly the same as for other cells in a population of organisms. Of course, this need only apply in the steady state of balanced cell growth. The hydrostatic pressure applied by the experimenter adds to the turgor pressure of the cells in leading to the collapse of the vesicles. Thus, if the turgor pressure, P_t, changes there will be a complementary change in C_a, the applied pressure [measured above atmospheric pressure] that causes collapse of a specified proportion of the vesicles.

Walsby mainly studied blue-green algae. But our interest is in the turgor pressure of *E. coli*, since this is the proper, and timely, study of a bacterial physiologist. Unfortunately, this organism has no gas vesicles. Nicole Tandeau de Marsac has cloned the vesicle protein gene into *E. coli* (42), but has not yet induced vesicle assembly in the new organism. While we are waiting for her to succeed, we have developed refined methods and tested them using the gram-negative heterotroph, *Ancylobacter aquaticus*. Our extensions and improvements of Walsby's methods are aimed in two directions. First, we have arranged our pressure delivery system in such a manner that we can watch under the phase microscope while we apply hydrostatic pressure, as Klebahn had done earlier (11). Second, we have developed a spectrofluorometer as a pressure nephelometer or turbidimeter. The advantage of the latter over previously available equipment is that ours responds mainly to the light scattered by the vesicles. The vesicles are small and have an index of refraction quite different from the cytoplasm and scatter light according to Rayleigh's law. On the other hand, the $90°$ light scattering signal responds only slightly to the cells themselves. This is because the cells are of a dimension comparable to the wavelength of light and have an index of refraction nearly equal to that of the surrounding medium and scatter light more nearly in the forward direction. Thus our signal to background level is much more favorable than with the previously used non-collimated systems. We have also introduced a number of mechanical and computer-linked features to improve the accuracy and speed of measurement.

With the microscope technique, we were able to demonstrate that the turgor pressure does not change systematically with cell size. In these studies, we recorded the applied pressure that first caused a detectable decrease in the refractility. We chose this measure because when a vesicle collapses there is an instantaneous reduction in the turgor pressure which will influence the collapse of other vesicles within the same cell. Therefore this is much better than measuring the pressure that gives rise to the collapse of 50% of the vesicles. However, this criterion of the first detectable decrease is a statistically poorer measure of turgor pressure because a particular cell may, by chance, have some vesicles that are weaker or stronger than does the typical cell. Thus our procedure confounds the turgor pressure measurement with the variability of the collapse pressure of the weakest vesicles that happen to be within a cell.

While we cannot avoid this problem, we can estimate its magnitude. There are roughly 100 vesicles in a cell and the human eye first responds when there is a decrease of about 5% of light intensity. A Monte Carlo simulation on a computer was carried out. Since the collapse pressure varies from vesicle to vesicle with a standard deviation (SD) of 90 kPa, the simulation showed that the SD of C_a from cell to cell due to the random sampling of vesicles within cells would be about 19 kPa with this 5% criterion. The SD of the mean of the collapse pressures within the cells similarly is about 9 kPa. Actually, we found that over the central two-fold range of cell sizes abstracted from our population of cells in balanced growth, the mean value of C_a varied only 7 kPa. The SD of all actual collapse pressure measurements of single cells was 42 kPa. This means that the turgor pressure varies quite a bit from cell to cell but not in a consistent way with cell growth. We will return to this issue below.

With the aid of our pressure nephelometer we can make fast, continuous measurements while the pressure is progressively increasing. The pressure nephelometer provides a way to estimate the turgor pressure, but the evaluation of P_t is dependent on knowing the value of C. With this knowledge, the turgor pressure can be calculated from:

$$P_t = C - C_a \qquad [1]$$

The problem however is that C is unknown. Walsby's solution to this dilemma was to repeat the measurement after the addition of 0.5 osmolal sucrose. The osmotic pressure of the sucrose (1260 kPa) is surely much greater than the turgor pressure of the cells. He assumed that this would lead to plasmolysis. Thus this method is based on the assumption that such an osmotic treatment will reduce the turgor pressure to exactly zero without changing C. We have spent much time and energy in trying to validate this methodology. A number of pitfalls are possible:

1). The cells may recover their turgor pressure by pumping in potassium ions, glycine, betaine, proline, etc. These processes commonly occur in bacteria, higher microorganisms and plants. The cells may have partially recovered their turgor pressure during the measurement procedure.

2). The pressure that collapses the vesicles may be altered by changes in the dehydrated environment around them as water is sucked out of the cell by the increased external osmotic pressure.

3). Some vesicles may show a delay in collapsing, instead of collapsing immediately when a sufficiently high pressure is applied.

4). Variability in the turgor pressure from cell to cell may be so great that the turgor pressure of all cells might not be decreased to zero by a given osmotic challenge.

5). After the external osmotic pressure is made greater than the internal pressure, the efflux of water will lead to an increase in the internal osmotic pressure. This would partially ameliorate the effect of added solute.

6). After the external osmotic pressure has caused the removal of water and the cell has shrunk to the point that the sacculus is no longer stretched, the turgor pressure may actually become negative. This could happen if the sacculus does not shrink below its stress-free state or plasmolysis does not occur. In fact, plasmolysis does ensue and the cell membrane retracts from the peptidoglycan layer, *but not instantaneously*. It may not happen soon or completely enough and thus cause an overestimation of the turgor pressure.

7). The cytoplasm of bacteria appears to be divided into two regions (45). One part contains the ribosomes and has a higher density of cell substance than the region of the cell that contains the chromosomal material. These regions may differ in their physical state and one or both may be gels and therefore have mechanical properties that may resist compression. It may be possible that a negative turgor pressure might remain in some regions even after the wall bears no stress and some visible plasmolysis has taken place.

8). The vesicles may be reformed or formed *de novo* while the measurements are being performed.

To estimate the importance of these eight possibilities, control studies in which the rate of pressure increase, time of treatment with sucrose, concentration of sucrose, use of other osmolytes, etc., led to the conclusion that some of our fears were unwarranted and all could be circumvented. We found that within a range of rate constants for pressure increase and within a range of sucrose concentrations, the estimate of C was approximately correct. However, most of the possibilities raised above actually occur and can be important under some circumstances. Thus while we believe that we have estimated C correctly in our case, the same procedure might not suffice for other physiological conditions with the same organism and might be totally inappropriate for another.

A critical factor in making this judgement is the result of experiments with a penicillin type antibiotic, ampicillin. With this antibiotic at the high concentration of 500 µg/ml, we found that the cells started to lyse after 20 min. The vesicles liberated thereby were soon destroyed and we presume that their physical properties changed as this took place. We therefore prepared collapse curves at intervals between 0 and 22 min. The pattern was clear. At intermediate times the curves were made up of two components in proportions that changed progressively. Component I was the only component at zero time and the value at the 50% point of the collapse curve was 290 kPa. As the proportion of Component I decreased, the value of C_a became progressively lower. This is consonant with the idea that the turgor pressure inside the cells was gradually increasing as further active transport and anabolism was leading to the accumulation of more substance under conditions where no new stress-bearing wall could be added. However, at some point the cell wall ruptured and the turgor pressure was reduced to zero. The vesicles in such cells, along with the vesicles subsequently released from such cells, exhibited a C_a of 478 kPa. These vesicles constitute Component II. This confirms our estimate of C since this value is similar to the 480 kPa estimated from our data fitted to experiments where the sucrose osmolality varied from 0 to 1 (see note in proof).

We intend to extend our measurements to other heterotrophic bacteria with gas vesicles and for each, choose a range of conditions with external osmolytes and compare the fit to the C predicted by this antibiotic procedure. But in summary for the work with *A. aquaticus*, we believe that the following description accounts for our observations: the wall is elastic and as the external osmotic pressure is raised the cell shrinks. After about a 43% decrease in volume, the strain on the sacculus is zero. With higher external osmolyte, the cells briefly have a negative turgor pressure and then plasmolyze with a return to zero hydrostatic pressure [above atmospheric]. The cells compensate fairly quickly by accumulating intracellular solutes, and this effect influences the observed value of C_a appreciably up to external concentrations of about 0.4 osmolal. Beyond 0.4 osmolal, the change in C_a is rate-limited by the solute accumulation process. At very high osmolyte concentrations, the dehydration of the cell contents also leads to an increase in C. We concluded that the relationship between the turgor pressure and the strain in a linear dimension of

82

the sacculus obeys Hooke's law. Consequently the volume of the cell is proportional to $(1 + K \cdot P_t)^3$ where K is related to the reciprocal of the Young's modulus of the murein.

THE STRAIN IN THE WALLS OF BACTERIA
DUE TO TURGOR

The bacterial cell wall, being a covalently closed network has physical properties resembling a knitted fabric or a sheet of rubber (34). It can be stretched, and resist stress in the two dimensions of a plane. It is elastic up to a point and then if stretched further must break. It has little ability to resist forces normal to their surface. Cellulose molecules on the other hand have excellent resistance to stretch in one dimension, and give only slightly in an elastic manner in this dimension. In other dimensions cellulose can be easily deformed. When a piece of dry wood is broken, only a few covalent bonds of the cellulose chains are ruptured. Rather it is hydrogen bonds between chains that break and reform, allowing sliding motions before rupture. Thus with tension in the direction of the chains, it is the breaking of the hydrogen bond holding non-overlapping chains that actually causes the more macroscopic breaks that one first measures as "creep" (see chapter by Cleland). Of course, once the structure is weakened, stress is transferred to adjacent material and consequently it is more likely that this adjacent material will rupture. This starts a chain reaction causing the creation of cracks and fissures. Consequently, ripping and tearing events are more likely to start where there is a defect in the starting material.

It takes several orders of magnitude more force to break covalent bonds mechanically than to break the collection of hydrogen bonds, even though the total binding energy may be the same. This is under the proviso that the stress is applied continuously and not as an impulse. Clearly this is important for the design of bacterial walls. The quite different construction of the bacterial wall means that certain phenomena present in plant and man-made material, such as 'creep', can not occur. Though it has never been properly measured, creep must be much less prominent and more rapidly and fully reversed in procaryotic saccular material than in higher plant walls. This could only be determined if the mechanical properties of a wall of a single cell could be examined.

Our approach has been to study the strain produced by the hydrostatic pressure within a single cell. The engineering of stress-strain relationships in closed containers of a variety of shapes is well understood (3). Of course, the pressure affects the wall equally in all regions of the cell. This tends to move the thin wall outward with respect to the dimensions of the cell, until sufficient stress develops within the plane of the wall to counter it. The stress depends on the curvature at any point. The important point for our purposes is that the stress is also proportional to the pressure. Consequently, although the fabric is stressed in the two dimensions differently depending on the morphological aspects of the situation, it is stressed in direct proportion to the turgor pressure. At low hydrostatic pressures the expansion in the two dimensions will be independent of each other and therefore the surface in each linear dimension will expand proportionately to the pressure [and Hooke's law will be obeyed]. This leads to the relationship stated above that the volume will rise as the third power of $(1 + K \cdot P_t)$. But like the Chinese finger puzzle, once the fabric is extended, the stress in one direction affects the stress in the other direction.

Molecular models show that by rotation around single bonds the peptidoglycan could extend over vastly different areas. A sheet made of fully cross-linked peptidoglycan, of the type produced by *E. coli* or *B. subtilis,* can expand in the direction of the peptide 'pleats' by four-fold. The degree of expansion in the direction of the carbohydrate chains is small. Because the native structure is not regular, nor completely cross-linked, nor entirely in the plane of the surface, a still larger expansion is possible. My analysis of the available analytical and electron microscopic data is that after fixation for electron microscopy the structure of the *E. coli* wall is approximately that of a single compact layer. This conclusion may be revised when certain studies now in progress are completed.

We utilized the rectangular capillary technique described above to measure the degree of stretch of the cell in growing bacteria. Basically the organisms attached themselves to a coating of polylysine on the glass surface. They adhered strongly enough so that we could flush the growth medium away and replace it with the same medium containing a detergent; e.g., dodecyl sulfate. When the cell membrane(s) were breached, the cellular constituents leaked out. This was signaled by a loss in phase contrast. As phase contrast was lost the cell wall structure shrank.

In order to measure the change in length accurately, a sufficiently large object was needed, i.e., several times the wavelength of the light employed. This could be easily achieved for *E. coli* by using temperature sensitive mutants that form long filaments without partitions between cellular units. We have also studied the large microorganism, *B. megaterium* utilizing cetyl trimethyl ammonium bromide as the membrane disrupting agent. In both systems the cells shrank about 17%. Although numerically the same, the implications are quite different. For the gram-positive cylinder, the degree of shrinking represents a compromise between the innermost and outermost portions of the wall which are almost stress free and the central part that should shrink a great deal in achieving a relaxed conformation.

This is an interesting result considering that scale models for peptidoglycan can expand up to 400% without breaking covalent bonds. Samples of wall fragments of gram-positive wall can be caused to expand more than 200% by altering the pH so that there are only charges of the same sign in the wall's molecular structure. Then the electrostatic repulsion between like charges causes expansion (30). Our studies with the sacculi of *E. coli* (Woeste and Koch, in preparation) gave similar results. It will remain for future studies to quantitate the actual stress-strain relationships in the walls of bacteria as they exist *in vivo*. A full understanding will assess the contribution of the inner (and outer) membranes. It may turn out that the gram-positive organism lays down its murein in a very compact form and expands it almost to the breaking point in the poles and past this point for sidewall growth, while the gram-negative only permits its wall to be expanded by about 20% before new wall is inserted.

SUMMARY

Successful life forms have a higher concentration of colligative particles in their cytoplasm than is present in their environment. Many ways of accomodating the osmotically-caused turgor pressure have been found and used by extant organisms. The bacteria, like plants, have a stress-resisting wall; however, unlike plants, bacteria do not have cytoskeleton elements within them. They rely on other mechanisms than do animal cells to maintain non-spherical shapes as they grow and divide. These approaches depend on the stable properties of mechanical systems designated 'shell' structures in engineering terminology. The cell selects the sites of wall formation, the circumstance of the application of stress to new wall, and formation, localization, and activation enzymes for cleavage of stress-bearing wall. Thus the cell arranges the biophysical circumstance to advance its own growth and division. This article reviews biophysics and microbial physiology of bacterial cell walls, as well as our current studies of the turgor pressure and strain in the walls of growing procaryotes.

Note added in proof: Recent experiments have shown that Component I cells are not precursors of Component II cells. Rather the latter represent cells that lose turgor pressure quickly and the former are cells that resist the ampicillin and develop and maintain a higher turgor pressure.

LITERATURE CITED

1. BURDETT IDJ, AL KOCH 1984 Shape of nascent and complete poles. J Gen Microbiol 130: 1711-1722
2. DOYLE RJ, AL KOCH 1987 The function of autolysins in the growth and division *Bacillus subtilis*. CRC Crit Rev Micro 15(2): 169-222
3. FLÜGGE W 1973 Stresses in Shells. Springer Verlag, Berlin
4. FOX SW, S YUYAMA 1963 Abiotic production of primitive protein and formed microparticles. Ann NY Acad Sci 108: 487-494
5. GAMOW RI 1980 Phycomyces: mechanical analysis of the living cell wall. J Expt Bot 123: 947-956
6. GREEN PB, FW STANTON 1967 Turgor pressure: direct manometric measurement in single cells of *Nitella*. Science 155: 1675-1676.
7. HALL JL 1983 Cells and their organization: current concepts. *In* FC Steward, ed, Plant Physiology, Vol. I Academic Press, NY, pp. 3-156
8. HOBOT JA, E CARLEMALM, V WERNER, E KELLENBERGER 1984 Periplasmic gel: new concept resulting from the reinvestigation of the bacterial cell envelope ultrastructure by new methods. J Bacteriol 160: 143-152
9. KING, J 1980 Regulation of structural protein interaction as revealed in phage morphogenesis. *In* R. F. Goldberger, eds., Biological Regulation and Development, Vol 2 Plenum, New York, pp. 101-132
10. KIRCHNER G, AL KOCH, RJ DOYLE 1984 Energized membrane regulates cell pole formation in *Bacillus subtilis*. FEMS Microbiol Letts 24: 143-147

11. KLEBAHN H 1922 Neue Untersuchungen über die Gasvakuolen. Jb Wiss Bot 61: 535-589
12. KOCH AL 1982 On the growth and form of *Escherichia coli*. J Gen Microbiol 128: 2527-2540
13. KOCH AL 1982 The shape of the hyphal tips of fungi. J Gen Microbiol 128: 947-951
14. KOCH AL 1983 The shapes of Gram-negative organisms: Variable T model. *In* R. Hakenbeck, J. Höltje, and H. Labischinski (eds.) The Target of Penicillin. deGruyter and Co., Berlin, pp. 99-104
15. KOCH AL 1983 The surface stress theory of microbial morphogenesis. Adv Microbial Physiol 24: 301-366
16. KOCH AL 1984 How bacteria get their shapes: the surface stress theory. Com Mol Cell Biophys 2: 179-196
17. KOCH AL 1984 Shrinkage of growing *Escherichia coli* cells through osmotic challenge. J Bacteriol 159: 914-924
18. KOCH, AL 1985 How bacteria grow and divide in spite of internal hydrostatic pressure. Can J Microbiol 31: 1071-1083
19. KOCH AL, IDJ BURDETT 1984 The variable T model for Gram-negative morphology. J Gen Microbiol 130: 2325-2338
20. KOCH, AL, IDJ BURDETT 1986 Normal pole formation during total inhibition of wall synthesis. J Gen Microbiol 132: 3441-3449
21. KOCH, AL, IDJ BURDETT 1986 Biophysics of pole formation of Gram-positive rods. J Gen Microbiol 132: 3451-3457
22. KOCH, AL, RJ DOYLE 1985 Inside-to-outside growth and the turnover of the Gram-positive rod. J Theoret Biol 117: 137-157
23. KOCH, AL, RJ DOYLE 1986 The growth strategy of the Gram-positive rod. FEMS Microbiol Rev 32: 247-254
24. KOCH, AL, ML HIGGINS 1984 Control of wall band splitting in *Streptococcus faecalis* ATCC 9790. J Gen Microbiol 130: 735-745
25. KOCH AL, ML HIGGINS, RJ DOYLE 1981 Surface tension-like forces determine bacterial shapes: *Streptococcus faecium*. J Gen Microbiol 123: 151-161
26. KOCH AL, ML HIGGINS, RJ DOYLE 1982 The role of surface stress in the morphology of microbes. J Gen Microbiol 128: 927-945
27. KOCH AL, G KIRCHNER, RJ DOYLE, IDJ BURDETT 1985 How does a *Bacillus* split its septum right down the middle? Ann Microbiol (Inst. Pasteur) 136A: 91-98
28. KOCH AL, HLT MOBLEY, RJ DOYLE, UN STREIPS 1981 The coupling of wall growth and chromosome replication in Gram-positive rods. FEMS Microbiology Letters 12: 201-208
29. KOCH AL, RWH VERWER, N NANNINGA 1982 Incorporation of diaminopimelic acid into the old poles of *Escherichia coli*. J Gen Microbiol 128: 2893-2898
30. MARQUIS RE, EL CARSTENSEN 1973 Electrical conductivity and internal osmolarity of intact bacterial cells, J Bacteriol 113: 1198-1206
31. MENDELSON MH, D FAVRE, JJ THWAITES 1984 Twisted states of *Bacillus subtilis* macrofibers reflect structural states of the cell wall. Proc Natl Acad Sci USA 81: 3562-3566

32. MITCHELL P, J MOYLE 1956 Osmotic function and structure of bacteria. Symp Soc Gen Micro 6: 150-180
33. MOBLEY HLT, AL KOCH, RJ DOYLE, UN STREIPS 1984 Insertion and fate of cell wall in *Bacillus subtilis*. J Bacteriol 158: 169-179
34. NEWMAN FH, VHL SEARLE 1957 The General Properties of Matter. Fifth ed. Edward Arnold, London
35. NOBEL PS 1983 Biophysical Plant Physiology and Ecology. Freeman, San Francisco
36. OPARIN A 1953 The Origin of Life. (republication of 1938 ed) Dover Publication, NY
37. PRESTON RD 1974 The Physical Biology of Plant Cell Walls. Chapman and Hall, London
38. ROELOFSEN PA 1951 Orientation of cellulose fibrils in the cell wall of growing cotton hairs and its bearing on the physiology of cell wall growth. Biochim Biophys Acta 7: 43-53
39. ROELOFSEN PA 1965 Ultrastructure of the wall in the growing cell and its relation to the direction of growth. Adv Bot Res 2: 69-148
40. ROELOFSEN PA, AL HOUWINK 1953 Architecture and growth of the primary cell wall in some plant hairs and in *Phycomyces* sporangiophores. Acta Bot Neerl 2: 218-225
41. ROGERS HJ, JR PERKINS, JB WARD 1980 Microbial Cell Walls and Membrane. Chapman Hall, London
42. TANDEAU DE MARSAC N, D MAZEL, DA BRYANT, J HOUMARD 1985 Molecular cloning and nucleotide sequence of a developmentally regulated gene from the cyanobacterium *Calothrix* PCC 76011: a gas vesicle protein gene. Nucl Acids Res 123: 7223-7236
43. SONNENFELD EM, TJ BEVERIDGE, AL KOCH, and RJ DOYLE 1985 Asymmetric distribution of charges on the cell wall of *Bacillus subtilis*. J Bacteriol 163: 1167-1171
44. SONNENFELD EM, AL KOCH, RJ DOYLE 1985 Cellular location of origin and terminus of replication in *Bacillus subtilis*. J Bacteriol 163: 895-899
45. VALKENBURG JAC, CL WOLDRINGH, GJ BRAKENHOFF, HTM van der VOORT, N NANNINGA 1985 Confocal scanning light microscopy of the *Escherichia coli* nucleoid: comparison with phase-contrast and electron microscope images. J. Bacteriol. 161: 478-483
46. WALSBY AE 1986 The pressure relationships of halophilic and non-halophilic prokaryotic cells determined by using gas vesicle as pressure probes. FEMS Microbial Rev 39: 45-49
47. ZIMMERMANN U 1978 Physics of turgor and osmoregulation. Annu Rev Pl Physiol 29: 121-148

Physiology of Cell Expansion During Plant Growth, *D.J. Cosgrove and D.P. Knievel* Eds.,
Copyright ©1987, The American Society of Plant Physiologists

LINKAGE OF WALL EXTENSION WITH WATER AND SOLUTE UPTAKE

DANIEL J. COSGROVE

*Department of Biology, Penn State University,
University Park, PA 16802*

INTRODUCTION

When a growing cell expands, its mass increases primarily by filling of the vacuole with water and solutes. This process requires close coordination between wall yielding and extension on the one hand and water uptake on the other; sustained growth also requires solute uptake or synthesis to prevent excessive dilution of intracellular solutes. This chapter will deal primarily with the linkage between the mechanical and transport processes that constitute cell expansion.

Mechanistically, we think of wall expansion, water uptake and solute absorption as distinct processes, so the question naturally arises, how are these processes coordinated? In this review, I want to argue for wall relaxation as the linchpin which links these separate but essential growth processes.

As a physical structure, the cell can be viewed as a pressurized, fluid-filled sack restrained from expansion by a tough yet flexible wall. The mechanical strength of the wall allows the cell to develop by osmosis a high turgor pressure, which in turn produces a tensile stress in the plane of the wall (Fig. 1).

Both water uptake and wall yielding (irreversible wall expansion) are needed for growth, and this raises the proverbial chicken-and-the-egg question, which comes first? Two opposing views have been advocated. One view (14) has been that growth starts with an initial uptake of water, which increases cell turgor and raises wall stress beyond the elastic limit of the wall. An elastic instability then results in a yielding of the wall and a consequent reduction in wall stress and cell turgor back to the starting point. An opposing view (24) holds that growth starts with a modification of the cell wall, leading to a reduction, or *relaxation*, of wall stress (see Fig. 1 for explanation). Because wall stress is the reactionary force to turgor pressure, stress relaxation of the wall lowers cell turgor pressure and water potential and thereby induces a water influx by osmosis.

Until recently there was little experimental evidence to test these opposing hypotheses. One critical distinction between them arises from the predicted behavior of a growing cell when it is deprived of water. According to the first theory, turgor will remain high because the mechanical instability of the wall requires a preceding water uptake and turgor increase. In contrast, the

88

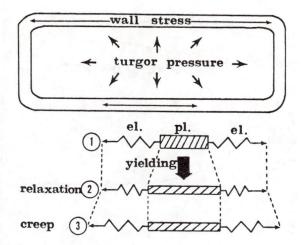

FIG. 1. TOP: Water absorption by a cell stretches the wall. In resisting such stretch, the wall compresses the protoplast, thus giving rise to turgor pressure, and at the same time developing an internal tensile stress in the plane of the wall. **BOTTOM**: In a growing cell wall, the stress is borne by elastic (el.) and plastic (pl.), or irreversibly deformable, elements in series (point 1). Yielding of the plastic elements gives rise either to a relaxation of stress if the wall dimensions are held constant (point 2) or to extension (creep) if the stress is maintained constant (point 2). Note that during stress relaxation, the elastic elements, modelled as springs, contract to the same extent that the plastic elements extend. Such contraction is the reason for reduction or relaxation in stress.

second theory predicts that wall relaxation will reduce cell turgor. That is, turgor of growing cells should decay when water uptake is prevented. How do plant cells actually behave under such circumstances? As shown in recent studies (5, 7, 10, 28), they exhibit a substantial relaxation in both turgor and wall stress, thus refuting the first hypothesis and confirming the second.

KINETICS OF STRESS RELAXATION *IN VIVO*

Some of the earliest observations of relaxation in growing cells were concerned with induced errors in water potential measurements (1, 11). More recently, *in vivo* stress relaxation[1] has been developed as a tool in its own right for studying the mechanism and control of plant cell growth (5, 6, 7).

In general terms, the rate and magnitude of relaxation provide quantitative ways to characterize the wall loosening process. As shown in Fig. 2, a growing pea stem shows a relaxation in turgor pressure of about 3 bar when water is withheld. Relaxation is initially rapid, but slows down and eventually ceases (or nearly ceases) when turgor reaches 2 to 3 bar. Such behavior accords with the usual physical view of wall extension in which the rate of extension is a partial function of wall stress (see chapters by Ray and by Cleland). Because

---

[1]The term *in vivo stress relaxation* was coined to distinguish it from stress relaxation of isolated wall specimens held in mechanical clamps at constant strain, as in the work by Masuda and coworkers (18). The important differences between these techniques are discussed by Cosgrove (5, 6).

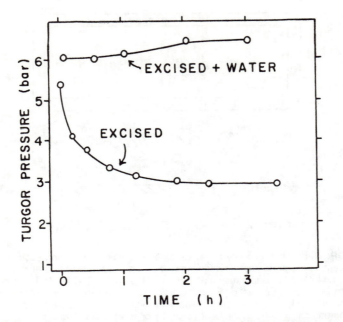

FIG. 2. Time course for stress relaxation *in vivo* of excised pea stems, as measured by the pressure probe technique (5). At time zero, a growing segment was excised, held under nontranspiring conditions, but deprived of external water. Each point represents the average pressure of 8-10 cells. The control was treated in the same fashion except that after excision a drop of water was placed on the cut surfaces.

wall stress is difficult to measure and is a complex function of cell geometry, it is common practice to describe wall extension in equivalent terms of turgor pressure (P, which is proportional to wall stress) with the relation (6, 24):

$$dV/Vdt = \emptyset \, (P - Y) \qquad \text{Eq. 1}$$

where dV/Vdt is relative growth rate in volume, $\emptyset$ is the apparent yield coefficient of the wall (often called extensibility), and Y is the minimum turgor for growth (the yield threshold). The conventional way to measure cell wall growth properties has been to vary P, measure growth rate, and calculate $\emptyset$ and Y by the equation above (3, 19, 27).

As depicted in Fig. 3, $\emptyset$ and Y may also be determined from the kinetics of relaxation. Assuming the wall properties in equation 1 remain constant, turgor should decrease with time according to:

$$P_t = (P_0 - Y) \exp(-\emptyset \varepsilon t) + Y \qquad \text{Eq. 2}$$

where P_0 and P_t are the turgor pressures at time zero and time t, and ε is the volumetric elastic modulus of the cell (5, 6). Equation 2 is related to equation 1, and says that turgor should decay exponentially to the yield threshold (Y) with a time constant of $-\emptyset \varepsilon$. As Fig. 2 illustrates, growing pea stems exhibit stress relaxation approximating the ideal behavior represented in equation 2.

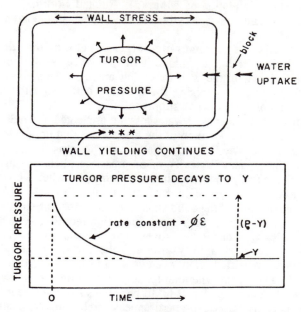

FIG. 3. Theoretical time course for ideal stress relaxation *in vivo*. Top panel illustrates that relaxation requires constant cell size, and thus some type of blockage of water uptake. Lower panel diagrams the exponential decay of turgor pressure to the yield threshold, with a time constant determined by both the elastic and the plastic characteristics of the cell wall (5).

Relaxation time courses, as shown in Fig. 2, are important for three reasons. First, they confirm our fundamental theory that wall modification and stress relaxation are the underlying causes of cell expansion. These results should put to rest the notion that water uptake is an initiating event for growth. They also indicate that water uptake and the associated driving forces are *consequences* of wall relaxation.

Second, because relaxation slows as turgor decays to the yield threshold, wall yielding is evidently a function of turgor in excess of a threshold. This is a major new confirmation of equation 1, but with some caveats. Certain details of the initial kinetics of relaxation show that equation 1 is an oversimplification of a more dynamic growth process (see below, also chapter by Hsiao). Moreover, because only a few growing tissues have been studied by this new technique, it may be premature to assume equation 1 is generally valid. In particular, it might be enlightening to examine the stress relaxation behavior of tissues reported to have little or no growth dependence on turgor pressure (16). It is also worth noting that Green et al. (13) found a somewhat nonlinear relationship between growth rate and turgor of rye coleoptiles.

Finally, relaxation time courses can be used to quantify the important biophysical characteristics of wall extension (ø and Y, where equation 1 is applicable), and thus to investigate the mechanism for growth alteration by hormones, light and other stimuli. For example, a recent study using a relaxation technique showed that an increase in ø could quantitatively account for the entire growth stimulation by auxin (5). Related work has indicated that Y may be altered by agents such as gibberellin, water stress and aging (unpublished data).

Methods of measuring stress relaxation *in vivo*. The essential condition for this technique is that cell size remain constant without directly interfering with the wall loosening process. This condition has been met most frequently by excising the growing cells from the plant, thereby depriving them of an external supply of water (2, 5, 10, 20, 28). When transpiration and other means of water loss are prevented, the reduction in turgor exactly matches stress relaxation of the wall. Note that under these conditions, turgor falls without water loss; the wall simply loosens its compression of the protoplast. Since water is nearly incompressible, there is only a negligible change in volume (very slight expansion, in principle) when the wall relaxes and turgor decreases. Readers should consult Cosgrove (5, 7) for discussion of technical problems in making valid stress relaxation measurements.

In vivo stress relaxation has been monitored directly with the pressure probe in growing pea, cucumber, soybean and zucchini stems (7). These four species exhibited qualitatively similar relaxation behavior, although the quantitative details varied. The average value for Y ranged from 1.3 bar for zucchini to 3 bar for pea. The total reduction in turgor pressure was typically 2 to 4 bar over the course of 1 to 2 h. In pea stems (5), there seemed to be fairly good agreement between wall properties measured by stress relaxation (Eq. 2) and those measured by varying turgor (Eq. 1).

Stress relaxation also reduces tissue water potential, as noted above, and has been monitored psychrometrically in pea and soybean stems (2, 10) and in bean leaves (28). Relaxation has likewise been followed in an oil-filled pressure chamber by measuring the time-dependent increase in balance pressure in excised soybean stems and castor bean leaves (7, 20). In most of these cases, little or no attempt was made to relate quantitatively the rate of relaxation with the wall properties. Nevertheless, in at least some cases it is possible to see from the data that faster-growing tissues relax at a faster rate, as one would expect if ø were increased in these tissues.

A new technique has recently been devised to permit *in vivo* stress relaxation in intact plants (7). Called the *pressure-block technique*, it uses a custom-made pressure chamber installed with an internal position transducer attached to the plant, as shown in Fig. 4. After the growing stem is sealed into the chamber, the chamber is pressurized so as to block water uptake by the growing cells. Water uptake is monitored continuously as an increase in stem length, and the essence of the method is to apply just sufficient pressure to keep length constant, without causing shrinkage. Initially, this pressure is low, but with time wall relaxation proceeds and lowers cell turgor pressure and water potential. Such relaxation would normally induce water uptake and cell expansion, but when the chamber pressure is increased to match the relaxation, the tissue water potential increases so the driving force for water uptake is cancelled out. Thus, stem length remains constant. The "trick" in this technique is that relaxation in turgor pressure is exactly and continuously matched by an increase in chamber pressure. The result is a constantly high water potential, no driving force for water uptake, and consequently a constant cell size (a necessary

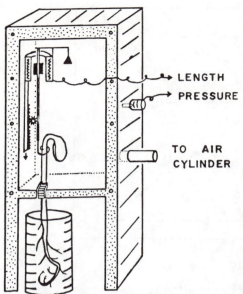

LENGTH

PRESSURE

TO AIR
CYLINDER

FIG. 4. Apparatus used for the pressure-block technique. The growing part of the stem is sealed into the custom-built pressure chamber with rapid-set epoxy and attached to an internal position transducer (LVDT). The front face of the chamber is sealed with a plate. Length and pressure are recorded continuously, and pressure may be adjusted via the pressure port connected to a gas cylinder (7).

condition of the stress relaxation technique). The chamber pressure is both the means to block water uptake and a quantitative measure of relaxation.

The pressure-block technique has numerous advantages over earlier methods of inducing and measuring stress relaxation. The growing tissue need not be excised from the plant, and so wound responses are avoided and the cells are supplied with the normal nutrients and hormones from the rest of the plant. Many of the technical problems associated with pressure probe or psychrometric measurements are avoided. The technique is relatively easy to use and affords high resolution of the relaxation time course.

Figure 5 shows the overall kinetics of relaxation in an intact pea stem, as measured by the pressure-block technique. The magnitude and general form of the relaxation are similar to that observed with the pressure-probe method, but the relaxation rate is usually faster, perhaps because the growing tissue is not excised or otherwise wounded during the pressure-block procedure. The inset shows that the data are approximately linear when plotted on a log plot, in accord with equation 2.

Because of the high resolution of the pressure-block technique, we can now discern that the initial kinetics of relaxation exhibit three phases (Fig. 6). In the first phase, the chamber pressure rises quickly (in 1 to 2 min) to about 0.5 bar. This first phase is caused by collapse of the water potential gradient within the tissue. This is the gradient that is created by wall relaxation and that sustains water movement from the xylem to the epidermis (4, 21, 25); from phase 1 we may estimate its size, as shown in Fig. 6. In phase 2, the chamber pressure slowly increases because of continued relaxation. The surprising part is phase 3, where the relaxation rate increases. Such behavior is not predicted by equation 2 and suggests that the plant responds to the growth blockage by accelerating the rate of wall loosening (7). This apparent reaction needs to be further characterized. At the very least, it suggests that wall properties are more

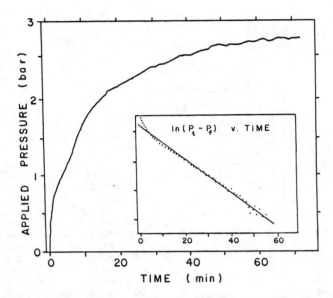

FIG. 5. Relaxation time course for the epicotyl of an intact pea seedling, as measured by the pressure-block technique. Inset shows a log plot of the same data, where P_f is final pressure and P_t is pressure at time t.

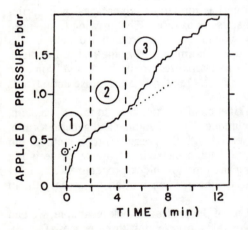

FIG. 6. Detailed kinetics of the initial relaxation of the epicotyl of a pea seedling, as measured by the pressure-block technique. Three phases are evident in this experiment, as explained in the text. Dotted line shows extrapolation of phase 2 relaxation to time zero. This is a measure of the internal water potential gradient sustaining growth.

dynamically controlled than equation 1 suggests. A similar conclusion was reached in the past, based on high-resolution measurements of dynamic changes in growth rate (17).

To summarize the conclusions so far, cell growth begins with a biochemical modification or rearrangement of the wall which results in relaxation of wall stress and cell turgor pressure. If the cell is prevented from absorbing water, then continued wall relaxation can lead to a substantial reduction in turgor pressure and water potential, and the kinetics of such relaxation provide us with a new way to examine the wall yielding process.

94

One important point should be noted before moving to the next topic: the wall yielding properties (ø and Y) are not intrinsic mechanical properties of the wall. Rather, they require sustained metabolic input, without which the wall quickly becomes relatively unyielding (see 6 for review). This means that purely mechanical measures of wall specimens will in general measure properties quite different from ø and Y, although there may be a useful correlation between the two types of parameters (see also chapters by Ray, by Cleland, and by Kutschera).

THE HYDRAULIC CONDUCTANCE OF GROWING STEMS IS LARGE

Normally, sustained relaxation of the type shown in Fig. 2 does not occur in a growing plant because water absorption tends to raise turgor and thereby counteract stress relaxation. At the onset of relaxation, the reduction in water potential generates a driving force for water uptake; at the same time the reduction in turgor pressure slows the rate of wall relaxation. At some point there is a stable crossover between these two processes. At this point turgor pressure is in dynamic balance between the counterposing processes. If the conductance of the water pathway is large, then only a minuscule relaxation will take place before water uptake becomes sufficiently large to match wall yielding and prevent further relaxation. The reduction in turgor will then be minuscule. On the other hand, if the conductance is small, then a large water potential drop is needed for water uptake (see also chapters by Ray and by Boyer). In this latter situation, turgor pressure may be depressed substantially below its full value ($\Delta \pi$) before wall yielding and water uptake come into balance. In this case, water transport may restrict the growth rate because wall yielding will be reduced by the lower turgor.

In single isolated cells, the main barrier to water uptake is the plasma membrane (in series with the tonoplast, in the case of vacuolar water uptake). It turns out that the hydraulic conductance of such membranes is so large that growing cells do not deviate substantially from water potential equilibrium, even at rapid growth rates. The reduction in turgor is in the range of 0.1% to 1% of full turgor (6).

In multicellular tissues, the water pathway is longer, more convoluted, and imposes a greater barrier to water flow (6, 21, 25). Because the resistance to water flow is diffuse, growing tissues will develop an internal gradient in water potential (15, 21, 25). Early psychrometric studies suggested that the size of this gradient was large, perhaps as large as 4 bar in young growing stems (21; see also chapter by Boyer). Such a large gradient would mean that growth was limited in large part by water transport through the growing tissue.

We have carried out studies to test this idea in several ways, but most of our results have contradicted this conclusion. Instead, our results indicate that the internal gradient sustaining water uptake into growing stems is about 0.5 bar and that it restricts growth by about 10%.

The major experiments leading to this conclusion are listed in Table I. First, we infiltrated growing pea stems with water. This short-circuits the normal hydraulic pathway, eliminating the gradient and raising turgor

Table I. *Evidence for large hydraulic conductance of growing stems.*

Numbers in parentheses are reference citations.

---

1. Infiltration Experiments: P increased by 0.5 bar (9)
2. Growth Alteration Experiments: P changed by 0.5 bar max (9)
3. Swelling/Shrinking is Faster than Relaxation (5)
4. Balance Pressure is Small (0.3 bar) in Excised, then Cooled Stems (7)
5. Initial Pressure to Block Growth is Small (Pressure-Block) (7)
6. Water Influx Starts/Stops Quickly when Wall Relaxation is Induced/Inhibited

---

correspondingly. With the pressure probe we measured an 0.5 bar increase in turgor upon infiltration (9). Second, we manipulated the growth rate (= water influx) of pea stems by procedures that alter wall yielding properties, then measured the passive change in turgor pressure. Growth rates were varied by a factor of 20, yet the maximum turgor change was 0.5 bar (9). Third, swelling kinetics in pea epicotyls were shown to be an order of magnitude faster than relaxation kinetics. By applying these kinetics to a theoretical model of water flow in multicellular tissues, we can calculate an internal gradient of about 0.4

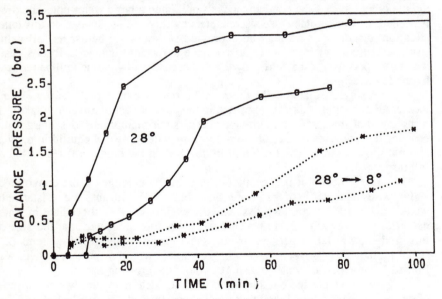

FIG. 7. Relaxation in soybean hypocotyls, as measured with an oil-filled pressure chamber. At time zero, the apical 1.5 cm of the seedling was excised and the balance pressure was measured in the usual fashion. In one set the tissue was kept at room temperature (28°C), in the second set the pressure chamber was lowered into an ice bath immediately after excision. Tissue temperature was measured to be about 8°C. At this temperature, growth and wall relaxation are greatly reduced. Redrawn after (7).

96

bar (5). Fourth, the balance pressure of soybean stems was measured in a pressure chamber. The stem was cooled immediately after excision to retard subsequent relaxation (Fig. 7). This procedure should give an estimate of the internal water potential gradient before excision, without relaxation artefacts. For soybean hypocotyls it averaged 0.3 bar (7). Fifth, the initial pressure required to stop growth was measured in the pressure-block technique (Fig. 6). It averaged 0.35 bar for soybean, 0.41 bar for pea, 0.51 bar for zucchini, and 0.56 bar for cucumber (7). Finally, water influx rapidly ceases when wall relaxation is inhibited, for example, by cyanide or cold treatments. Likewise, water influx increases greatly when walls are made more extensible by auxin, fusicoccin and other agents. If hydraulic conductance was low, such transitions in growth rate would be much slower (4).

Thus, six different approaches to the question yielded a consistent answer: the driving force for water uptake into growing stems is about 0.5 bar. This conclusion might seem at odds with numerous water potential measurements showing that well-hydrated growing tissues are 2 to 4 bar below, water potential equilibrium (21, reviewed in 6). In part such low water potentials might be an artefact of wall relaxation after excision (10). In addition, substantial apoplastic solutes have been found in growing and nongrowing tissues (8 and refs cited therein; also see poster abstract by Meinzer and Moore in this volume). Such solutes will reduce the water potential of tissues independently of wall relaxation, but will not induce significant water flow because the reflection coefficient of the wall is near zero (23).

LINKAGE OF SOLUTE TRANSPORT
WITH CELL GROWTH

As described above, water uptake by growing cells is a consequence of wall relaxation, and appears to restrict stem growth rate slightly. Long term cell expansion, if not otherwise compensated, would dilute cell solutes, and so inevitably come to a standstill because of inadequate osmotic pressure, turgor pressure and the wall stress necessary for wall yielding and expansion. Solutes are also needed as substrates for respiration and for synthesis of wall polymers and other structural parts of the cell. Thus, solute import is required to meet both the physical and the biochemical needs of sustained cell expansion.

A quick calculation can put the physical need for solute import into perspective. Take as an example an isolated cell growing at a fast rate of 10% per h and having an osmotic pressure of 6 bar, turgor of 6 bar, and yield threshold of 3 bar. Assuming no solute import, it is easy to see that growth will stop when cell size doubles and osmotic pressure and turgor thus drop to the yield threshold. If the cell continued to grow at 10% per hour, this would take $\ln(2)/0.1$ h, or 6.9 h. However, as the cell expands, turgor should fall and the growth rate decline. It can be shown (unpublished results) that the cell will asymptotically approach its final size with a time constant given by $-\phi Y$. The final size will be larger than the initial size by the factor π_0/Y, where π_0 is the initial osmotic pressure. In the case given above, growth rate will decrease to 5% per h after about 4 h, and to 2.5% per h after 9.1 h. The point is, for growth of plant cells, dilution is important over the scale of several hours, but not substantial on the scale of minutes. However, if turgor pressure were very close to the yield threshold, then dilution would be correspondingly more significant.

It is clear from a number of studies that solute import compensates at least in part, for dilution during growth of intact tissues. In contrast, when growing regions are excised and floated on dilute media, osmotic pressure commonly decreases as the cells expand (26). By supplying absorbable solutes to the tissues, osmotic pressure increases and long term growth rate is enhanced. Thus, at least in this artificial situation, solute import can be partially limiting for growth.

It is not yet clear, however, whether such limitation also exists in the intact plant. From a purely physical standpoint, higher solute concentration would entail higher turgor and wall stress, and thus higher growth rate. In this sense, solute import always limits growth. However, growth is likely to be more finely regulated. For example, Marshall and Penny (17) applied a tensile force to a lupin hypocotyl to increase the wall stress and elongation rate. However, there was only a temporary increase in elongation rate before the growth rate was adjusted downward to the original value. From these and similar results (12), it seems likely that plants may be able to adjust wall properties, within limits, to compensate for physical disturbances in growth.

A number of very basic questions about the mechanisms of solute import, the regulation of transport rates, and the coupling of transport to cell expansion remain unanswered (see chapter by Spanswick). Solute import and accumulation also play an apparently adaptive role during water stress (see chapter by Hsiao and Jing). Although some marine algae have mechanisms to regulate osmotic pressure and turgor pressure, there remains some debate about whether higher plants behave similarly (22). This area is ripe for careful, quantitative studies.

SUMMARY

In this chapter I have discussed the process of wall relaxation, to show that relaxation procedures can test the basic tenets of our theories of how plant cells expand, and to illustrate some new ways to measure stress relaxation *in vivo*. The fact that plant cells do undergo relaxation demonstrates that growth is initiated by a modification and relaxation of the cell wall. *In vivo* stress relaxation provides a new way to characterize the wall properties that govern cell expansion. Water uptake, although it is essential for cell expansion, is a consequence of relaxation and does not appear to impose a large restraint on the expansion rate of young stems. Solute uptake is needed for long-term growth, but the exact relationship between the two processes is not well worked out. The relationship will no doubt prove complicated, in part because there are both physical and biochemical roles for imported solutes. Better understanding of plant growth is likely to come from in-depth studies of the characteristics and mechanisms of wall relaxation, and from manipulative experiments to work out the linkages between growth and solute influxes.

Acknowledgments--Supported by grants from the U.S. National Science Foundation and from the Department of Energy.

LITERATURE CITED

1. BAUGHN JW, CB TANNER 1976 Excision effects of leaf water potential of five herbaceous species. Crop Sci 16: 184-190
2. BOYER JS, AJ CAVALIERI, E SCHULZE 1985 Control of the rate of cell enlargement: Excision, wall relaxation, and growth-induced water potentials. Planta 163: 527-543
3. CLELAND RE 1976 The control of cell enlargement. Integration of activity in the higher plant. Symp Soc Exp Biol 31: 101-115
4. COSGROVE DJ 1981 Analysis of the dynamic and steady-state responses of growth rate and turgor pressure to changes in cell parameters. Plant Physiol 68: 1439-1446
5. COSGROVE DJ 1985 Cell wall yield properties of growing tissues. Evaluation by *in vivo* stress relaxation. Plant Physiol 78: 347-356
6. COSGROVE DJ 1986 Biophysical control of plant cell growth. Ann Rev Plant Physiol 37: 377-405
7. COSGROVE DJ 1987 Wall relaxation in growing stems: Comparison of four species and assessment of measurement techniques. Planta 171: 266-278
8. COSGROVE DJ, RE CLELAND 1983a Solutes in the free space of growing stem tissues. Plant Physiol 72: 326-331
9. COSGROVE DJ, RE CLELAND 1983b Osmotic properties of pea stem internodes in relation to growth and auxin action. Plant Physiol 72: 332-338
10. COSGROVE DJ, E VAN VOLKENBURGH, RE CLELAND 1984 Stress relaxation of cell walls and the yield threshold for growth: Demonstration and measurement by micro-pressure probe and psychrometer techniques. Planta 162: 46-52
11. GANDAR PW, CB TANNER 1975 Comparison of methods for measuring leaf and tuber water potentials in potatoes. Am Potato J 52: 387-396
12. GREEN PB, K BAUER, WR CUMMINS 1977 Biophysical model for plant cell growth: auxin effects. *In* AM Jungreis, TK Hodges, A Kleinzeller, SG Schultz eds, Water Relations in Membrane Transport in Plants, Academic Press, New York, pp 30-45
13. HAYASHI T, G MACLACHLAN 1984 Pea xyloglucan and cellulose. I. Macromolecular organization. Plant Physiol 75: 596-604
14. HETTIARATCHI DRP, JR O'CALLAGHAN 1974 A membrane model of plant cell extension. J Theor Biol 45: 459-465
15. HSIAO TC, WK SILK, J JING 1985 Leaf growth and water deficits: biophysical effects. *In* NR Baker, WD Davis, C Ong eds, Soc Exp Bio Seminar Series 27: Control of Leaf Growth, Cambridge University Press, Cambridge, pp 239-266
16. KUZMANOFF KM, ML EVANS 1981 Kinetics of adaptation to osmotic stress in lentil (*Lens culinaris* Med.) roots. Plant Physiol 68: 224-247
17. MARSHALL DC, D PENNY 1976 High-resolution measurements of transient changes in the growth rate of intact lupin seedlings. Austr J Plant Physiol 3: 237-246

18. MASUDA Y 1978 Auxin-induced cell wall loosening. Bot Mag Tokyo Spec Issue 1: 103-123
19. MATTHEWS MA, E VAN VOLKENBURGH, JS BOYER 1984 Acclimation of leaf growth to low water potentials in sunflower. Plant Cell Environ 7: 199-206
20. MILBURN JA 1979 Water Flow in Plants. Longman, London and New York
21. MOLZ FM, JS BOYER 1978 Growth-induced water potentials in plant cells and tissues. Plant Physiol 62: 423-429
22. MUNNS R, A TERMAAT 1986 Whole-plant responses to salinity. Austr J Plant Physiol 13: 143-160
23. PASSIOURA JB 1984 Hydraulic resistance of plants. I. Constant or variable?. Aust J Plant Physiol 11: 333-339
24. RAY PM, PB GREEN, RE CLELAND 1972 Role of turgor in plant cell growth. Nature 239: 163-164
25. SILK WK, KK WAGNER 1980 Growth-sustaining water potential distributions in the primary corn root. A noncompartmented continuum model. Plant Physiol 66: 859-863
26. STEVENSON TT, RE CLELAND 1981 Osmoregulation in the *Avena* coleoptile in relation to auxin and growth. Plant Physiol 67: 749-753
27. STUART DA, RL JONES 1977 Roles of extensibility and turgor in gibberellin- and dark-stimulated growth. Plant Physiol 59: 61-68
28. VAN VOLKENBURGH E, RE CLELAND 1986 Wall yield threshold and effective turgor in growing bean leaves. Planta 167: 37-43

Physiology of Cell Expansion During Plant Growth, D.J. Cosgrove and D.P. Knievel Eds.,
Copyright ©1987, The American Society of Plant Physiologists

PROPERTIES OF PROTON-PUMPING ATPases AND THEIR INVOLVEMENT IN GROWTH

ROGER M. SPANSWICK

Section of Plant Biology
Division of Biological Sciences
Cornell University
Ithaca, NY 14853

INTRODUCTION

The upsurge of interest in the acid growth hypothesis for the action of auxin on cell wall extension (10, 16) coincided closely with the realization that the electrogenic ion pump in the plasma membrane of plant cells transports hydrogen ions (21, 36, 37). It was also realized that an electrogenic hydrogen ion pump would fit into a chemiosmotic scheme (24) as the primary active transport system that would provide the energy for secondary active transport systems (cotransport systems or symports) for ions (35) or sugars and amino acids (27, 34). Thus it became possible, in principle, to begin considering the relationship between auxin action and the transport of metabolic substrates and ions.

I shall consider, first, recent work on the properties of the plasma membrane H^+-ATPase and its consequences for the acid growth hypothesis. Then I shall comment on the use of metabolic control analysis as a framework within which to examine the events leading to cell wall extension (38).

LOCALIZATION AND CHARACTERIZATION OF TRANSPORT ATPases

The difficulties inherent in measurement of unidirectional H^+ fluxes, resulting from the exchange of H^+ with H_2O and the high permeability of membranes to water, and of net H^+ fluxes, due to the return pathways for H^+ movement via cotransport systems and the effects of CO_2 movements on pH, made it necessary to look for a system simpler than an intact tissue with which to elucidate the H^+ transport systems in plant cells. The solution, which made it possible to take advantage of methods worked out earlier for bacterial and plastid membranes, was to use isolated membrane vesicles. The development of this work has been reviewed in detail by Sze (39). ATP-dependent H^+ transport in isolated membrane vesicles was demonstrated almost simultaneously by a number of groups, though the first paper giving a direct demonstration of H^+ transport, apparently unnoticed for some time by the other groups, was published by Hager and his coworkers (17).

101

The tonoplast ATPase. There was some initial confusion concerning the interpretation of this work because the properties of the ATPase and proton transport associated with the first "microsomal" preparations that demonstrated H^+ transport were unlike those of the ATPase associated with earlier plasma membrane preparations. In particular, it was found that the H^+ transport and ATPase activity was stimulated by Cl^- rather than K^+, inhibited by NO_3^-, and unaffected by vanadate. This contrasted with the ATPase associated with plasma membranes which was stimulated by K^+, indifferent to the anion, and inhibited by vanadate. Isolation of the activity on sucrose density gradients soon demonstrated that the membrane vesicles associated with this activity had a low density typical of that associated with the tonoplast (5). Furthermore, an anion-sensitive, vanadate-resistant ATPase activity was found to be associated with isolated vacuoles from red beets (42) and with the membranes of the vacuolysosomal organelles from the latex of *Hevea brasiliensis* (1). The vacuolar origin was demonstrated directly by isolation of vacuoles from red beets, the formation of membrane vesicles and the localization on a sucrose density gradient at a density associated with the anion-stimulated activity obtained from microsomal preparations. Furthermore, the ATPase activity associated with vesicles obtained in this way was almost completely abolished by NO_3^- (4).

Evidence from several laboratories indicates that the tonoplast ATPase has a holoenzyme molecular mass in the range 300-600 kD, contains several subunits, and does not form a phosphorylated intermediate (e.g., 23, 28).

The plasma membrane ATPase. Isolation of plasma membrane vesicles with the capability of ATP-driven H^+ transport proved more difficult than for the tonoplast. However, several groups have now been able to demonstrate H^+ transport in plasma membranes from a variety of plants (39). As expected, the H^+ transport associated with these membrane vesicles is sensitive to vanadate (4, 25). Consistent with other ATPases that are inhibited by vanadate, the plasma membrane H^+-ATPase from plants also forms a phosphorylated intermediate (6). Unlike the tonoplast ATPase, it appears to consist of a single subunit having a molecular mass in the 90-100 kD range. Radiation inactivation data suggest that the molecular mass of the holoenzyme is 228 kD (7). This is consistent with cross-linking studies on the purified enzyme from tomato roots which suggest that it can exist as dimers and trimers (1).

There is also evidence for H^+-ATPases associated with the endoplasmic reticulum and Golgi apparatus (39), but our main concern here is with the plasma membrane.

REGULATION OF THE PLASMA MEMBRANE ATPase

There is considerable controversy surrounding the way in which IAA and fusicoccin produce an increase in H^+ efflux. In this section, I shall attempt to assess the relevance of what we have learned about the properties of the plasma membrane ATPase to the effects of these substances on intact cells. It is important to keep in mind that with tightly sealed membrane vesicles the ATPase activity and proton transport properties will reflect the activity of the fraction of vesicles having an inside-out orientation, since ATP will not have access to the catalytic site of sealed right-side-out vesicles.

Regulation by pH. We have found that the pH optimum is close to 6.5 for the plasma membrane ATPase activity from a number of tissues and for H^+ transport in native vesicles from red beets (4) and corn (11) and in the reconstituted ATPase from red beets (25; Fig. 1). The same pH optimum is also

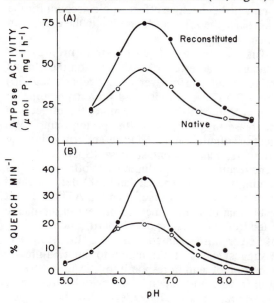

FIG. 1. Effect of pH on ATPase activity and H^+ transport associated with red beet plasma membrane (o) and reconstituted H^+-ATPase(•). Methods are given in ref. 25. (From ref. 25, with permission.)

observed for the purified plasma membrane ATPase from tomato roots (1). Since the cytoplasmic pH of plant cells tends to be in the range 7.0-7.5 (see ref. 13 and references therein), this means that a decrease in the pH of the cytoplasm may be expected to result in a significant increase in proton pump activity. This is reflected in a hyperpolarization of the membrane potential as weak acids lower the cytoplasmic pH towards the pH optimum for the ATPase and stimulate its function as an electrogenic pump (13).

Regulation by metabolic substrates and products. There is evidence that plant cells under non-stressed conditions maintain high ratios of ATP/ADP (15, 31). Thus, even though we have found that ADP inhibits the plasma membrane ATPase of red beets (I_{50} = 2.5 mM; 3), the ADP level in the cytoplasm may not exert control over the enzyme except under conditions in which ATP regeneration is inhibited. The substrate for the ATPase appears to be MgATP, and the enzyme is inhibited by free ATP. However, it seems probable that most ATP in the cytoplasm is normally in the MgATP form since the cytoplasmic concentration estimated from flux measurements (22), is 6.5 meq of Mg per l.

The K_m for MgATP is in the range 0.3 to 0.5 mM (3, 39), which may not be far below the cytosolic concentration of ATP in the cytoplasm (32).

CONSEQUENCES FOR HYPOTHESES CONCERNING THE ACTION OF AUXIN AND FUSICOCCIN

Direct effects of IAA on the plasma membrane ATPase. Two sets of results indicate that IAA may have a direct effect on the plasma membrane ATPase which manifests itself as a decrease in the Km for ATP (14, 33). However, one might question whether such an effect would account for the observed increase in acid secretion since the ATP concentration in the cytoplasm is apparently somewhat greater than the K_m.

Direct effects of fusicoccin on the plasma membrane ATPase. Early attempts to demonstrate an effect of fusicoccin on the plasma membrane ATPase produced mixed results. However, recent reports (29, 30) demonstrate a direct stimulation of ATPase activity and proton pumping in plasma membrane vesicles isolated from radish. Furthermore, the effect was demonstrated to represent an increase in the V_{max} for the proton pump. We subsequently attempted to detect a stimulation of ATPase activity using the purified ATPase from tomato roots, but without success (Spanswick and Anthon, unpublished results). The results are too preliminary to offer a rigorous interpretation, but it is worth noting that Stout and Cleland (38) demonstrated binding of fusicoccin to a membrane component other than the ATPase in oat root membranes. It is conceivable that this component is present in the native membrane vesicles but is separated from the ATPase during purification.

Indirect effects of IAA and fusicoccin on the ATPase. Brummer and Parish (8) have suggested, as an alternative to the acid-growth hypothesis, that the observed secretion of acid in response to IAA or fusicoccin may be a secondary response following acidification of the cytoplasm by these substances. In a more detailed elaboration of the hypothesis (26), they suggest that auxin may trigger the release of calcium into the cytoplasm and that this, in turn may produce a decrease in pH that will stimulate the proton pump. Evidence in favor of this hypothesis is that an observed decrease in cytosolic pH, as a result of IAA or fusicoccin treatment, appears to precede the membrane hyperpolarization (9). However, this does not take into account the more recent evidence suggesting a direct effect at the membrane (29, 30). It is evident that considerable work may be required to separate cause and effect in this area.

INTEGRATION OF PROTON FLUXES WITH GROWTH

The fact that cells grow and cease growth in a well-defined and ordered manner suggests that growth is controlled in a much more complicated way than our current, relatively primitive hypotheses could account for. Historically, plant physiologists have been tempted to try and identify a single "rate-limiting" factor for each process investigated. Thus it might be expected that an advocate of the acid-growth hypothesis would suggest that, at least in the short term, the efflux of H^+ might be a limiting factor in growth rate. It could also be pointed out that an increase in the activity of the electrogenic proton pump in the plasma membrane would lead to an increase in the driving force on proton cotransport systems that could be involved in the import of materials into the cell, thus providing for an integration of import with cell expansion. This, however, ignores the possibility that the import of substrates by the rapidly growing regions of developing roots and leaves may well be via the symplasm (12, 41).

It is now generally accepted, particularly in the long term, that a number of biosynthetic reactions related to wall synthesis must be involved in cell growth, in addition to the processes necessary for the maintenance of the osmotic potential at a value sufficiently negative to produce the turgor to drive cell extension. A recent candidate for a rate-limiting role in IAA-induced cell elongation is the level of UDP-glucose in the cytoplasm (19). While not disputing this as a possibility, I do wish to draw attention to a recent article by Trewavas (40) in which he points out that the sheer complexity of biological systems may make it futile to attempt to identify rate-limiting steps in a complex process. He draws a parallel with metabolic networks which are increasingly being analysed using the "metabolic control analysis" invented by Kacser and Burns (20) and independently by Heinrich and Rapaport (18). It is not possible to explain this approach in detail here. However, an important aspect of this theory is the fact that "flux control coefficients," defined as the fractional change in the flux through the pathway divided by the fractional change in enzyme level producing the change in flux, may be defined for each enzyme in a pathway. It was demonstrated theoretically that the flux control coefficients in a pathway must sum to 1.0. Each flux control coefficient may be regarded as a measure of the relative contribution of that step in the pathway to the overall control of the flux. In those biochemical pathways that have been examined in detail, it is found that control is generally distributed over several steps. Thus the idea of a "rate-limiting step" becomes obsolete and is replaced by the flux control coefficient as a measure of the contribution of the step in the pathway to control of the overall flux through the pathway. I am currently developing applications of this approach to the study of partitioning. Application of metabolic control analysis to the study of growth will be even more complicated, but attention to the ideas advanced by Trewavas (40) may point us in a direction that will avoid much pointless controversy.

Acknowledgments--The author acknowledges support from the National Science Foundation for work on membrane transport, including NSF grant DBM-8415035.

LITERATURE CITED

1. ANTHON GE, RM SPANSWICK 1986 Purification and properties of the H+-translocating ATPase from the plasma membrane of tomato roots. Plant Physiol 81: 1080-1085
2. D'AUZAC J 1977 ATPase membranaire de vacuoles lysosomales: les lutoides du latex d'*Hevea brasiliensis*. Phytochemistry 16: 1881-1885
3. BENNETT AB, SD O'NEILL, M EILMANN, RM SPANSWICK 1985 H+-ATPase activity from storage tissue of *Beta vulgaris*. III. Modulation of ATPase activity by reaction substrates and products. Plant Physiol 78: 495-499
4. BENNETT AB, SD O'NEILL, RM SPANSWICK 1984 H+-ATPase activity from storage tissue of *Beta vulgaris*. I. Identification and

characterization of an anion-sensitive H^+-ATPase. Plant Physiol 74: 538-544

5. BRISKIN DP, RT LEONARD 1980 Isolation of tonoplast vesicles from tobacco protoplasts. Plant Physiol 66: 684-687

6. BRISKIN DP, RT LEONARD 1982 Partial characterization of a phosphorylated intermediate associated with the plasma membrane ATPase of corn roots. Proc Natl Acad Sci USA 79: 6922-6926

7. BRISKIN DP, WR THORNLEY, JL ROTI-ROTI 1985 Target molecular size of the red beet plasma membrane ATPase. Plant Physiol 78: 642-644

8. BRUMMER B, RW PARISH 1983 Mechanisms of auxin-induced plant cell elongation. FEBS Lett 161: 9-13

9. BRUMMER B, A BERTL, I POTRYKUS, H FELLE, RW PARISH 1985 Evidence that fusicoccin and indole-3-acetic acid induce cytosolic acidification of *Zea mays* cells. FEBS Lett 189: 109-114

10. CLELAND R 1971 Cell wall extension. Ann Rev Plant Physiol 22: 197-222

11. DE MICHELIS MI, RM SPANSWICK 1986 H^+-pumping driven by the vanadate-sensitive ATPase in membrane vesicles from corn roots. Plant Physiol 81: 542-547

12. DICK PS, ap REES, T 1975 The pathway of sugar transport in roots of *Pisum sativum*. J. Exp Bot 26: 305-314

13. FELLE H 1987 Proton transport and pH control in *Sinapsis alba* root hairs: a study carried out with double-barrelled pH micro-electrodes. J Exp Bot 38: 340-354

14. GABATHULER R, MM MOLONEY, PE PILET 1983 Possible regulation by IAA of an ATP-dependent electrogenic proton uptake into maize and pea root membrane vesicles. Plant Physiol 72S: 116

15. GOLLER M, R HAMPP, H ZIEGLER 1982 Regulation of the cytosolic adenylate ratio as determined by rapid fractionation of mesophyll protoplasts of oat. Effect of electron transfer inhibitors and uncouplers. Planta 156: 255-263

16. HAGER A, H MENZEL, A KRAUSS 1971 Versuche und Hypothese zur Primärwirkung des Auxins beim Streckungswachstum. Planta 100: 47-75

17. HAGER A, R FRENZEL, D LAIBLE 1980 ATP-dependent proton transport into vesicles of microsomal membranes of *Zea mays* coleoptiles. Z Naturforsch Sect C 35: 783-793

18. HEINRICH R, TA RAPAPORT 1974 A linear steady-state treatment of enzymatic chains. General properties, control and effector strength. Eur J Biochem 42: 89-95

19. INOUHE M, R YAMAMOTO, Y MASUDA 1987 UDP-glucose level as a limiting factor for IAA-induced cell elongation in *Avena* coleoptile segments. Physiol Plant 69: 49-54

20. KACSER H, JA BURNS 1973 The control of flux. Symp Soc Exp Biol 27: 65-104

21. KITASATO H 1968 The influence of H^+ on the membrane potential and ion fluxes of *Nitella*. J Gen Physiol 52: 60-87

22. MACKLON AES, SIM A 1976 Cortical cell fluxes and transport to the stele in excised root segments of *Allium cepa* L. III. Magnesium. Planta 128: 5-9
23. MANDALA S, L TAIZ 1985 Partial purification of a tonoplast ATPase from corn coleoptiles. Plant Physiol 78: 327-333
24. MITCHELL P 1961 Coupling of phosphorylation to electron and hydrogen transfer by a chemi-osmotic mechanism. Nature (Lond.) 191: 144-148
25. O'NEILL SD, RM SPANSWICK 1984 Characterization of native and reconstituted plasma membrane H^+-ATPase from the plasma membrane of *Beta vulgaris*. J Membr Biol 79: 245-256
26. PARISH RW, H FELLE, B BRUMMER 1986 Evidence for a mechanism by which auxins and fusicoccin may induce elongation growth. *In* AJ Trewavas, ed, Molecular and Cellular Aspects of Calcium in Plant Development, Plenum Press, New York, pp 301-308
27. POOLE RJ 1978 Energy coupling for membrane transport. Ann Rev Plant Physiol 29: 437-460
28. RANDALL SK, H SZE 1986 Properties of the partially purified tonoplast H^+-pumping ATPase from oat roots. J Biol Chem 261: 1364-1371
29. RASI-CALDOGNO F, MI DE MICHELIS, MC PUGLIARELLO, E MARRÈ 1986 H^+-pumping driven by the plasma membrane ATPase in membrane vesicles from radish: stimulation by fusicoccin. Plant Physiol 82: 121-125
30. RASI-CALDOGNO F, MC PUGLIARELLO 1985 Fusicoccin stimulates the H^+-ATPase of plasmalemma in isolated membrane vesicles from radish. Biochem Biophys Res Commun 133: 280-285
31. ROBERTS JKM, O JARDETZKY 1981 Monitoring of cellular metabolism by NMR. Biochim Biophys Acta 639: 53-76
32. ROBERTS JKM, AN LANE, RA CLARK, RH NIEMAN 1985 Relationships between the rate of synthesis of ATP and the concentrations of reactants and products of ATP hydrolysis in maize root tips, determined by ^{31}P nuclear magnetic resonance. Arch Biochem Biophys 240: 712-722
33. SCHERER GFE 1984 Stimulation of ATPase activity by auxin is dependent on ATP concentration. Planta 161: 394-397
34. SLAYMAN CL, CW SLAYMAN 1974 Depolarization of the plasma membrane of *Neurospora* during active transport of glucose: evidence for a proton-dependent cotransport system. Proc Natl Acad Sci USA 71: 1935-1939
35. SMITH FA 1970 The mechanism of chloride transport in Characean cells. New Phytol 69: 903-917
36. SPANSWICK RM 1972 Evidence for an electrogenic ion pump in *Nitella translucens*. I. The effects of pH, K^+, Na^+, light and temperature on the membrane potential and resistance. Biochim Biophys Acta 288: 73-89
37. SPANSWICK RM 1981 Electrogenic ion pumps. Ann Rev Plant Physiol 32: 267-289
38. STOUT RG, RE CLELAND 1980 Partial characterization of fusicoccin binding to receptor sites on oat root membranes. Plant Physiol 66: 353-359

39. SZE H 1985 H$^+$-translocating ATPases: advances using membrane vesicles. Ann Rev Plant Physiol 36: 175-208
40. TREWAVAS A 1986 Understanding the control of plant development and the role of growth substances. Aust J Plant Physiol 13: 447-457
41. TURGEON R 1986 The import-export transition in dicotyledonous leaves. *In* J. Cronshaw, WJ Lucas, RT Giaquinta, eds, Phloem Transport, Alan R. Liss, New York, pp 285-291
42. WALKER RR, RA LEIGH 1981 Characterization of a salt-stimulated ATPase activity associated with vacuoles isolated from storage roots of red beet (*Beta vulgaris* L.). Planta 153: 140-149

Physiology of Cell Expansion During Plant Growth, D.J. Cosgrove and D.P. Knievel Eds.,
Copyright ©1987, The American Society of Plant Physiologists

HYDRAULICS, WALL EXTENSIBILITY AND WALL PROTEINS

JOHN S. BOYER

Department of Soil and Crop Sciences
Texas A&M University,
College Station, TX 77843, USA

INTRODUCTION

Our understanding of cell enlargement has increased recently because of new methods for studying how water participates in the process. Most previous studies used tissue segments submerged in solutions where the environment can be conveniently and precisely controlled. However, this approach has some disadvantages. Factors requiring intactness cannot be studied and it is difficult to measure the water status of the tissue independently of the surrounding solution.

The new methods allow these problems to be avoided. Thermocouple psychrometry plays a particularly important role because the water status of the enlarging tissue can be determined continuously while growth occurs in completely intact plants (7). The miniature pressure probe is similarly useful although it must penetrate individual cells (21). Thus, it is possible for the first time to determine both the driving force for water uptake and the driving force for wall extension in enlarging cells of intact plants. The preservation of the intact state allows forces in the xylem and surrounding tissues to be evaluated as well.

These methods confirm that turgor drives cell enlargement by extending the walls of the cells irreversibly. Abundant evidence shows that turgor is the force for this extension (20, 34, 35) and that turgor must be above a threshold level before the force is sufficient to cause irreversible effects on the wall (12, 35). Above this level, termed the yield threshold, the wall is deformed and begins to flow much as a plastic changes shape when subjected to sufficient force (14, 35).

However, the newer approaches show that, although turgor drives cell wall extension, several other components are equally necessary. Water uptake is essential and can alter the rate of the growth process (7). Changes in wall extensibility (25), solute transport and use (25, 26, 27, 28) and the yield threshold turgor (20, 25) can change the growth rate without a change in turgor. Moreover, changes in the water status of the xylem have immediate effects on growth rates (7).

Nowhere are these additional effects more important than in the growth of plants at low water potentials. If these conditions are severe enough, turgor can decrease and inhibit growth. However, it is increasingly clear that solutes accumulate under these conditions, particularly in localized growing regions. The accumulation acts to maintain turgor (26, 31). Despite turgor maintenance,

109

growth is inhibited (18, 27, 28). Why this occurs is a central question for plant physiologists. It is such a frequent problem in natural situations that the answer has important consequences, especially for agriculture.

The following review highlights some of the principles to have emerged from this work. For additional information, the reader may wish to see reviews of cell enlargement (2, 14, 35) and water transport (6).

WATER UPTAKE BY ENLARGING TISSUE

Plants grow in size mostly by increasing in cellular water content. Rapidly growing tissues typically add 3-6% to their water volume every hour and, in stems, this can produce linear growth rates of 0.25-0.75 μm·s^{-1}. At the cellular level, the small cells continually produced in the stem apex can double in length every 4 h on average. After about 15 h, they have enlarged 12-fold, or almost 4 doublings, and have reached their final size. Within the cells, the vacuole enlarges many-fold and it is the growth of this compartment that accounts for most of the increased water content of the cells. About half of the volume increase in the tissue is attributable to vacuolar water. Much of the remainder is air in the intercellular spaces, which increase in volume as the tissue enlarges.

The water flux to feed this process requires that the cells closest to the xylem transmit water at many times their own needs (6). The water must pass through many intervening cells and cell walls in the enlarging tissue and these form a frictional resistance to water movement (6, 33). As a consequence, the water potential inside each enlarging cell must be lower than in the water source, usually the xylem, by an amount that depends on the resistance. Cells farthest from the vascular system pull water through the largest resistance and can have a water potential several bars below that in the vascular system. For soybean hypocotyl tissue on average, the water potential is about 2 bars lower than in the xylem (6).

This water potential, unique to the enlarging tissue, has been termed a growth-induced water potential because of its association with the growth process (30). It is not present in the mature tissue (5, 7, 8, 11, 28, 30, 37, 38), and it provides a force that will extract water not only from the soil via the vascular system but also from the mature tissue. This effect is particularly dramatic when plant organs are detached so that there can be no water entering from the soil. Growth often continues, albeit at a slow rate, and the mature tissue dehydrates as it gives up water to the enlarging tissue (7, 11). Natural examples are the leaf growth that continues on a tree felled in the spring, and the flowering and seed set that occur in certain weedy species removed from the soil.

It has been suggested that growth-induced water potentials result from solutes in the cell wall solution, i.e., the free space or apoplast (17). However, more recent measurements show that although small amounts of solute are present (33), most of the potential arises from a low water potential in the protoplasts that is transmitted to the cell walls and causes a tension in the wall solution (33). The water potentials in the tissue thus form a potential field around the vascular system with the highest potentials in the xylem and lowest potentials in the epidermal cells at the outer surface of the tissue (33). Water moves in a direction normal to the equipotential surfaces in the field. For stems, this is a radial movement from the xylem outward.

A consequence of the field is that the xylem plays an important role in determining growth rate. Boyer et al. (7) observed an abrupt decline in growth (within 1 min) when the water supply was decreased by excising all tissues other than mature tissue at the base of stems of soybean seedlings. The water potential of the enlarging tissue did not change significantly until at least 1 h later. Others have also observed rapid decreases in growth rate when transpiration rates increase or soil water potentials rapidly decrease (1, 23, 24).

This behavior is most easily explained if the xylem water potential decreases, changing the potential field so that water entry decreases in the enlarging tissue. Because growth is almost completely dependent on water entry, its rate also would decrease. The water potential of most of the enlarging cells would not change immediately, however. Their water potential already is below that in the xylem, and a decrease in xylem potential would serve only to invert the potential field for those cells closest to the xylem. This inversion may be the first event that slows or prevents water entry into the enlarging tissue. All the cells would be deprived of water because the cells next to the xylem must supply all the outlying cells. Therefore, the entire tissue experiences a decrease in growth (7).

Gradually the water potential of the enlarging tissue should adjust to the lowered potential of the xylem. Cavalieri and Boyer (11) showed that the growth-induced water potential had reformed at lower potentials several days after the water supply to the roots was restricted.

Turgor shows a particularly interesting behavior because it may remain completely constant throughout these events. At first, the localization of the effect in the xylem causes no change in the turgor (7). Just as the potential field begins to adjust to the changes in the xylem, turgor would ordinarily decrease but solutes begin to accumulate in the enlarging cells (27). This has the effect of maintaining turgor (23, 24, 26, 27, 28, 31) despite the decrease in water potential that is occurring in the cells.

Eventually, of course, these changes result in a high turgor, a high concentration of solute most of which is photosynthate, and a reestablishment of the growth-induced water potential at low tissue water potentials. With so many conditions favoring rapid growth, why don't the cells grow?

CELL WALL EXTENSION

Sometime after these changes take place, the extensibility of the cell wall declines. Matthews et al. (25) observed a decreased wall extensibility in leaves that were exposed to low water potentials for several days. This implies that, even if turgor were maintained and the growth-induced water potential had recovered, cell enlargement would be limited by the characteristics of the wall. Apparently, the conditions necessary for rapid wall extension were lost.

It is worth exploring the significance of this change in the light of the changes that already would have taken place. What began as a response of the cells to a change in hydraulics resulted in a changed wall that must have a biochemical basis. Evidently, the hydraulic changes eventually influenced cell metabolism.

111

The exact way in which this occurs is unknown. However, there may be some information in the way solute accumulation occurs. There is considerable evidence that the accumulation is a consequence of the reduced growth and not a cause (27, 32). If we view the osmotic pool as the sum of the influx and the efflux of solute, it must reflect the dynamics of these fluxes (27). A few hours after low water potentials are imposed on the xylem, the use of solute diminishes (27). Because influx continues rapidly, solutes begin to pile up unused (27, also chapter by Hsiao).

However, this osmotic adjustment is accompanied by some other changes. The rate of cell division decreases to the same extent as cell enlargement (26). The rate of maturation of the cells decreases (27). There is a general decrease in the rate of dry matter incorporation into newly matured tissues which is diagnostic for the decreased rate of biosynthesis generally (27). The decrease in wall extensibility may be a manifestation of these coincident effects.

If the decreased extensibility of the walls is associated with these changes, are components of the wall unable to be used and piling up in the wall or cytoplasm? On the other hand, is the transport of structural material or wall loosening factors to the wall being decreased, in effect starving the wall for components needed for extension?

WALL PROTEINS

There is evidence that both types of phenomena are occurring. The ability of the cell to acidify the cell wall decreases at low water potential (36). Because protons act as wall looseners (14), the delivery of wall loosening factors appears to have been reduced. Thus, the wall may be starved for protons. In addition, components of the wall are changing. For example, the protein extractable from the wall increases (9). Figure 1 shows that walls of the dividing and elongating cells contain much more salt-extractable protein than walls of the mature cells. Within 24 h after low water potentials were imposed, the extractable proteins increased. Assuming the proteins are associated with wall extension or biosynthesis, this result implies that they are not being used as rapidly as they are being transported to the wall or, less likely, that there is an increased biosynthesis and transport of wall components. The unlikelihood of increased biosynthesis comes from evidence that rates of protein synthesis usually decrease in enlarging tissue at low water potentials (3, 19, 23). Therefore, proteins for the wall may be piling up unused (not incorporated into wall structure? not degraded at the usual rate?) and showing up as an increased amount in the extractable pool.

Interestingly, not all the wall proteins show the same pattern. Figure 2 shows that a 28 kD protein accumulates particularly strongly among the proteins extractable from the wall (9). As yet, no role can be assigned for this protein but its appearance in the dividing and elongating tissue and almost undetectable amount in the mature tissue suggests that it is involved in the growth process. The specific accumulation of this protein relative to others in the wall raises the possibility that there are specific effects of low water potentials on cell wall metabolism and that, despite the general accumulation of cell wall protein, there

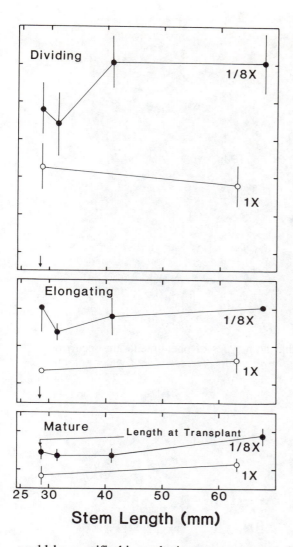

FIG. 1. Amount of protein extractable from the cell walls of the dividing, elongating, and mature regions of dark-grown soybean stems after transplanting the seedlings to vermiculite having high (1X) or low (1/8X) water contents. The 1X and 1/8X vermiculite had a water potential of -0.1 bar and -3 bars, respectively. Arrow indicates the length of the stems at transplant. Stems of 1/8X plants did not grow for first 24 h after transplant and thus had same length as 1X plants at transplant. Thereafter, each succeeding data point was measured 24 h after the previous one. Proteins were extracted from isolated cell walls in 0.5 M $CaCl_2$ at pH 7.2 and assayed according to Bradford (10). Data represent means ± 1 SD. See Bozarth et al. (9) for details.

could be specific biosynthetic or transport or utilization steps that have been altered. More work is needed to explore these possibilities.

ORIGIN OF GROWTH—INDUCED WATER POTENTIALS

The extensibility of the walls is important in an entirely different context. We have seen that a significant amount of force is necessary to overcome the frictional resistances to water movement in enlarging tissue. The force originates within the growing tissue and therefore the growth process itself must act in some way to generate the force. It is increasingly likely that the extensibility of the wall plays a role in the development of this force (see chapters by Ray and by Cosgrove).

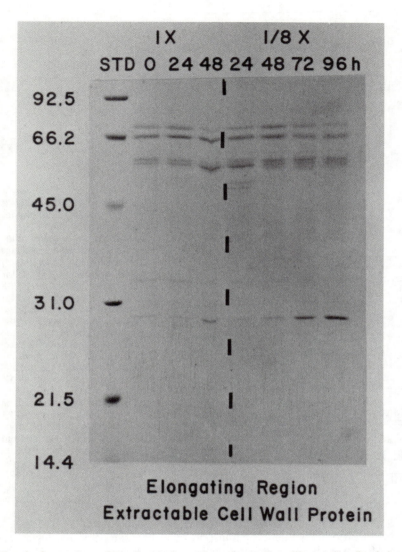

FIG. 2. Comparison of banding patterns of proteins extracted from cell walls of elongating regions of dark-grown soybean stems at 24 h intervals after transplanting the seedlings to vermiculite having high (1X) or low (1/8X) water contents. Protein was extracted as in Fig. 1. SDS-PAGE gel was loaded with 20 µg protein/lane and stained with Coomassie Blue. Note appearance of 28 kD protein in 1/8X tissue at 72 h and 96 h. See Bozarth et al. (9) for details.

One may imagine the following sequence of events. As water enters cells in the enlarging region in response to the osmotic potential of the protoplast solution, turgor increases. After reaching the yield threshold, growth begins as turgor causes the wall to yield. However, this yielding prevents turgor from increasing further. The inability of turgor to reach its maximum causes the

114

water potential to be below zero in the enlarging cells, thus generating the growth-induced water potential (4).

There is evidence that the walls can yield rapidly to turgor. Boyer et al. (7) showed a rapid relaxation (within 5 min) of cell walls when the water supply was totally removed from stem enlarging tissue by excising the other parts of the plant. The relaxation occurred because the extension of the wall was not matched by water inflow, and turgor decreased as the walls extended without a water supply. The rapidity of the effect indicates that the wall must have been yielding rapidly in the intact plant.

Cosgrove and coworkers (15, 16) also observed wall relaxation in a similar experiment but it was much slower, requiring up to 5 h for completion. The reasons for this discrepancy are unknown.

ANALYSIS OF GROWTH

This evidence suggests that the turgor must be high enough to extend the walls irreversibly but low enough to generate the growth-induced water potential that brings water in to enlarge the cell (4, 7). Thus, wall extension creates a demand for water but at the same time generates the force that allows water to be supplied. The two acts are inevitably linked and both need to be included before an accurate understanding of growth is possible.

Recently, a method was described for measuring all of the factors affecting cell enlargement in the intact plant (7). The analysis, based on the theory of Lockhart (22), considers steady tissue enlargement (G) to result from the extension of the cell walls by turgor thus creating a demand for water. The relationship depends on the average cell turgor (ψ_p) above the yield threshold (Y) and on the average wall extensibility (ϕ, $s^{-1}bar^{-1}$) according to:

$$G = \phi(\psi_p - Y) \tag{1}$$

The demand for water is satisfied when water moves from the xylem into the cells in response to the water potential field created by the yielding of the walls to turgor. Thus, the average water potential of the enlarging cells (ψ_w) is below that of the water supply (ψ_o, usually the xylem) and the tissue has an average hydraulic conductance (L, $s^{-1}bar^{-1}$) for water:

$$G = L(\psi_o - \psi_w) \tag{2}$$

where G is the relative rate of increase in water volume ($dV/dt \cdot 1/V$). The G of Eqs. 1 and 2 are numerically equal because the volume of the intercellular spaces, though contributing to G in Eq. 1, cancels in the differential form of the expression to give the G of Eq. 2.

The cell solves Eqs. 1 and 2 simultaneously. Because the potentials driving wall extension and water uptake are additive:

$$G = \frac{\phi L}{(\phi + L)} (\psi_o - \psi_w + \psi_p - Y) \tag{3}$$

115

and, because $\psi_w - \psi_p = \psi_s$:

$$G = \frac{\emptyset L}{(\emptyset + L)} \ (\psi_0 - \psi_s - Y) \qquad (4)$$

Equation 4 governs the rate of tissue enlargement and shows that the rate is determined by a coefficient containing wall properties and water flow properties. If L is large compared to $\emptyset$, the coefficient approximates $\emptyset$ and Eq. 4 reduces to Eq. 1 (see Eq. 3). If $\emptyset$ is large compared to L, the reverse is true and Eq. 4 reduces to Eq. 2 (see Eq. 3). The force generating cell enlargement is the total osmotic force diminished by the water potential of the xylem and the yield threshold turgor. If the water potential of the xylem (ψ_0) decreases, the force decreases. Likewise, if the yield threshold increases, the force decreases. The force is affected by osmotic adjustment, which alters ψ_s.

Each of these components is measured in a simple fashion (7). The method uses a newly developed guillotine psychrometer (7) to detect the vapor pressure of the solution in the apoplast of the growing tissue. No transpiration occurs during the measurement, which assures that all water movement and tissue potentials reflect only the growth process. Because the apoplast is in local equilibrium with the protoplasts (6, 29) and the plants are intact and growing during the measurement, the water potential of the enlarging tissue (ψ_w) is measured without disturbing the growing system. The ψ_0 is determined separately, usually on mature tissue close to the enlarging region. The mature tissue reflects the water potential of the xylem when no transpiration is occurring (11, 37).

After ψ_0 and ψ_w have been evaluated, the enlarging tissue is excised with the guillotine, which makes a cut on the outside of the vapor pressure chamber. This removes the water supply, and turgor decreases as the cell wall relaxes. Growth ceases abruptly under these conditions (7), and the new turgor represents Y. The amount of change in the tissue water potential is caused entirely by this turgor change. The water potential change is thus a direct measure of $(\psi_p - Y)$.

Because the water potential change is completed in 5 min or less (7), $(\psi_p - Y)$ must be similar to that in the intact plant before excision. G is determined in the same plants during the psychrometer measurements. The ψ_s also is measured in the same tissue after the experiment is completed. Thus, all the terms in Eqs. 1-4 are known except $\emptyset$ and L which can be calculated.

These equations and principles of measurement are shown diagrammatically in Fig. 3. Figure 3A represents the extensibility expression (Eq. 1), Fig. 3B the conductance expression (Eq. 2), and 3C the combined expression from 3A and 3B (Eqs. 3 and 4). Note that the combined expression defines the growth-induced water potential at the intersection of the lines (Fig. 3C) from Fig. 3A and 3B. The relaxation of the cell wall occurs after the tissue is excised. The relaxation shifts the conductance expression to the right (ψ_0 moves to the right in Fig. 3D). This collapses the potential difference that drives cell enlargement ($\psi_0 - \psi_s - Y$ approaches zero) and identifies Y. In soybean stems, this analysis indicated that ($\psi_0 - \psi_w$) was about 2 bars and ($\psi_p - Y$) was about 1 bar when the tissue was growing rapidly with adequate water (7). The

116

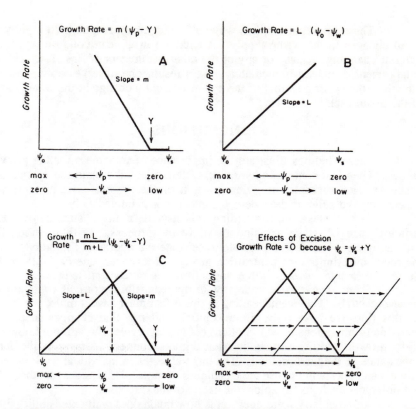

FIG. 3. Diagrammatic representation of Equations 1 to 4 showing effects of A) wall extensibility ø, B) water conductance L, C) both wall extensibility and water conductance, and D) effects of depletion of water supply by plant tissue that is enlarging steadily. In D, depletion causes the water potential of the water source (ψ_o) to move lower (to the right) until $\psi_o = \psi_w = \psi_s + Y$, where wall extension ceases. This change can be used to measure Y. Among the factors that cause depletion are a) the use of water in the soil without replenishment or b) the use of water in the xylem after excision of the tissue in air. Y is most accurately measured when depletion is rapid, as in b). From (6).

coefficients ø (0.95×10^{-5} $s^{-1} bar^{-1}$) and L (0.78×10^{-5} $s^{-1} bar^{-1}$) were the same order of magnitude. Therefore, cell enlargement was co-limited by wall extensibility and the conductance of the enlarging stem tissue for water (7).

The analysis also illustrates that turgor is a dynamic property of enlarging cells and that cell enlargement can be affected by several factors in addition to the turgor. Some of them are hydraulic (ψ_w, ψ_p), some are determined mostly by metabolism (ø, Y, L) and some are external to the enlarging tissue (ψ_o and sometimes L). If any are disrupted, enlargement is disrupted and each factor must readjust until the steady state is again achieved. This requires considerable time because complex metabolically-determined factors must change as well as hydraulic ones. Thus, the return to the steady state may take hours (see 7 for an example). During this time, Eqs. 1-4 cannot be applied because they rely on the steady state. However, after the steady state resumes, the analysis may be used again.

These principles explain why cell enlargement responds so rapidly to small changes in the water supply. A decrease in ψ_0 can collapse $(\psi_0 - \psi_w)$ without changing ψ_w, ψ_p, or any of the other parameters of Eqs. 1-4. Tissue enlargement decreases immediately. As a result, one observes a response of growth to the water potential of the xylem without a change in the water status of the tissue itself.

CONCLUSIONS

These findings illustrate the importance of intactness for the growth process. The plant contains a hydraulic system oriented around the xylem, and tissue excision disrupts the xylem portion of the system. Because growth depends on hydraulics in the xylem, the rate responds immediately.

Nevertheless, understanding this aspect of the system allows the judicious use of tissue excision to study the process. With methods for evaluating Eqs. 1 to 4, it is possible to separate the hydraulic limitations from the metabolic limitations controlling growth. Moreover, one can follow the changes in each factor as growth changes from one steady state to the next.

There appear to be decreases in the extensibility of cell walls at low water potentials. The events leading to this effect must have a molecular basis. At this time we have only circumstantial evidence that changes in proton secretion or alterations in wall proteins could play a role. We do not know the early growth-limiting events or what signals connect them with the later appearance of the wall proteins. The large changes in individual wall proteins provide an opportunity to use the techniques of molecular genetics together with physiological methods to explore the nature of the response.

A particularly large question is how much does wall extensibility limit cell enlargement in comparison with other factors that could be equally limiting, particularly at low water potentials? The answer to this question could contribute a great deal to our understanding of when molecular events in the wall affect the growth process. Such an analysis also would be useful for understanding not only how plants cope with low water potentials but how they acclimate to other unfavorable environments, altered hormonal relations, changes in light intensities, and so on.

Considerable progress has already been made. It is now possible to account for instances where tissue enlargement is affected by small changes in water supply without changes in the water status of the enlarging cells. The rates can change too rapidly (within 1 min) to involve growth regulators. The most likely explanation lies in effects on the xylem water potential. At the same time, this concept allows us to understand the enigma of how enlarging tissue can grow at the expense of water in other tissues.

In the future we hope it will be possible to link the events controlling plant growth more fully. Perhaps it will be possible to alter growth rates in various tissues and organs after more of the mechanisms are known. When one compares the contrasting effect of low water potentials on the growth of roots and shoots (see 38 and the chapter by Hsiao for examples), it is clear that reduced growth is not inevitable at low water potentials. The genetic and metabolic apparatus already exists to alter the growth response. We only must find it.

LITERATURE CITED

1. ACEVEDO E, TC HSIAO, DW HENDERSON 1971 Immediate and subsequent growth responses of maize leaves to changes in water status. Plant Physiol. 48: 631-636
2. BARLOW EWR 1986 Water relations of expanding leaves. Aust J Plant Physiol 13: 45-58
3. BEWLEY JD, KM LARSEN 1982 Differences in the responses to water stress of growing and non—growing regions of maize mesocotyls: protein synthesis on total, free and membrane—bound polyribosome fractions. J Exp Bot 33: 406-415
4. BOYER JS 1968 Relationship of water potential to growth of leaves. Plant Physiol 43: 1056-1062
5. BOYER JS 1974 Water transport in plants: mechanism of apparent change in resistance during absorption. Planta 117: 187-207
6. BOYER JS 1985 Water transport. Ann Rev Plant Physiol 36: 473-516
7. BOYER JS, AJ CAVALIERI, ED SCHULZE 1985 Control of the rate of cell enlargement: excision, wall relaxation, and growth-induced water potentials. Planta 163: 527-543
8. BOYER JS, G WU 1978 Auxin increases the hydraulic conductivity of auxin-sensitive hypocotyl tissue. Planta 139: 227-237
9. BOZARTH CS, JE MULLET, JS BOYER 1987 Proteins in cell walls at low water potentials. Plant Physiol: in press
10. BRADFORD MM 1976 A rapid and sensitive method for the quantitation of microgram quantities of protein utilizing the principle of protein - dye binding. Anal Biochem 72: 248-254
11. CAVALIERI AJ, JS BOYER 1982 Water potentials induced by growth in soybean hypocotyls. Plant Physiol 69: 492-496
12. CLELAND RE 1959 Effect of osmotic concentration on auxin - action and on irreversible and reversible expansion of the *Avena* coleoptile. Physiol Plant 12: 809-825
13. CLELAND RE 1976 The control of cell enlargement. Symp Soc Exp Biol 31: 101-115
14. CLELAND RE 1981 Wall extensibility: hormones and wall extension. *In* Encyclopedia of Plant Physiology (NS) Vol 13B: Plant Carbohydrates. II. Extracellular Carbohydrates. Ed. W Tanner, FA Loewus, Springer-Verlag, New York, pp 255-276
15. COSGROVE DJ 1985 Cell wall yield properties of growing tissue: evaluation by *in vivo* stress relaxation. Plant Physiol 78: 347-356
16. COSGROVE DJ, E VAN VOLKENBURGH, RE CLELAND 1984 Stress relaxation of cell walls and the yield threshold for growth: demonstration and measurement by micro-pressure probe and psychrometer techniques. Planta 162: 46-54
17. COSGROVE DJ, RE CLELAND 1983 Solutes in the free space of growing stem tissues. Plant Physiol 72: 326-331

18. CUTLER JM, PL STEPONKUS, MJ WACH, KW SHAHAN 1980 Dynamic aspects and enhancement of leaf elongation in rice. Plant Physiol 66: 147-152
19. DASGUPTA J, JD BEWLEY 1984 Variations in protein synthesis in different regions of greening leaves of barley seedlings and effects of imposed water stress. J Exp Bot 35: 1450-1459
20. GREEN PB, RO ERICKSON, J BUGGY 1971 Metabolic and physical control of cell elongation rate: *in vivo* studies in *Nitella*. Plant Physiol 47: 423-430
21. HÜSKEN D, E STEUDLE, U ZIMMERMANN 1978 Pressure probe technique for measuring water relations of cells in higher plants. Plant Physiol 61: 158-163
22. LOCKHART JA 1965 An analysis of irreversible plant cell elongation. J Theor Biol 8: 264-275
23. MASON HS, K MATSUDA 1985 Polyribosome metabolism, growth and water status in the growing tissue of osmotically stressed plant seedlings. Physiol Plant 64: 95-104
24. MATSUDA K, A RIAZI 1981 Stress-induced osmotic adjustment in growing regions of barley leaves. Plant Physiol 68: 571-576
25. MATTHEWS MA, E VAN VOLKENBURGH, JS BOYER 1984 Acclimation of leaf growth to low water potentials in sunflower. Plant Cell Environ 7: 199-206
26. MEYER RF, JS BOYER 1972 Sensitivity of cell division and cell elongation to low water potentials in soybean hypocotyls. Planta 108: 77-87
27. MEYER RF, JS BOYER 1981 Osmoregulation, solute distribution, and growth in soybean seedlings having low water potentials. Planta 151: 482-489
28. MICHELENA VA, JS BOYER 1982 Complete turgor maintenance at low water potentials in the elongating region of maize leaves. Plant Physiol 69: 1145-1149
29. MOLZ FJ, E IKENBERRY 1974 Water transport through plant cells and cell walls: theoretical development. Soil Sci Soc Amer Proc 38: 699-704
30. MOLZ FJ, JS BOYER 1978 Growth-induced water potentials in plant cells and tissues. Plant Physiol 62: 423-429
31. MORGAN JM 1984 Osmoregulation and water stress in higher plants. Ann Rev Plant Physiol 35: 299-319
32. MUNNS R, J BRADY, EWR BARLOW 1979 Solute accumulation in the apex and leaves of wheat during water stress. Aust J Plant Physiol 6: 379-389
33. NONAMI H, JS BOYER 1987 Origin of growth-induced water potential: solute concentration is low in apoplast of enlarging tissues. Plant Physiol 83: 596-601
34. PROBINE MC, RD PRESTON 1962 Cell growth and the structure and mechanical properties of the wall in internodal cells of *Nitella opaca*. J Exp Bot 13: 111-127
35. TAIZ L 1984 Plant cell expansion: regulation of cell wall mechanical properties. Ann Rev Plant Physiol 35: 585-657

36. VAN VOLKENBURGH E, JS BOYER 1985 Inhibitory effects of water deficit on maize leaf elongation. Plant Physiol 77: 190-194
37. WESTGATE ME, JS BOYER 1984 Transpiration- and growth-induced water potentials in maize. Plant Physiol 74: 882-889
38. WESTGATE ME, JS BOYER 1985 Osmotic adjustment and the inhibition of leaf, root, stem and silk growth at low water potentials in maize. Planta 164: 540-549

Physiology of Cell Expansion During Plant Growth, *D.J. Cosgrove and D.P. Knievel* Eds.,
Copyright ©1987, The American Society of Plant Physiologists

INHIBITION OF ELONGATION BY ANTIBODIES SPECIFIC FOR WALL PROTEINS

D.J. NEVINS, R. HATFIELD[1] , T. HOSON[2] AND M. INOUHE

*Department of Vegetable Crops, University of California,
Davis, CA 95616*

INTRODUCTION

Growth of plant cells is accommodated by the uptake of water coupled with a rearrangement of structural elements in the wall. The reorganization of wall components, normally stabilized by intra- or inter-molecular bonds, is closely integrated with *de novo* synthesis and insertion of new polymers (1, 4, 23). The precise nature of the changes in structure leading to initial shifts in wall compliance is unknown but it is highly probable that the state of the wall is predisposed by metabolic events governing adjustments in polymer interactions. These events may allow walls to respond to stimuli mediated by phytohormones at appropriate temporal and spatial stages in development. Studies on composition and an emerging knowledge of the structural changes which occur in cell walls during growth support the concept of a dynamic yet highly ordered matrix (14, 18). As a result of the sequence of events and induced changes in the matrix, irreversible cell extension occurs, a process highly coordinated with other cells in the tissue. Substantial evidence supports the concept that elongation is, however, fundamentally regulated by responses of cells constituting the epidermis (16, 17, 20, 22, 26).

The background for understanding the nature of cell extension has been augmented in recent years by new techniques which provide significant insights into wall polysaccharide and protein structure. Nevertheless progress toward the goal of resolving the mechanisms will require the development of appropriate specific molecular probes to disclose regulatory events within the sequence

---

[1]Current address: USDA-ARS, US Dairy Forage Research Center, Madison, WI 53706.

[2]Current address: Dept. of Biology, Faculty of Science, Osaka City University, Sumiyoshi-ku, Osaka 558, Japan.

Abbreviations: Prot-I through Prot-IV represent protein antigens active in generating growth inhibitory antibodies at increasing stages during purification. Anti-I through Anti-III represent antibodies derived from Prot-I through Prot-III. Prot-exoglu is the exoglucanase and Prot-endoglu is the endoglucanase derived from maize walls.

leading to cell elongation. We therefore became intrigued with the possibility of developing specific antibodies against putative regulatory proteins within the wall. We recognize that there are inherent dangers involved in interpreting systems by dissociating specific events. Certainly progress toward this end has the potential to develop probes for influencing the action of specific gene expression products thereby affording a powerful new approach to contribute to an understanding of the extension process.

WALL PROTEINS

There is no need to develop in detail a case in this chapter for the existence and proposed function of cell wall proteins. Cell walls have numerous enzymatic proteins potentially responsible for mediating structural changes (10, 14, 15). Wall proteins are also widely recognized for their putative structural roles (13, 14, 15). Arabinose containing hydroxyproline-rich glycoproteins (extensins) have been the subject of numerous studies. The ongoing molecular characterization and molecular biology of extensin and extensin precursors continue to contribute to our understanding of an important class of wall structural proteins (13, 14, 15). It remains uncertain what precise functional role these proteins serve; however, quantitative changes in extensin content indicate that it is most abundant after elongation has ceased (8). The generally accepted interpretation of this observation is that structural protein serves to stabilize the wall matrix following extension. The role of isodityrosine as an intramolecular crosslinking mechanism is consistent with this view (15, 23).

Less established as wall components but perhaps of great significance are lectins. Lectins are carbohydrate binding proteins with a curious diversity in plants but their functional role remains obscure. Quantitative changes in lectins during development suggest an association with growth (11) and the dissociation of wall components by certain treatments has been attributed to a modification of a lectin-like function (3). A better understanding of polysaccharide structure, their interactions, metabolism and the identification of certain functions will serve to resolve some of these important issues.

In any case, current evidence suggests that an interaction between these proteins contributes to the regulation of the function of complex carbohydrates. Included in this series of interactions would be covalent binding as in the formation of glycoproteins, a myriad of enzymatic processes that serve to modify polysaccharide structure, and the possibility of regulation of polymer interactions by noncovalent binding. Because of affinities within the wall matrix, it only reluctantly yields individual components and often these components are extensively modified to effect solubilization. Recurring questions about the integrity of the liberated fractions are appropriate.

PROTEINS SOLUBILIZED BY LITHIUM CHLORIDE

During the course of an investigation of enzymes responsible for maize cell wall autohydrolysis (degradation of insoluble wall components by wall bound enzymes) we observed that the process was arrested following a treatment with molar concentrations of LiCl (9, 10). The basis for the suppression of autolysis by appropriate LiCl treatments was attributed to the solubilization and removal of selected enzymes. Since the enzymes were not denatured by extraction or by subjecting them to high salt, they could be studied in detail. Simulated autolytic activity was observed when an appropriate balance of selected solubilized enzymes was added to inactivated wall material (7, 10).

THE SYSTEM

The extracted protein fraction described above forms the basis for the observations reported herein. Proteins were derived from walls of shoots of 5 day old maize seedlings (*Zea mays* L. hybrid B73 x Mo17) (9). Frozen tissue was homogenized in an ice slurry containing 50 mM NaCl. The insoluble material retained by Miracloth was reextracted with 50 mM NaCl and treated with acetone (-20°C) followed by extensive rinsing with acetone and a final treatment with 50 mM NaCl. Free liquid was expressed from the wall by squeezing the retained material confined within the Miracloth. The insoluble walls were then extracted with 3 M LiCl at 2°C for 48 hrs (9). The LiCl extract was dialyzed, concentrated and used as a select source of wall proteins (Prot-I) for the experiments described.

The LiCl extracted proteins were of primary interest for several reasons. First, that fraction was comprised of enzymes responsible for cell wall glucan autohydrolysis. In subsequent experiments it was found that this fraction, when employed to immunize rabbits, generated antibodies with the capacity to suppress auxin induced elongation (9). A conventional protocol for antibody production was employed utilizing female New Zealand white rabbits subcutaneously immunized on a weekly basis with increasing dosages of nondenatured protein. The maximun amount of protein injected ranged up to 2000 μg in the case of crude protein and to 60 μg in more refined fractions. Nonterminal bleeding was initiated after 12 weeks and continued thereafter at two week intervals. It should be noted that over the course of these experiments many different rabbits have been used with consistant results. The IgG fraction from the serum was purified by Protein A chromatography.

To determine the inhibitory response the coleoptiles were gently abraded with carborundum. For most experiments 10 mm abraded sections were pretreated for 2 hr with the antibody preparation followed by rinsing and recutting of a 5 mm segment from the central region (Fig. 1). Uptake of antibody through the cut ends of the section tended to cause a lack of uniformity in the growth response. However, some inhibition of growth was observed in unabraded sections when the ends were not recut prior to auxin treatment. Treatment of sections with preimmune serum had no effect on elongation responses.

EVIDENCE FOR SPECIFICITY OF INHIBITION

Huber and Nevins (9) initially demonstrated that the antibody preparation (Anti-I) (Fig. 2) generated in response to LiCl extracted proteins (Prot-I) would inhibit the elongation of maize coleoptiles and that it would also suppress autolytic reactions of isolated walls. The initial approach did not distinguish between a nonspecific effect caused by diverse antibodies or a specific effect. To address this issue LiCl extracted proteins (Prot-I) were subjected to

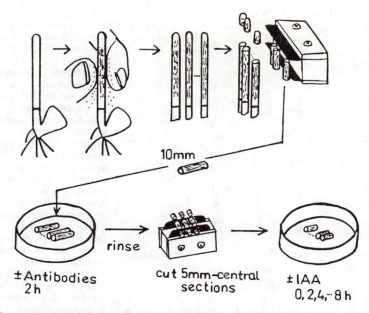

FIG.1. The protocol used to prepare tissue for evaluating the effect of antibodies on auxin-induced cell elongation of maize coleoptile segments is illustrated. The seedlings were grown for 5 days in the dark at 25°C. The surface of the coleoptiles was abraded longitudinally with carborundum. The abraded coleoptiles were washed with distilled water and 10 mm sections were cut from regions 3-13 mm below the tip. The sections were treated with 10 mM K-citrate phosphate buffer (pH 6.5) containing 100 to 1000 μg serum protein. Controls were prepared for each treatment. After preincubation, the segments were rinsed with distilled water and a 5 mm segment cut from the center. The 2.5 mm ends were discarded. The 5 mm segments were treated with or without IAA (10 μM) and elongation determined with the aid of binocular microscope equipped with an ocular micrometer.

fractionation (Fig. 2). SP-Sephadex columns were effective in resolving an unbound fraction (Prot-II) from major enzymatic components responsible for autolytic reactions. The enzymes within the elution profile included an exoglucanase (Prot-exoglu) (10) and an endoglucanase (Prot-endoglu) (6, 7). Both enzymes were capable of generating antibodies which suppress autolysis and to a variable extent could also inhibit auxin induced elongation. These responses will be the basis for subsequent reports.

The fraction not retained by the SP-Sephadex column (Prot-II) has proved to be especially effective in generation of antibodies (Anti-II) that inhibit auxin induced elongation. However, Anti-II has little or no effect on autolysis. Protein-II was subjected to Bio-Gel P-150 gel exclusion chromatography. Various major fractions resolved in the profile were evaluated and one (Prot-III) was found to generate antibodies (Anti-III) responsible for the suppression of growth. A relationship between the concentration of antibody used in the treatment and the extent to which inhibition occurred was observed (Fig. 3).

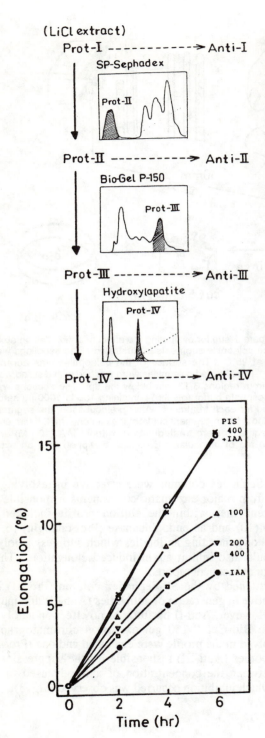

(LiCl extract)

Prot-I ----------→ Anti-I

SP-Sephadex

Prot-II

Prot-II ----------→ Anti-II

Bio-Gel P-150

Prot-III

Prot-III ----------→ Anti-III

Hydroxylapatite

Prot-IV

Prot-IV ----------→ Anti-IV

FIG. 2. Scheme for the fractionation of the LiCl extracted cell wall proteins. The protein (Prot-I) represents the crude LiCl extract used to generate the antibody (Anti-I). Subsequently Prot-I was subjected to SP-Sephadex chromatography to obtain Prot-II (shaded region of the profile). Prot-II was used to generate Anti-II. Prot-II was subjected to Bio-Gel P-150 gel exclusion chromatography to obtain Prot-III (shaded portion). Prot-III was used to generate Anti-III. Prot-III was fractionated on a hydroxylapatite column to generate Prot-IV (shaded portion). The generation of antibodies from Prot-IV is underway. Except for shaded regions most of the proteins resolved were ineffective in generating growth inhibitory antibodies. Purity of fractions was determined by SDS-PAGE. Details concerning the purification process are in preparation for publication.

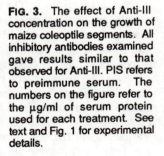

FIG. 3. The effect of Anti-III concentration on the growth of maize coleoptile segments. All inhibitory antibodies examined gave results similar to that observed for Anti-III. PIS refers to preimmune serum. The numbers on the figure refer to the μg/ml of serum protein used for each treatment. See text and Fig. 1 for experimental details.

126

DIAGNOSTICS FOR EVALUATING ACTIVE FRACTIONS

Initially fractions from various separation procedures were injected into rabbits using standard protocols. However, a convenient method was devised to facilitate progress. Antibody preparations containing inhibitory IgG components were treated with appropriate quantities of protein to effect binding and precipitation (Fig. 4). The aggregate complex was removed by centrifugation and serum components remaining in solution are evaluated in the bioassay system. Rather pure protein preparations have been shown to reverse inhibitory effects of some rather crude antibody preparations. Only a small amount of the total protein is removed from the serum by this procedure. In the appropriate combination highly refined fractions are defined which effectively reverse the inhibitory effect (Fig. 4). The results of this approach can be reconciled with those obtained using antibodies generated by injection of more highly purified proteins.

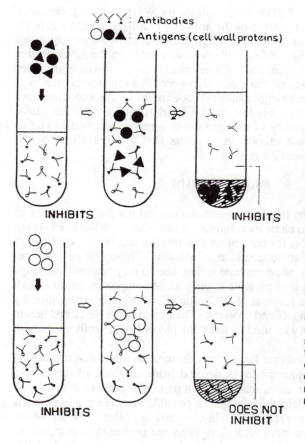

INHIBITS   INHIBITS

INHIBITS   DOES NOT INHIBIT

FIG. 4. Immunoprecipitation of antibodies from IgG preparations after Protein A purification. Top row illustrates the precipitation of inactive antibodies by wall proteins not containing active antigens. After precipitation the supernatant of the sample illustrated at the top right retains the inhibitory antibody. The bottom row illustrates the effect of adding one of the active antigens to an antibody preparation which may be rather crude. The supernatant of the tube at the bottom right is no longer capable of inhibiting elongation even though most serum components remain.

127

ADDITIONAL FRACTIONATION STEPS

Because of the effectiveness of the precipitation approach described above, we have progressed one step further. The additional step for fractionation of the active protein (Prot-III) employed hydroxylapatite. In this step two protein fractions were recovered and evaluated by the precipitation technique. One fraction did not reverse the growth inhibitory effects of an active antibody preparation (Anti-III) but the other (Prot-IV) did (See Fig. 2). The production of antibodies by Prot-IV is underway. It is of interest that at this stage, amino acid analysis of Prot-IV indicates that it is low in hydroxyproline. However, the results thus far do not exclude the possibility that the protein might be a glycoprotein. Immunoaffinity chromatography represents an additional fractionation method currently being explored.

WESTERN BLOTS

Antibody probes of SDS-PAGE plates by Western blot procedures (immuno blots) revealed association of the antibody (Anti-III) with the Prot-III band (2, 5, 12). Binding was mostly at a single band but some minor impurities remained. The binding to the SDS-PAGE zones derived from the SP-Sephadex void fraction (Prot-II) and from the LiCl extract (Prot-I) showed significant binding to the same region of the gel. However no such association was observed on gels of purified exoglucanase (Prot-exoglu) or the endoglucanase (Prot-endoglu). The specific binding to the band (Prot-III) and the direct relationship between the ability of this protein to generate antibodies and to reverse the inhibitory effect support the concept that the interaction of the antibody with the wall is a specific reaction.

A PROSPECTUS

There remain many intriguing questions concerning the significance of the proteins which generate elongation inhibitory antibodies. What is emerging from these results is support for the notion that interference with specific wall proteins by binding leads to suppression of growth. Coleoptile cells do not respond to IgG components of preimmune serum, nor do they respond to a range of antibodies generated by wall proteins in general. An important control in all of these experiments is the reversal of inhibition of a particular preparation by selective precipitation with isolated proteins. The experiments described herein illustrate that bioassays may be used to track the physiological activity through a purification scheme.

In studies conducted thus far we have focused on three proteins and the results presented here emphasize those derived from only one of these. No enzymatic activity has been associated with that particular protein (Prot-III). If enzyme activity could be detected it might be possible to develop a hypothesis for the biological effect of protein-III. In contrast, the other proteins which generate antibodies that interfere with elongation are produced in response to proteins associated with exoglucanase and endoglucanase activities. In the latter two cases, we may consider the possibility that elongation suppression is due to enzymatic inhibition (Fig. 5). In addition the fact that several specific antibodies inhibit growth allows speculation that several proteins may cooperate in controlling elongation. Critical experiments are planned to test these hypotheses.

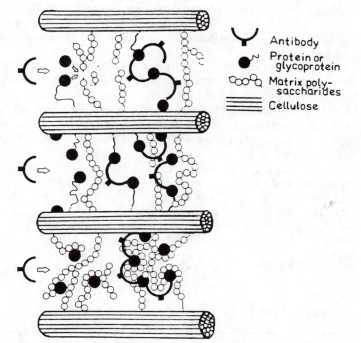

FIG. 5. Some possible mechanisms that might account for the inhibitory effect of the antibody preparation. The effects may be direct or indirect. Cellulose microfibrils serve to delimit three speculative concepts. **Top**: Enzymatic inhibition. At the top, left is an illustration of a hydrolytic process (the effect may not be limited to this type of enzymatic process). Antibodies bind and inactivate a hypothetical enzyme. **Center**: Growth may require functional structural proteins which interact with themselves or with other wall components to regulate compliance. Binding of antibodies to these proteins may alter their function perhaps by a mechanism as simple as increasing their bulk, thereby restricting their mobility. **Bottom**: Growth may be fine tuned by interactions with wall lectins. Growth regulation may be directly or indirectly related. Binding of antibodies to lectins may alter their function perhaps by aggregation. There are other ideas which could account for the observed effects. At this point little evidence supports any one of these concepts but extension of the current work should provide a good base for testing appropriate hypotheses.

Prot-III, the primary fraction considered in this chapter, is a candidate for several potential mechanisms. One possibility is that the antibody (Anti-III) interferes with a protein necessary for transient structural modifications within the wall. A preliminary experiment conducted in conjunction with Professor Y. Masuda has demonstrated that the degree of suppression of elongation is associated with a longer T_0 parameter as determined by stress relaxation measurements (Table I). A shorter T_0 value has typically been taken as evidence for modification of a wall mechanical property associated with auxin induced elongation (19). Moreover the antibody effect on T_0 is maintained through the preparation of tissues for Instron analysis, suggesting that physical displacement of the antibody at that point does not reverse its impact. The explanation may ultimately be linked to blockage of some yet undisclosed enzymatic activity. Perhaps the effect is the consequence of binding to a lectin (Fig. 5).

Table I. *Effect of Antibody-III on elongation and T_o values of maize coleoptiles.*
The protocol includes a 2 hr preincubation period with 400 µg/ml serum protein followed by a 2 hr treatment period. After appropriate tissue preparation the T_0 value was determined by stress relaxation methods in the laboratory of Professor Y Masuda. PIS refers to preimmune serum, Anti-III refers to the antibody preparation generated in response to Prot-III and Anti-III ppt Prot-III refers to Anti-III precipitated with Prot-III prior to the pretreatment period. ± refers to the std error.

Pretreatment of tissue	Treatment of tissue	Elongation % (2 hrs)	T_0 (msec)
Buffer	Buffer	4.0±0.5	17.7±0.7
Buffer	IAA	10.5±0.5	13.5±0.7
PIS	IAA	10.0±0.7	15.0±0.7
Anti-III	IAA	6.0±0.5	16.0±1.2
Anti-III ppt Prot-III	IAA	9.0±0.5	13.7±1.0

Further consideration of the system will have an important bearing on the usefulness of antibodies as probes. In order to explore elongation inhibition, abrasion of the cuticle is necessary. If, however, the activity of the epidermal cells govern the rate of elongation then one would expect a maximum expression of inhibition in the system as described. This spatial arrangement would also be less affected by the molecular dimensions of the antibody. The preparation and testing of FAB fragments generated from active antibodies is currently underway in an effort to enhance penetration. Suppression of auxin induced elongation by antibody mediated inhibition of auxin transport proteins also takes advantage of the regulatory influence of the epidermal cells (16, 17). Ultimately, labeled antibodies will be instrumental in determining the location of binding sites. Finally the antibodies offer attractive probes for identifying gene expression products responsible for auxin induced elongation.

Acknowledgments--The authors acknowledge the contributions of DJ Huber, E Labrador, J Koch and Professor Y Masuda. This work was supported by National Science Foundation grants PCM7818588 and DMB8505901.

LITERATURE CITED

1. BAKER DB, PM RAY 1965 Direct and indirect effects of auxin on cell wall synthesis in oat coleoptile tissue. Plant Physiol 40: 345-352

2. BURNETTE WN 1981 "Western blotting": Electrophoretic transfer of proteins from sodium dodecyl sulfate-polyacrylamide gels to unmodified nitrocellulose and radioactive protein A. Anal Biochem 112: 195-203

3. BATES GW, PM RAY 1981 pH Dependent interactions between pea cell wall polymers possibly involved in wall deposition and growth. Plant Physiol 68: 158-164

4. CLELAND RE 1971 Cell wall extension. Ann Rev Plant Physiol 22: 197-222

5. DELLAPENNA D, RE CHRISTOFFERSON, AB BENNETT 1986 Biotinylated proteins as molecular weight standards on Western blots. Anal Biochem 152:329-332

6. HATFIELD R, DJ NEVINS 1986 Purification and properties of an endoglucanase isolated from the cell walls of *Zea mays* seedlings. Carbohydr Res 148: 265-278

7. HATFIELD R, DJ NEVINS 1987 Hydrolytic activity and substrate specificity of an endoglucanase from *Zea mays* seedling cell wall. Plant Physiol 83: 203-207

8. HOSON T, S WADA 1980 Role of hydroxyproline-rich cell wall protein in growth regulation of rice coleoptiles grown on or under water. Plant Cell Physiol 21: 511-524

9. HUBER DJ, DJ NEVINS 1981 Wall protein antibodies as inhibitors of growth and of autolytic reactions of isolated cell wall. Physiol Plant 53: 533-539

10. HUBER DJ, DJ NEVINS 1982 Exoglucanases from *Zea mays*, L. seedlings: their role in β-D-glucan hydrolysis and their potential role in extension growth. Planta 155: 467-472

11. KAUSS H, C GLASER 1974 Carbohydrate-binding proteins from plant cell walls and their possible involvement in extension growth. FEBS Lett 45: 304-307

12. LAEMMLI UK 1970 Cleavage of structural proteins during the assembly of the head of bacteriophage T4. Nature 227: 680-685

13. LAMPORT DTA 1980 Structure and function of plant glycoproteins. *In* Biochemistry of Plants, ed J Preiss, 3: 501-541

14. LAMPORT DTA 1981 Glycoproteins and enzymes of the cell wall. Plant Carbohydrates II: Extracellular Carbohydrates, eds W Tanner, F Loewus. Springer-Verlag, NY, pp 133-167

15. LAMPORT DTA, L EPSTEIN 1983 A new model for the primary cell wall: A concatenated extension-cellulose network. Curr Top Plant Biochem Physiol, Proc Ann Plant Biochem Physiol Symp, ed DD Randall et al. 2:73-83. Columbia, Univ Missouri

16. LOBLER M, D KLAMBT 1985 Auxin-binding protein from coleoptile membranes of corn (*Zea mays* L.). I. Purification by immunological methods and characterization. J Biol Chem 260: 9848-9853

17. LOBLER M, D KLAMBT 1985 Auxin-binding protein from coleoptile membranes of corn (*Zea mays* L.). II. Localization of a putative auxin receptor. J Biol Chem 260: 9854-9859

18. LUTTENEGGER D, DJ NEVINS 1985 Transient nature of a $(1{\rightarrow}3),(1{\rightarrow}4)$β-D-glucan in *Zea mays* coleoptile cell walls. Plant Physiol 77: 175-178

19. MASUDA Y 1978 Auxin-induced cell wall loosening. Bot Mag Tokyo. (Special issue) 1: 103-123
20. MASUDA Y, R YAMAMOTO 1972 Control of auxin induced stem elongation by the epidermis. Physiol Plant 27: 109-115
21. NEVINS DJ, R HATFIELD, Y KATO 1984 Depolymerization of matrix polysaccharides by endogenous wall enzymes. *In* Structure, function and biosynthesis of plant cell walls, eds WM Dugger and S Bartnicki-Garcia Am Soc Plant Physiol, Rockville MD, pp 167-184
22. POPE DG 1982 Effect of peeling on IAA-induced growth in *Avena* coleoptiles. Ann Bot 49: 493-501
23. TAIZ L 1984 Plant cell expansion: Regulation of cell wall mechanical properties. Ann Rev Plant Physiol 35: 585-657
24. THIMANN KV, CL SCHNEIDER 1937 The role of salts, hydrogen ion concentration and agar in the response of *Avena* coleoptiles to auxins. Am J Bot 25: 270-280

Physiology of Cell Expansion During Plant Growth, *D.J. Cosgrove and D.P. Knievel* Eds.,
Copyright ©1987, The American Society of Plant Physiologists

POSSIBLE LINK BETWEEN AUXIN REGULATED GENE EXPRESSION, H+ SECRETION, AND CELL ELONGATION: A HYPOTHESIS

ATHANASIOS THEOLOGIS

*Plant Gene Expression Center, ARS, USDA, Albany, CA 94710
and Molecular Plant Biology Department,
University of California at Berkeley,
Berkeley, CA 94720*

INTRODUCTION

The primary mechanism of action of auxin is poorly understood. The classic effect of auxins is to promote cell elongation, a process that requires extension of the cell wall (1, 11). The rapidity of the initiation of cell elongation has made this biological phenomenon the most attractive for studying the primary mechanism of auxin (2). Throughout the years, two major views have emerged in explaining auxin induced cell growth. In the early 1960's the gene activation hypothesis suggested that the hormone regulates the synthesis of specific mRNAs coding for proteins involved in the growth process (4). Subsequently, during the early 1970's the acid growth theory was proposed suggesting that plant cell growth is regulated by the auxin induced proton secretion (12).

Recent advances in techniques of gene cloning have led to the isolation of DNA sequences complementary to rapidly induced mRNAs in soybean and pea tissue (19). This made it possible to examine whether there is a relationship between the two theories. Herein, I will present the experimental evidence against any role of proton secretion in the rapid mRNA induction and I will present a model uniting auxin-induced H+ secretion, mRNA induction and cell enlargement.

CLONING AUXIN-REGULATED mRNA SEQUENCES

Analysis of *in vitro* translation products of mRNAs from auxin treated soybean and pea tissue by 2-dimensional gel electrophoresis (17, 21) showed that the hormone specifically protentiates certain mRNAs rapidly. This potentiation was attributed to an increase in the amount of translatable mRNA which is due to the activation of transcription, post transcriptional processing or mRNA stabilization. Furthermore, the possibility existed that the hormone altered the translatability of preexisting mRNAs by polyadenylation, capping or internal methylation. To answer these questions, it was necessary to isolate cDNA clones to auxin-regulated mRNAs.

133

In the last few years, various laboratories were able to isolate complementary DNA sequences to some of the auxin-regulated mRNAs from soybean and pea tissue (19). Differential plaque filter hybridization of cDNA libraries constructed in the vector λgt10 allowed the isolation of two cDNA clones, pIAA4/5 and pIAA6, from pea tissue. They correspond to mRNAs that have been previously identified by *in vitro* translation to code for polypeptide 4, 5, and 6 (18). Clone pIAA4/5 (Figure 1A) hybridizes to two mRNAs that are

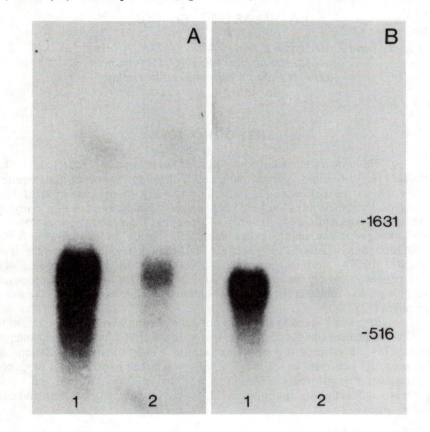

FIG. 1. IAA regulated mRNAs in pea tissue. Autoradiograms of RNA filter paper hybridized successively with ^{32}P-labeled pIAA4/5(A) and pIAA6(B) plasmid DNAs. mRNA was isolated from pea segments treated for 2 hr with 20 μM IAA (lane 1) or without hormone (lane 2).

greatly induced after 2 hr of incubation in the presence of 20 μM IAA. Similarly, clone pIAA6 hybridizes to a smaller size mRNA which is also induced after a similar treatment.

Isolation of the cDNA clones allowed a detailed characterization of the hormonal response in pea tissue. A significant problem in the study of the initial steps in hormone action at the gene level is establishing whether the mRNA induction represents a primary action of the hormone. There are three criteria to be fulfilled: the induction of the mRNA should be rapid, specific and insensitive to protein synthesis inhibition. The auxin mediated mRNA accumulation in pea tissue fulfills these criteria (18).

The induction of the pIAA4/5 and pIAA6 mRNAs depends on transcription because α-amanitin, an inhibitor of RNA polymerase II, completely abolishes the mRNA accumulation (Figure 2). Recent experimental

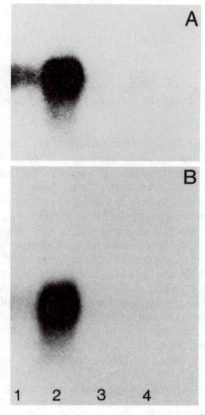

FIG. 2. Effect of α-amanitin on the expression of the IAA inducible mRNAs in pea tissue. Autoradiograms of RNA filter paper hybridized successively with ^{32}P-labeled pIAA4/5(A) and pIAA6(B) plasmid DNAs. Poly A$^+$-RNA was isolated from abraded pea segments incubated for 2 hr. Lane 1, control; lane 2, 20 μM IAA; lane 3, 5 μM α-amanitin and lane 4, IAA + α-amanitin.

evidence obtained with nuclear runoff assays and from determining the effect of the hormone on the half lives of the mRNAs indicates that auxin acts at the transcription initiation level and does not have any post-transcriptional effect (5). Furthermore the transcriptional inhibitor abolishes the auxin induced cell elongation and H$^+$ secretion in pea tissue (Table I). This indicates that transcription is a prerequisite for the plant cell growth as well as for H$^+$-secretion. Whether the protein coded by the auxin regulated mRNAs are involved in these two processes is not known at the present time.

IS IAA-INDUCED H$^+$ SECRETION THE DRIVING FORCE FOR mRNA INDUCTION?

The isolation of cDNA clones to some auxin regulated mRNAs made it possible to ask whether the auxin induced H$^+$ secretion is the driving force for

Table I. *Effect of α-amanitin on the IAA-induced cell elongation and H+ secretion.*

Inhibitor	ΔElongation	ΔpH
	mm	
Control	4	-0.18
IAA 20 μM	18	-0.57
α-amanitin 5 μM	2	-0.10
IAA + α-amanitin	1	-0.02

Abraded pea segments (100 per sample, ~2 g fr wt) were prepared and depleted as described in Ref. (18)

the transcriptional activation via an alkalinization of the cytoplasmic pH. The experimental approach to answer this question was to inhibit the auxin induced H^+ secretion with metabolic inhibitors, or to stimulate H^+ secretion by the fungal toxin fusicoccin (6) and monitor by RNA hybridization analysis whether the induction of the genes takes place. Firstly, cycloheximide, an inhibitor of protein synthesis, inhibits the auxin induced cell growth and H^+ secretion (Table II). However, the inhibitor alone induces the pIAA4/5 and pIAA6 mRNA to same level as auxin does (Table II). IAA in the presence of cycloheximide induces the genes, although H^+ secretion has been completely abolished. Secondly, vanadate, an inhibitor of H^+-ATPases (13, 15), inhibits the auxin induced H^+ secretion and cell elongation; however, the inhibitor is unable to prevent the mRNA accumulation (Table II). Thirdly, cerulenin, an inhibitor of fatty acid biosynthesis (18), also inhibits the IAA induced H^+ secretion and cell elongation and is ineffective in inhibiting the expression of the IAA inducible mRNAs. Lastly, mannitol at 0.4 M prevents both cell elongation and H^+ secretion but is unable to prevent the IAA induced mRNA accumulation (18). On the other hand, stimulation of proton secretion with fusicoccin (Table III) does not cause induction of the auxin-regulated mRNAs in pea segments (Fig. 3). Studies with [^{31}P] NMR have shown that treatment of pea epicotyl segments with IAA or fusicoccin does change the cytoplasmic or vacuolar pH (16). Consequently, the experimental evidence overwhelmingly rejects any role of auxin induced H^+ secretion in mediating the rapid mRNA induction. It can be confidently stated that pH changes cannot be the driving force for the mRNA induction in pea tissue.

RELATIONSHIP BETWEEN EARLY AUXIN REGULATED mRNA INDUCTION, H+ SECRETION AND CELL ELONGATION

Induction of specific mRNAs leads *in vivo* to changes in the synthesis of their corresponding polypeptides (9). The question then arises as to the possible function of the polypeptides coded by the auxin-regulated mRNAs in IAA induced H^+ secretion and cell elongation. The inability of H^+ secretion to be responsible for mRNA accumulation (Table II) offers two possible

Table II. *Effect of various inhibitors on the IAA-induced cell elongation, H^+ secretion and mRNA induction in pea tissue.*

Inhibitor	ΔElongation	ΔpH	mRNA Induction Peak Area	
			pIAA4/5	pIAA6
	mm		*(arbitrary units)*	
A. Cycloheximide				
Control	7	-0.11	1	1
IAA 20 µM	16	-0.29	80	120
Cycloheximide 20 µM	1	-0.31	90	130
IAA + Cycloheximide	1	+0.26	85	126
B. Vanadate				
Control	4	-0.18	1	0.1
IAA 20 µM	14	-0.57	100	95
VO^3 1 mM	2	-0.05	2	1
IAA + VO^{+3}	4	-0.08	105	103
C. Cerulenin				
Control	8	-0.28	0.5	0.5
IAA 20 µM	16	-0.40	6	80
Cerulenin 1 mM	1	-0.00	5	4
IAA + Cerulenin	1	+0.10	65	75

Tissue: abraded pea epicotyl segments. Time of incubation: 2 hr. For more experimental details, see Ref. (18).

Table III. *Effect of cycloheximide on the IAA and Fusicoccin induced cell elongation and H^+ secretion.*

Inhibitor	ΔElongation	ΔpH
	mm	
Control	7	-0.18
IAA 20 µM	16	-0.29
Cycloheximide 20 µM	1	+0.31
IAA + CH	1	+0.26
Fusicoccin 5 µM	21	-0.49
FC + CH	14	-0.25

ΔElongation = L_{final}-$L_{depleted}$; ΔpH = pH_{final}-$pH_{initial}$; Abraded segments. Incubation time 2 hours.

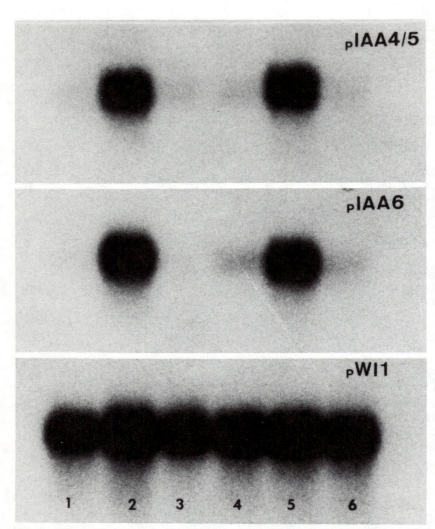

FIG. 3. Effect of fusicoccin and kinetin on the expression of the IAA inducible mRNAs in pea tissue. Autoradiograms of RNA filter paper hybridized successively with ^{32}P-labeled pIAA4/5, pIAA6 and pWI1 plasmid DNAs. Poly A+-RNA was isolated from non-abraded pea segments incubated for 2 hr. Lane 1, control; lane 2, IAA 20 µM; lane 3, fusicoccin 5 µM; lane 4, kinetin 100 µM; lane 5, IAA + kinetin and lane 6, fusicoccin + kinetin. pWI1 is a non-differential clone.

explanations for the mechanism of auxin action. On one hand, the hormone may have two independent primary mechanisms of action, one activates H+ secretion and the other causes mRNA induction. On the other hand, the rapid mRNA accumulation may be the primary response to the hormone, with H+ secretion a secondary consequence (see also chapter by Ray).

Prior to the discovery of rapid gene regulation by auxin, it had been proposed that the hormone has two primary mechanisms of action in initiating cell growth. To accommodate the widely accepted view that auxins do not act rapidly at the gene level (20) the two responses were proposed to have different latent periods. Cell wall acidification occurs with a lag period of approximately 15 min in soybean and was thought to initiate cell growth (phase I) without the participation of biosynthetic events. Furthermore, without experimental verification auxin-regulated cell wall material deposition was proposed to have a lag period of 50 min and to involve activation of transcriptional and translational machinery (20), leading to phase II of cell growth. In view of the recent finding that the latent period for the mRNA induction in abraded pea tissue is only 5 min (Koshiba and Theologis, unpublished) the model for two primary mechanisms of action is untenable on two accounts. First, in pea segments cell elongation and H^+ secretion are abolished by cycloheximide and α-amanitin soon after the application of the hormone, indicating an absolute dependence of the early cell elongation and initiation of H^+ secretion on protein and RNA synthesis. Second, a direct effect of auxin on a H^+-ATPase has not been demonstrated and the proposed cell wall "loosening enzymes" have not been identified yet.

In view of the above observations, the acid growth theory of auxin action needs to be re-evaluated. As I mentioned earlier, the mRNA accumulation is detected 5 min after auxin application in pea tissue, 10 to 15 min earlier than the induced H^+ secretion. Furthermore, the mRNA accumulation occurs 15 to 20 min earlier than the initiation of cell growth (3). The accumulated experimental evidence supports the following sequence of events in response to auxin.

mRNA induction -----> H^+ secretion -----> cell elongation

This model is shown in Figure 4. H^+ secretion is viewed as a consequence of the rapid mRNA induction which represents the primary response to auxin. H^+ secretion is proposed to be an expression of the growth process rather than its cause. Plant cell growth is fundamentally a biosynthetic event, because synthesis of cell wall materials is required during plant cell growth in order to avoid cell breakage. Growth proceeds by the intussusception of cell wall materials which are transported in secretory vesicles into the pre-existing cell wall (10). The flow of cell wall materials into the wall results in a decrease of the cell's turgor pressure with the concomitant stimulation of water uptake leading to cell expansion (see chapters by Cosgrove and by Ray). It is possible that the hormone regulates the secretory process responsible for cell wall formation by regulating the availability of specific mRNAs coding for polypeptides that facilitate the exocytosis (fusion-fission) of secretory vesicles filled with cell wall materials.

The view presented in Figure 4 may be considered a complex metabolic pathway with auxin regulating the head of the pathway (gene transcription), with the secretory process at the tail. The auxin induced protein secretion shown in Figure 4 is a consequence of the enhanced biosynthetic activity induced by the hormone. This view predicts the presence of two H^+ extrusion systems: one is activated indirectly by auxin and is linked to the secretory process, so when the pathway is activated by auxin so is the H^+ secretion system. The second system is activated directly by fusicoccin, is localized on the cell membrane (14), and is independent of the secretory process. The first type of H^+ extrusion system

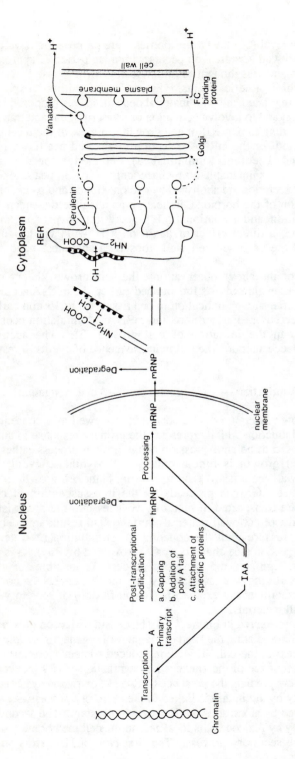

FIG. 4. Model linking auxin induced mRNA accumulation, H⁺ secretion and cell growth.

140

accounts for why auxin induced H^+ secretion is abolished by cycloheximide (Table II) and the fatty acid biosynthesis inhibitor cerulenin (Table II). Inhibition of synthesis of the components of the secretory vesicles (proteins and prospholipids) the secretory process is itself inhibited, and consequently so are auxin induced H^+ secretion and cell elongation.

H^+-ATPases are widespread in plant tissues (13, 15). Because vanadate inhibits auxin induced H^+ secretion (Table II), it is proposed that a H^+-ATPase is associated with the secretory vesicles, or may be more extensively associated with the endomembrane system from the Golgi to the plasma membrane. Vanadate sensitive H^+-ATPases associated with the plasma membrane have been widely studied (13, 15). It will be of great interest to see whether a vanadate-sensitive enzyme is present in plant secretory vesicles. Since fusicoccin activates a vanadate sensitive H^+-ATPase and since the fusicoccin H^+ secretion is insensitive to cycloheximide (Table III), it is proposed that the fusicoccin induced H^+ secretion is independent of the secretory process. Fusicoccin-induced cell elongation is viewed to be biochemically different from that induced by IAA. First, it is transient and does not depend on protein synthesis (Table III). Second, whereas the auxin induced cell elongation is inhibited by kinetin, the fusicoccin one is unaffected by the hormone (Figure 5). Thirdly, fusicoccin is unable to induce ethylene production in a wide range of concentration, whereas auxin always enhances ethylene production at concentrations higher of those required for cell growth (Figure 6). Consequently, I would like to suggest that the fusicoccin induced cell growth is mediated by exocytosis of preexisting secretory vesicles. I would like also to point out that kinetin does not prevent the auxin-induced mRNA accumulation (Figure 3). If the auxin regulated mRNAs are involved in the growth process, its inhibition by kinetin should be at a post transcriptional level.

The role of the secretory-vesicle-associated H^+-ATPase, indirectly activated by auxin, is shown in Figure 7. The intussusception of a secretory vesicle (sv) filled with cell wall material into the cell wall is thermodynamically an unfavorable process because the chemical potential μ_j of species j inside the vesicle μ_j^{sw} is much smaller than that in the cell wall μ_j^{cw}. This is because the activity (a_j) and the electrochemical potential (E) of species j in the secretory vesicle are smaller than they are in the cell membrane-cell wall complex. According to the view presented in Figure 7 the $\Delta\mu_j$ is satisfied by an electrochemical gradient generated by the electrogenic H^+-ATPase localized in the secretory route (Figure 4). Proton electrochemical gradients generated by H^+-ATPases serve as the driving force for active transport of inorganic and organic cations, anions and sugars in various tissues (13, 15). The auxin induced H^+ secretion shown in Figure 4 is proposed to be the expression of a H^+-ATPase whose electrochemical potential is the driving force for the transport of secretory vesicles responsible for the IAA induced cell growth. It is of interest that in mammalian systems newly endocytosed materials are exposed to low pH soon after internalization (7), and acidification of intracellular compartments are mechanistically involved in processes that normally occur soon after adsorptive or receptor-mediated endocytosis. Similarly, it may be that acidification of the cell wall is a prerequisite of exocytosis in plants.

It has been stated by George Palade (8) that "Transport operations are dominated by specific membrane interactions which lead to membrane fusion-

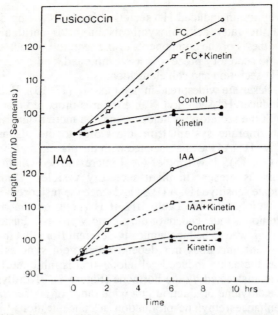

FIG. 5. Effect of kinetin on the IAA and fusicoccin induced cell elongation in pea epicotyl segments. IAA 20 μM; fusicoccin 5 μM; incubation medium pipes/tris pH 6.0.

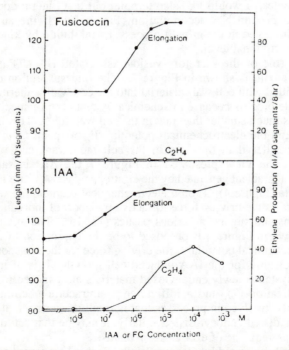

FIG. 6. Effect of fusicoccin and IAA on cell elongation and ethylene production in pea epicotyl segments.

fission and eventually to continuity established between the interacting compartments. Somewhere in this chain of reaction energy is required..." Auxin-induced H^+ secretion in plants may be the expression of the energy requirement to which Professor Palade has alluded.

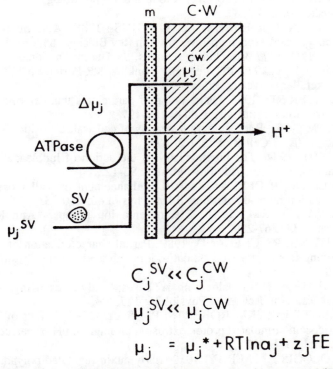

$$C_j^{SV} \ll C_j^{CW}$$

$$\mu_j^{SV} \ll \mu_j^{CW}$$

$$\mu_j = \mu_j^* + RT \ln a_j + z_j FE$$

FIG. 7. An electrogenic H^+-ATPase involved in the transport of secretory vesicles to the cell wall. sv = secretory vesicle; m = cell membrane; cw = cell wall; μ_j = chemical potential of species j; μ_j = standard chemical potential; a_j = activity of species j which is equal to $\gamma \cdot c_j$; where γ_j = activity coefficient and c_j = concentration; R = gas constant; T = temperature; z_j = charge number of species j; F = Farady constant; E = electrical potential.

Acknowledgments--The support of the National Institutes of Health (GM35447) and National Science Foundation (DCB 84-21147) is gratefully acknowledged.

LITERATURE CITED

1. CLELAND R 1971 Cell wall extension. Ann Rev Plant Physiol 22: 197-222
2. EVANS M, PM RAY 1969 Timing of the auxin response in coleoptiles and its implications regarding auxin action. J Gen Physiol 53: 1-20
3. JACOBS M, PM RAY 1976 Rapid auxin-induced decrease in free space pH and its relationship to auxin-induced growth in maize and pea. Plant Physiol 58: 203-209

4. KEY JL 1969 Hormones and nucleic acid metabolism. Ann Rev Plant Physiol 20: 449-474
5. KOSHIBA T, A THEOLOGIS 1987 Regulation of gene expression by IAA and protein synthesis inhibitors. Plant Physiol 83: S-920
6. MARRÉ E 1979 Fusicoccin: A tool in plant physiology. Ann Rev Plant Physiol 30: 273-288
7. MELLMAN J, R FUCHS, A HELENIUS 1986 Acidification of the endocytic and exocytic pathways. Ann Rev Biochem 55: 663-700
8. PALADE G 1977 Concluding remarks. *In* The International Congress on Cell Biology, 1976-1977, eds BR Brinkley, KR Porter, pp 337-340. The Rockefeller University Press
9. PALMITER RD 1975 Quantitation of parameters that determine the rate of ovalbumin synthesis. Cell 4: 189-197
10. RAY PM 1967 Radioautographic study of cell wall deposition in growing plant cells. J Cell Biol 35: 659-674
11. RAY PM 1974 The biochemistry of the action of indoleacetic acid on plant growth. Rec Adv Phytochem 7: 93-122
12. RAYLE DL, R CLELAND 1970 Enhancement of wall loosening and elongation by acid solutions. Plant Physiol 46: 250-253
13. SPANSWICK RM 1981 Electrogenic ion pumps. Ann Rev Plant Physiol 32: 267-289
14. STOUT RG, RE CLELAND 1980 Partial characterization of fusicoccin binding to receptor sites on oat root membranes. Plant Physiol 66: 353-359
15. SZE H 1985 H^+-translocating ATPases: Advances using membrane vesicles. Ann Rev Plant Physiol 36: 175-208
16. TALBOTT LD, JKM ROBERTS, PM RAY 1984 Effect of IAA- and fusicoccin-stimulated proton extrusion on internal pH of pea cells. Plant Physiol 75: S-232
17. THEOLOGIS A, PM RAY 1982 Early auxin-regulated polyadenylylated mRNA sequences in pea stem tissue. Proc Natl Acad Sci USA 79: 418-421
18. THEOLOGIS A, TV HUYNH, RW DAVIS 1985 Rapid induction of specific mRNAs by auxin in pea epicotyl tissue. J Mol Biol 183: 53-68
19. THEOLOGIS A 1986 Rapid gene regulation by auxin. Ann Rev Plant Physiol 37: 407-438
20. VANDERHOEF LN, RR DUTE 1981 Auxin-regulated wall loosening and sustained growth in elongation. Plant Physiol 67: 146-149
21. ZURFLUH LL, TJ GUILFOYLE 1982 Auxin-induced changes in the population of translatable messenger RNA in elongating sections of soybean hypocotyl. Plant Physiol 69: 332-337

Physiology of Cell Expansion During Plant Growth, *D.J. Cosgrove and D.P. Knievel* Eds.,
Copyright ©1987, The American Society of Plant Physiologists

REGULATION OF NUCLEAR ENZYMES IN PLANTS

S. J. ROUX

*The Department of Botany, The University of Texas at Austin
Austin, Texas 78713*

INTRODUCTION

Fifteen years ago it would have seemed strange to talk about the regulation of nuclear enzymes in the context of a symposium that was focused on cell growth. At that time, discussions of growth regulation centered around post-translational events such as wall acidification. However, results during the last decade or so have revealed a closer link between nuclear activities and growth. It is now known, for example, that new proteins must be synthesized for auxin to promote growth, and that auxin may induce a change in gene expression before it induces growth changes (9; see also chapter by Theologis). These findings indicate that metabolic changes in the nucleus are probably important for growth, even during the first minutes of an induced change in growth rate. In this context, a discussion of how nuclear enzymes are regulated does not seem so remote from the topic of growth physiology.

Many of the earlier studies on the regulation of nuclear enzymes by growth-affecting factors emphasized the effects of auxin on nuclear enzymes. These studies which are still ongoing have been recently reviewed by Guilfoyle (9). This chapter will mainly discuss the regulation of nuclear enzymes by two other growth affecting factors, light and Ca^{2+}. More specifically, we will review recent data from our laboratory that indicate Ca^{2+} may help to mediate the photoreversible effects of red light on certain nuclear enzymes. The photoreceptor for the responses, phytochrome, also rapidly modulates growth rates (13) and gene expression (20). There is as yet no evidence that there is a causal connection between phytochrome effects on RNA synthesis and its effects on growth, but the lag time for light induced growth changes (8-10 minutes) is certainly long enough for altered gene expression to play a causative role.

The light and Ca^{2+} regulated nuclear enzymes we have studied thus far are a chromatin-associated nucleoside triphosphatase (NTPase) and protein kinases. In the process of studying their regulation, we found that they were also stimulated by another growth regulator, the polyamines, with spermine being the most effective. The specific nuclear protein kinase affected by spermine appears to be different from the one(s) affected by light, but we will review evidence that the effect of spermine on nuclear metabolism may overlap with effects that are stimulated by light. Because phytochrome strongly influences polyamine metabolism (19), polyamines could also help to mediate light affects on growth. All of the nuclear enzyme studies were carried out using nuclei from pea plumules, where phytochrome is known to regulate both growth and the activity of a key enzyme in the polyamine biosynthetic pathway, arginine decarboxylase (19).

REGULATION OF NUCLEAR NTPase ACTIVITY BY PHYTOCHROME AND CALCIUM

The first report of phytochrome regulated ATPase activity in pea nuclei was by Wagle and Jaffe (21). They found that red-light treatments given to pea plumules could stimulate by more than 50% the ATPase activity that was detectable in nuclei isolated from the etiolated tissue. Far-red light given immediately after the red light could reverse its effects, resulting in nuclear ATPase activity levels similar to those of the dark controls. These results were particularly interesting to us because Matsumoto et al (12) had reported Ca^{2+} and calmodulin-stimulated ATPase activity associated with the chromatin of pea nuclei, and we had a long-standing interest in the role of Ca^{2+} in regulating phytochrome responses. The obvious question to test was whether Pfr stimulated ATPase activity in pea nuclei through the mediation of Ca^{2+}.

After confirming the results of Wagle, Y-R. Chen tested whether nuclear ATPase activities could be modified by irradiations given *in vitro* to isolated nuclei. Phytochrome co-purifies with pea nuclei (Kim and Roux, unpublished), though whether it has a nuclear locale *in vivo* has not been established. Phytochrome that co-purifies with mitochondria can regulate Ca^{2+} release and Ca^{2+}-regulated enzymes in these organelles (18), so there was some precedent for believing that phytochrome could regulate Ca^{2+}-dependent functions in purified organelles.

A 2 min red-light treatment given to highly purified pea nuclei could stimulate ATPase activity in these nuclei by over 50%. This stimulation was reversible by far-red light (2). Although this enzyme activity was highest with ATP as the substrate, it had significantly high levels also with GTP, UTP and CTP as substrates, so it was more appropriately termed a nucleoside triphosphatase (NTPase). Of special interest was the finding that the red-light stimulation of NTPase activity was blocked by calcium chelation (EGTA) and by low concentrations of calmodulin antagonists, including the naturally-occurring antagonist quercetin (Table I). Confidence that this NTPase was *nuclear* was increased by the observation that it was insensitive to vanadate, nitrate, and oligomycin (Table I) and could not be stimulated by K^+, which distinguished it from well characterized ATPases associated with the plasma membrane, vacuoles and mitochondria (2).

PURIFICATION OF NUCLEAR NTPase AND DEMONSTRATION OF ITS CALMODULIN SENSITIVITY

Even the basal (dark) activity of the pea nuclear NTPase could be stimulated by micromolar Ca^{2+} (Fig. 1). Such low levels of Ca^{2+} usually require the mediation of Ca^{2+}-binding proteins to stimulate enzymes, and immunocytochemical evidence indicated that purified nuclei contained significant levels of calmodulin (7), the best characterized Ca^{2+}-binding protein in plants. These results, taken together with the inhibitor studies, were supportive of the

Table I. *Effect of Inhibitors on NTPase Activity of Pea Nuclei*

Inhibitor	Concentration	Relative Activity
		% of control
Control		100
Vanadate	15 μM	103
	30 μM	116
Nitrate	50 mM	112
Oligomycin	2 μg/ml	106
EGTA	0.2 mM	60
Quercetin	5 μg/ml	65
	10 μg/ml	40
	20 μg/ml	10
	25 μg/ml	Trace
Compound 48/80	1 μg/ml	50

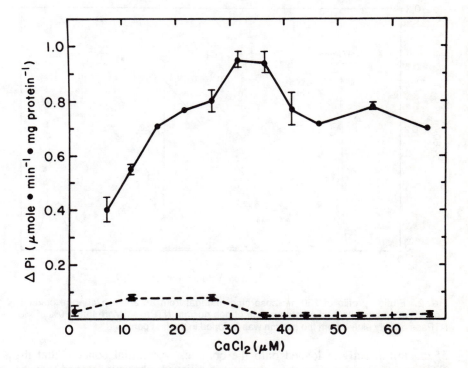

FIG.1. Effect of Ca^{2+} concentration on NTPase activity in the presence (●——●) or absence (●- - - -●) of 6.0 mM Mg^{2+}. (from 2)

conclusion of Matsumoto et al. (12) that pea nuclei contained an ATPase that was regulated by Ca^{2+} and calmodulin. This regulation could have been *direct*, by the binding of Ca^{2+}-activated calmodulin to the NTPase, or as in the case of quinate:NAD oxidoreductase (15), *indirect*, by calmodulin controlling another factor (e.g., a protein kinase) that could stimulate the NTPase. To investigate this question, we attempted to fractionate the nuclear NTPases and test whether any of them were Ca^{2+}/calmodulin sensitive.

In the initial nuclear fractionation, more that 80% of the NTPase activity remained associated with a chromatin-enriched fraction, and more than 80% of this activity could be released from the chromatin by a simple salt extraction with 0.3M NaCl. After further purification by two ammonium sulfate fractionation steps, still more than 50% of the total activity that had been measurable in the crude nuclear preparation was present in a single fraction. This activity was purified to >90% homogeneity by one pass through an anion-exchange HPLC column. The purified NTPase had a molecular weight of 47,000 and an isoelectric point of 6.7 (1).

Although virtually all of the NTPase activity that eluted from the column was present in an early-eluting peak (Fig. 2), this activity was only 30%

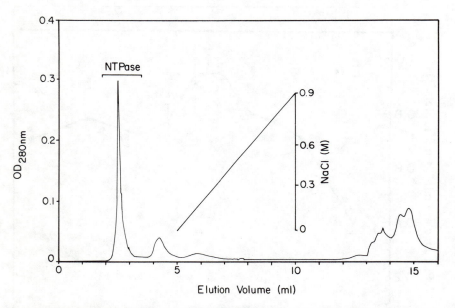

FIG. 2. Elution profile of 280 nm-absorbing material from an anion exchange column loaded with a partially purified preparation of pea nuclear NTPase. More than 90% of the NTPase activity eluted from the column was included in the first peak.

of the initial activity loaded onto the column. An initial concern that the NTPase was denatured on the column was relieved when we learned that the "lost" activity could be restored by simply adding Ca^{2+}-activated calmodulin to the preparation. Endogenous calmodulin from the nuclei that co-purified with the NTPase until the anion-exchange column would have been separated from it on that column, because calmodulin is very acidic and binds tightly to DEAE columns under the conditions used.

The calmodulin sensitivity of the purified NTPase was significant (1). When tested with oat calmodulin, the enzyme activity showed half-maximal stimulation at 30 nM calmodulin and a 2.5 - 3.5X stimulation at 60 nM (Fig. 3). The stimulation was due to an increase in V_{max} rather than K_M for ATP.

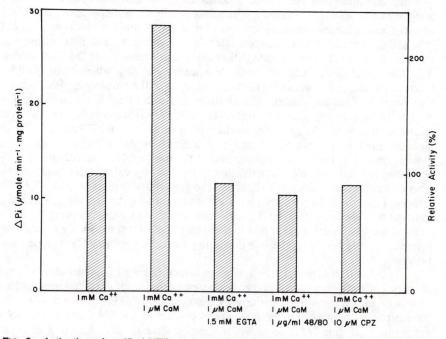

FIG. 3. Activation of purified NTPase by purified bovine brain calmodulin (CaM), and reversal of the activation by EGTA and by calmodulin antagonists. Oat calmodulin stimulates the NTPase by 3.5 fold (1).

If, as preliminary evidence indicates, this enzyme is phosphorylated, it will be important to determine the phosphorylation state of the enzyme before finally assessing its sensitivity to calmodulin activation, since phosphorylation is known in some cases to decrease the binding affinity of an enzyme for calmodulin (17). A completely de-phosphorylated NTPase preparation could show an even greater sensitivity to an activation by calmodulin. It is not likely that pea calmodulin would be any more effective than oat calmodulin in stimulating the pea NTPase, since all angiosperm calmodulins tested appear to have virtually identical compositions and sequences (17).

There is precedent for one Ca^{2+}-binding protein substituting quantitatively for another in activating the same enzyme (17). Thus, it is possible that though the purified NTPase is activated by purified calmodulin *in vitro*, *in vivo* it may be activated by some other Ca^{2+}-binding protein. We have identified another Ca^{2+}- binding protein, a calcimedin, in plant tissue (8), but this protein is present at very low levels, if at all, in pea nuclei, whereas calmodulin appears to be highly concentrated in pea nuclei (7). Calmodulin is the most likely activator of the NTPase *in vivo*, based on available evidence so far.

We have raised a library of monoclonal antibodies against the purified NTPase and have found that one of these recognized only the 47 kilodalton NTPase when tested against a crude preparation of pea nuclear proteins by Western blot analysis (Y-R. Chen, unpublished). We have used this monoclonal antibody to localize the NTPase by immunocytochemical methods. Preliminary results (both immunofluorescence and immunogold surveys) indicate that the NTPase is, indeed, highly concentrated in the nucleus, and that within the nucleus, it has a preferential distribution along chromatin fibrils (M. Dauwalder, Y.R. Chen, S.J. Roux, unpublished). We are now testing whether the NTPase binds to protein or to nucleotide (DNA) sequences in the chromatin. We propose that this chromatin-associated, calmodulin-regulated NTPase is the same one that is photoreversibly stimulated by light. The evidence is indirect but, we believe, strong nonetheless. As indicated earlier, Pfr stimulation of NTPase activity in isolated nuclei requires Ca^{2+} and is inhibited by calmodulin antagonists. Correspondingly, the *in vitro* stimulation of the purified NTPase requires Ca^{2+}-activated calmodulin, and its activation is blocked by calmodulin antagonists. This correspondence does not eliminate the possibility that some other NTPase fraction could also be regulated by Ca^{2+} and calmodulin. However, since the purified NTPase appears to be the major one in the nucleus, accounting for up to 50% of the total units measurable in the intact nuclei, it must surely rank as the prime candidate for being the one that is regulated by light in a Ca^{2+}-dependent fashion.

Although at present there is no evidence linking Pfr-stimulated NTPase activity with Pfr-regulated changes in the transcription of certain genes, such a link is certainly possible. At the very least, nuclear NTPases could influence transcription by controlling the pool size of precursors for DNA-dependent RNA synthesis. If the NTPase were inhomogeneously distributed along the chromatin, changes in its activity could conceivably have a more localized impact on the supply of nucleotides, and thus have a selective impact on the synthesis of only certain genes. Beyond these theoretical considerations, the only rationale for postulating that this NTPase could influence gene expression is its localization on the chromatin and its regulation by an environmental cue (red light) that also regulates transcription rates. Certainly, other functions that are only distantly related to gene expression are equally plausible. Needed data which should help to elucidate its function would include determining whether it selectively associates with only certain DNA sequences or with only certain protein components of the chromatin.

LIGHT AND Ca^{2+}-REGULATED PROTEIN PHOSPHORYLATION

A well characterized NTPase in animal nuclei is regulated by phosphorylation (14). This consideration led us to initiate a study of whether light and Ca^{2+} could influence nuclear protein phosphorylation under the same *in vitro* conditions that it stimulates NTPase activity. Our initial results (3) indicated that red light photoreversibly stimulated the phosphorylation of several nuclear proteins including prominently, a 47 kilodalton protein. This effect was blocked by both EGTA and by calmodulin antagonists, indicating that Ca^{2+}-activated calmodulin could be an important mediator of nuclear protein phosphorylation. There are precedent data on Ca^{2+}-regulated kinases in plant nuclei (17) and there are precedent data on enzymes being regulated both by phosphorylation and by Ca^{2+}-activated calmodulin (17). We are still testing whether the NTPase we have purified is a phosphoprotein whose phosphorylation state can influence its function. Concurrently we have begun to fractionate the protein kinases in pea nuclei and test their sensitivity to Ca^{2+} and calmodulin.

Phytochrome could regulate nuclear protein phosphorylation by changing the activity of either protein kinases or phosphoprotein phosphatases in the nuclei. Interest in the regulation of protein kinases by phytochrome has been stimulated by the recent finding that phytochrome itself may have kinase activity (22). General interest in the regulation of nuclear kinases has been stimulated by reports that phosphorylation regulates the activity of DNA-dependent RNA polymerases, and that induced changes in transcriptional activity are coincident with induced changes in the phosphorylation level both of nuclear enzymes and of chromatin-associated proteins (10). These considerations and our awareness of the role of Ca^{2+} in regulating nuclear enzymes have led us to begin fractionating the protein kinases in pea nuclei and testing their sensitivity to Ca^{2+} and calmodulin. That there are Ca^{2+} regulated protein kinases in plant nuclei seems certain (reviewed in 17), but purification and analysis of such kinases will be required to ascertain the mechanism of this regulation. Our initial fractionation results indicate that among the protein kinases in pea nuclei, Ca^{2+}-regulated kinases are not so dominant as the Ca^{2+}-regulated NTPase is among the NTPases in pea nuclei (Datta, unpublished data).

POLYAMINE-STIMULATED NUCLEAR PROTEIN PHOSPHORYLATION

Prominent in the literature on the regulation of nuclear kinases are reports on the effects of polyamines (19). Polyamines induce changes in the endogenous protein phosphorylation pattern of chromatin-associated proteins and stimulate specific protein kinases in the nuclei of various plant and animal cells. Because phytochrome can regulate polyamine biosynthesis in peas through its effects on arginine decarboxylase (19), it was reasonable to test whether polyamines could mediate the effects of light on protein phosphorylation in pea nuclei.

The initial studies by Datta et al. (4) tested whether polyamines, like light, could stimulate the phosphorylation of nuclear proteins. They found that spermine increased the general protein phosphorylation by about 15% but increased the phosphorylation of a 47 kilodalton polypeptide by 150%. Other polyamines were far less effective. Spermine could stimulate the phosphorylation of the 47 kilodalton polypeptide even in the presence of NaF, which is an effective inhibitor of protein phosphatase activity (Fig. 4). This result indicated that spermine was probably activating a nuclear protein kinase. In contrast to the regulation of protein phosphorylation by light, the effects of spermine were not blocked by EGTA or by the potent calmodulin antagonist,

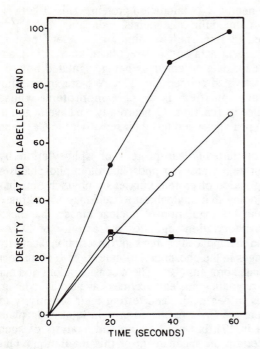

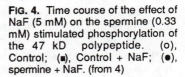

FIG. 4. Time course of the effect of NaF (5 mM) on the spermine (0.33 mM) stimulated phosphorylation of the 47 kD polypeptide. (o), Control; (■), Control + NaF; (●), spermine + NaF. (from 4)

Compound 48/80. This result rendered improbable any role for Ca^{2+} or calmodulin in mediating the phosphorylation response to spermine treatment. It could indicate that spermine and Pfr affect different nuclear kinases by different mechanisms. Alternatively, since both Pfr and spermine stimulate the phosphorylation of a 47 kilodalton polypeptide, they could both be affecting the same kinase, which would happen, for example, if Pfr stimulated an increased level of spermine in nuclei. We have recently detected ornithine decarboxylase activity in isolated pea nuclei (Roux and Schell, unpublished) and are currently testing whether this enzyme, which controls spermine biosynthesis, is regulated by phytochrome in pea nuclei. The effects of light on spermine levels in pea nuclei will also need to be tested.

As indicated in an earlier section, we have begun fractionating protein kinases from pea nuclei. Although our primary interest has been to purify and analyze Ca^{2+}-stimulated protein kinases, we have also tested the effects of spermine on the various kinase fractions we have separated. Recently Datta et al. (5) have partially purified from a chromatin fraction of pea nuclei a cAMP-independent casein kinase that is stimulated over 5-fold by spermine. The effect of spermine is substrate specific; with casein as substrate, spermine stimulates the kinase activity; with phosvitin as substrate, spermine completely inhibits the activity. This kinase appears to be an NII kinase; that is, it is inhibited by heparin and can use GTP as a phosphate donor (NI kinases have neither characteristic). Using [γ-^{32}P] GTP as the phosphate donor, Datta et al. find that spermine promotes the phosphorylation of a 47 kilodalton polypeptide in isolated nuclei; thus, this stimulation must be mediated, at least in part, by an NII kinase. Purification of this kinase will be necessary to further characterize

NII kinase. Purification of this kinase will be necessary to further characterize its activation by spermine. Since spermine can also stimulate NTPase activity in isolated nuclei (Chen, unpublished) it will also be of interest to determine whether the NII kinase can phosphorylate the 47 kilodalton NTPase, and, if so, whether phosphorylation alters its activity.

We have treated the activation of the NII kinase by spermine as independent of any Ca^{2+} influence, but this assumption has not been tested, and, in the light of recent findings by Kauss and Jablick (11), it should be. Those authors found that the plasma-membrane located 1,3-β-D glucan synthase of suspension-cultured soybean cells is stimulated by spermine in the absence of Ca^{2+} ($<10^{-8}$M), but that the addition of as little as 0.8 μM Ca^{2+} to the spermine further stimulates this enzyme by more than 2-fold. It will be interesting to learn whether this example of Ca^{2+}/spermine synergism is an isolated occurrence or happens more generally in plant cells.

SUMMARY AND SIGNIFICANCE

The transduction chain leading from a hormonal or environmental stimulus to changes in the growth rate of plants is surely multipartite and complex. Current evidence favors the involvement of altered gene expression at some step in the transduction chain, either to sustain the growth change, or possibly even to initiate it. The mechanisms that convert environmental and biochemical stimuli into changes in gene expression are currently unknown. Given this void, it will be helpful to learn as much as possible about how nuclear metabolism is regulated.

Light simultaneously induces changes in both growth and gene expression. To test whether Ca^{2+} and calmodulin, which almost certainly mediate some photomorphogenic responses (18), also control light-induced changes in the activity of important nuclear enzymes, we have used isolated pea nuclei as a model system. The demonstration that Ca^{2+}/calmodulin regulate a chromatin-associated NTPase, which is also modulated by phytochrome, represents the most definitive progress on this project to date. It is a significant finding, because it shows clearly for the first time how Ca^{2+} and calmodulin play a role in controlling nuclear metabolism. There is a stronger rationale now for investigating whether this potent "second messenger" regulates other nuclear functions, especially given the evidence that Ca^{2+} can induce specific changes in gene expression in animal cells (16). In this regard it will be of interest to learn whether the chromatin-bound NTPase has any specific interactions with DNA.

The evidence that light and Ca^{2+} regulate nuclear protein phosphorylation is not as complete as the NTPase data, but the preliminary findings certainly warrant an effort to identify and purify specific Ca^{2+}-regulated kinases in nuclei. Our discovery of an NII kinase that is stimulated by spermine is not evidently related to the calcium studies. However, it could prove to have a calcium connection if this NII kinase, when purified, exhibits a synergistic stimulation by Ca^{2+} and spermine, as does the only spermine-binding enzyme thus far characterized in plants (11). Learning about how nuclear kinases are regulated has a certain intrinsic interest, but the ultimate significance of this information will be discovered only when we can identify the protein/enzyme targets of the kinase action and specify how the phosphorylation of these target molecules affects nuclear metabolism, including gene expression.

Research on the regulation of nuclear enzymes in plants is at a very early stage. The preliminary findings are encouraging, for they reveal that at least some nuclear enzymes can be separated from their association with chromatin, purified, and still retain regulatory characteristics they had *in situ*. If future dissections of nuclear activities are equally successful they will provide valuable insights into how cells regulate their nuclear functions.

Acknowledgments--The author acknowledges support from the National Science Foundation (PCM 8402526), from the United States Department of Agriculture (86-CRCR-1-2001) and from the National Aeronautics and Space Administration (NSG 7480), for the research cited from his laboratory.

LITERATURE CITED

1. CHEN Y-R, M DAUWALDER, SJ ROUX 1987 Characterization of a calmodulin-regulated nucleoside triphosphatase localized in the matrix of pea nuclei. Plant Physiol 83 (Suppl): 546
2. CHEN Y-R, SJ ROUX 1986 Characterization of nucleoside triphosphatase activity in isolated pea nuclei and its photoreversible regulation by light. Plant Physiol 81: 609-613
3. DATTA N, Y-R CHEN, SJ ROUX 1985 Phytochrome and calcium stimulation of protein phosphorylation in isolated pea nuclei. Biochem Biophys Res Commun 128: 1403-1408
4. DATTA N, LK HARDISON, SJ ROUX 1986 Polyamine stimulation of protein phosphorylation in isolated pea nuclei. Plant Physiol 82: 681-684
5. DATTA N, MB SCHELL, SJ ROUX 1987 Spermine stimulation of a nuclear NII kinase. Plant Physiol 84: 1397-1401
6. DATTA N, SJ ROUX 1986 Regulation of enzymes in isolated pea nuclei. BioEssays 5: 120-123
7. DAUWALDER M, SJ ROUX, L HARDISON 1986 Distribution of calmodulin in pea seedlings: immunocytochemical localization in plumules and root spices. Planta 168: 461-470
8. DAUWALDER M, SJ ROUX, LK HARDISON, JR DEDMAN 1986 Localization of calcimedin in pea seedlings by immunocytochemistry. J Cell Biol 103: 453a
9. GUILFOYLE T Auxin-regulated gene expression in higher plants. CRC Crit Rev Plant Sci 4: 247-276
10. HOCHHAUSER SJ, JL STEIN, GS STEIN 1981 Gene expression and cell cycle regulation. Int Rev Cytol 71: 95-243
11. KAUSS H, W JEBLICK 1987 Solubilization, affinity chromatography and Ca^{2+}/polyamine activation of the plasma membrane-located 1,3-B-D-glucan synthase. Plant Science 48: 63-69
12. MATSUMOTO H, T YAMAYA, M TANIGAWA 1984 Activation of ATPase activity in the chromatin fraction of pea nuclei by Ca^{2+} and calmodulin. Plant Cell Physiol. 25: 191-195

13. MORGAN DC, H SMITH 1978 Simulated sunflecks have large, rapid effects on plant stem extension. Nature 273: 534-536
14. PURRELLO F, DB BURNHAM, ID GOLDFINE 1983 Insulin regulation of protein phosphorylation in isolated rat liver nuclear envelopes: potential relationship to mRNA metabolism. Proc Natl Acad Sci USA 80: 1189-1193
15. RANJEVA R, G REFENO, AM BOUDET, D MARME 1983 Activation of plant quinate:NAD+ 3-oxidoreductase by Ca^{2+} and calmodulin. Proc Natl Acad Sci USA 80: 5222-5224
16. RESENDEZ JR, E, J TING, KS KIM, SK WOODEN, AS LEE 1986 Calcium ionophore A23187 as a regulator of gene expression in mammalian cells. J Cell Biol 103: 2145-2152
17. ROBERTS DM, TJ LUCAS, DM WATTERSON 1986 Structure, function, and mechanism of action of calmodulin. CRC Crit Rev Plant Sci 4: 311-339
18. ROUX SJ, RO WAYNE, N DATTA 1986 Role of calcium ions in phytochrome responses: an update. Physiol Plant 66: 344-348
19. SLOCUM RD, R KAUR-SAWHNEY, AW GALSTON 1984 The physiology and biochemistry of polyamines in plants. Arch Biochem Biophys 235: 283-303
20. TOBIN EM, J SILVERTHORNE 1985 Light regulation of gene expression in higher plants. Annu Rev Plant Physiol 36: 569-594
21. WAGLE J, MJ JAFFE 1980 The association and function of phytochrome in pea nuclei. Plant Physiol 65 (Suppl): 3
22. WONG, Y-S, H-C CHENG, DA WALSH, JC LAGARIAS 1986 Phosphorylation of *Avena* phytochrome *in vitro* as a probe of light-induced conformational change. J Biol Chem 261: 12089-12097

BIOCHEMICAL GENETICS AND THE REGULATION OF STEM ELONGATION BY GIBBERELLINS

JAKE MACMILLAN AND BERNARD O. PHINNEY

School of Chemistry (J.M.), University of Bristol, Bristol BS8 1TS, U.K., and Department of Biology (B.O.P.), University of California, Los Angeles 90024, U.S.A.

INTRODUCTION

This paper reviews the current state of knowledge on the role of gibberellins (GAs) in the control of stem elongation in *Zea mays* L. and *Pisum sativum* L. This subject provides an excellent example of an interdisciplinary approach in plant science. It is therefore instructive to recall the main events in the development of this study.

In the mid-1950s gibberellic acid (GA_3) was known only as a plant growth promoting metabolite of the fungus, *Gibberella fujikuroi*. In the course of exploring the range of growth promoting effects of GA_3, Brian and Hemming (2) discovered that GA_3 restored normal plant growth to the dwarf cultivar, Meteor, of *Pisum sativum*. Shortly afterwards Phinney (33) showed that GA_3 did likewise when applied to a series of dwarf mutants of *Zea mays*. These observations led directly to the discovery that GA-like substances occurred naturally in *Zea mays* (37, 54) and *Pisum sativum* (42) and indirectly to the first isolation of identified GAs from seeds of *Phaseolus* species (30, 55). In 1961 Phinney (34) recognised the significance of these findings, in terms of biochemical genetics, for the investigation of GA-dependent stem elongation in *Zea mays*. His analysis was based on earlier evidence (9, 13) that a mutant gene negates a specific step in a biochemical pathway to a product required for growth and that non-allelic mutant genes in the same pathway block different metabolic steps in that pathway. On this basis the following predictions (34) were made:

(a) intermediates in the GA-pathway should be active or inactive for a particular dwarf mutant, depending on the position of the mutant gene in the pathway;

(b) any GA that is active for a dwarf mutant should be much less active for the normal plant, if native GAs are limiting for stem elongation;

(c) GAs producing a differential response for a mutant should not produce a differential response in normal plants since normal plants would to able to convert a particular intermediate to the GA(s) necessary for growth;

156

(d) GAs should be present in normal plants and absent, or present in reduced amounts, in the mutants. However, quantitation of the native GAs by bio-assay would be difficult either if there is more than one active GA or its inactive precursors can be converted into active GAs. In both cases the choice of the mutant for bio-assay is critical and the dwarf with a mutant gene furthest back in the pathway would be the best choice.

It has taken about 25 years to confirm these predictions and to define the role of GAs in stem elongation of maize and pea. The main reason for this hiatus has been the technical problem of identifying and establishing the metabolic pathway of native GAs because of their low levels in the vegetative tissue of normal and dwarf mutants. For example the levels of GA_1 are now known to be as low as 100 picogrammes per gramme fresh weight in normal maize and pea seedlings. The required sensitive methods, particularly capillary GC-MS for identification (27), combined with the use of stable isotopic labelling for metabolic and quantification studies (51), have only been available during the present decade.

In the intervening 1960s and 1970s much progress was made on the identification and metabolic pathways of GAs occurring naturally in cultures of *Gibberella fujikuroi* and developing seeds of higher plants. These studies, made possible by the relatively high levels of GAs, established the common pathway (Fig. 1) from mevalonic acid to GA_{12}-aldehyde, the first intermediate with the GA ring structure. These studies, reviewed in various chapters of the book edited by Crozier (6), also showed the existence of different metabolic pathways from GA_{12}-aldehyde in the fungus and seeds of different plant species. Of particular relevance to the present theme was the work on seeds of *P. sativum*. In these seeds, the following GAs were identified by GC-MS (for leading references see ref. 17): GA_9, GA_{17}, GA_{19}, GA_{20}, GA_{29}, GA_{44}, GA_{51}, GA_{29}-catabolite and GA_{51}-catabolite; and two metabolic pathways from GA_{12}-aldehyde were established both in intact growing seeds (see ref. 49 and references therein) and in cell-free enzyme preparations from embryos (see ref. 20 and references therein). Of these two pathways the major one, shown in Fig. 2, is usually referred to as the early 13-hydroxylation pathway from GA_{12}-aldehyde. Although the native GAs and the GA-metabolic pathways are not necessarily the same for seeds and vegetative tissue of the same species (17), this early 13-hydroxylation pathway (Fig. 2) for seeds of *P. sativum* has provided the basis for the subsequent studies on the biochemical genetics of GA-regulated stem elongation in *P. sativum* and *Zea mays*.

STEM LENGTH MUTANTS

It is not surprising that the reversal of genetic dwarfism by GA_3 was first observed for pea and maize mutants since the genetics of both species had been intensively studied for many years. The pea cultivar, *Meteor*, used by Brian and Hemming (2), is homozygous for the *le*-gene (56), a dwarfing gene that has been known since the time of Mendel. The single gene maize mutants, used by Phinney (33), came from random isolates, collected by maize geneticists since 1912. Stem length mutants of many plant species are now known. They comprise three main groups:

157

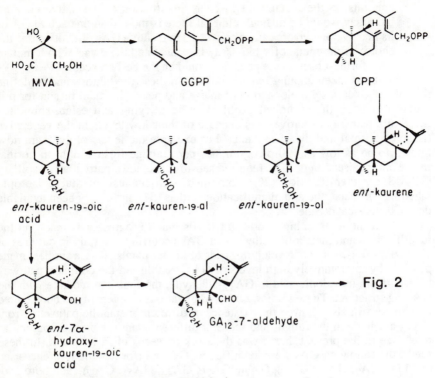

FIG. 1. Biosynthetic pathway to GA$_{12}$-7-aldehyde.

(a) dwarf phenotypes which respond to applied GAs or GA precursors by normal phenotypic growth;
(b) dwarf phenotypes which do not respond to GA-treatment, some of which may accumulate native GAs;
(c) slender or giga mutants which appear to be independent of the levels of native, or applied GAs.

To date the biochemical genetics of stem elongation has concentrated mainly on the hormonal role of GAs using GA-responding dwarf mutants of maize and pea. These studies comprise the main theme of this review. Where appropriate stem-length mutants of types (b) and (c) are briefly discussed.

MAIZE

Stem-length mutants. Of the 1,000 or so known mutants of maize more than 30 involve the height of the plant (5,8). Of these stem-length mutants, *dwarf*-1, *dwarf*-2, *dwarf*-3, *dwarf*-5 and *anther ear*-1 have been shown to be GA-responders and *dwarf*-na$_1$, *dwarf*-na$_2$, *dwarf*-pe$_1$, *dwarf*-mi$_2$ and dominant *Dwarf*-8 give no, or little, response to applied GA$_3$ at the early seedling stage (33,36,37). Dominant *Dwarf*-8 has recently been shown to accumulate native

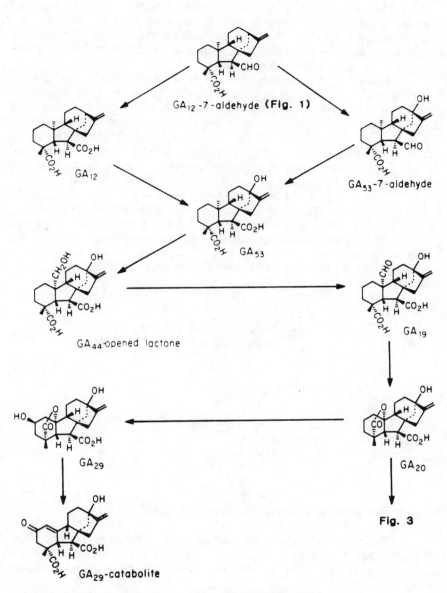

GA$_{12}$-7-aldehyde (Fig. 1)

GA$_{12}$

GA$_{53}$-7-aldehyde

GA$_{53}$

GA$_{44}$-opened lactone

GA$_{19}$

GA$_{29}$

GA$_{20}$

Fig. 3

GA$_{29}$-catabolite

FIG. 2. Early 13-hydroxylation pathway from GA$_{12}$-aldehyde to GA$_{20}$.

GAs (Phinney et al., unpublished). In addition to these random isolates from field-grown stock, Phinney et al. (38) have recently described a number of GA-responding mutants originating from the transposon line, Robertson Mutator (*Mu1*) (46). So far five mutants, (*Mu*) *d*A, *d*B, *d*C, *d*D and *d*E have been isolated from F$_2$ progeny of selfed lines carrying multiple copies of *Mu1*.

159

The most detailed studies on the biochemical genetics of GA-regulated stem elongation of maize have so far concentrated on the GA-responding *dwarf*-1, *dwarf*-2, *dwarf*-3 and *dwarf*-5 mutants. These mutants are non-allelic simple recessives and their dwarf phenotypes are expressed in the light and dark from the young seedling stage to the mature plant.

Native GAs in maize shoots and their metabolic relationship. Using capillary GC-MS, the following GAs have been identified (52, Phinney et al., unpublished) from full scan MS and Kovats Retention Indices in extracts of 21 day-old seedlings of normal maize: GA_{53}, GA_{44}, GA_{19}, GA_{17}, GA_{20}, GA_{29}, GA_1 and GA_8. The qualitative and quantitative GA-content of seedlings of normal, *dwarf*-1, *dwarf*-2, *dwarf*-3 and *dwarf*-5 is discussed below. It is interesting to note that the young tassels of normal maize contain the same GAs as the vegetative shoots (11, 12).

The GAs, identified in the vegetative tissue of normal seedings, are all members of the early 13-hydroxylation pathway which originates from GA_{12}-aldehyde. This pathway was first established for seeds of pea. With this precedent and the established pathway (Fig. 1) from mevalonic acid the GA_{12}-aldehyde, the pathway, shown in Fig. 1 → Fig. 2 → Fig. 3, was constructed as a working hypothesis for the metabolism of GAs in maize seedlings.

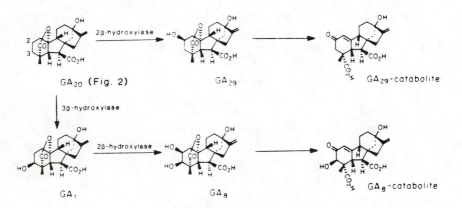

FIG. 3. Steps in the early 13-hydroxylation pathway after GA_{20}.

Three types of evidence for this pathway have been sought using *dwarf*-1, *dwarf*-2, *dwarf*-3 and *dwarf*-5 seedlings. These are bio-assay data, endogenous levels and metabolic studies and they are discussed in the following three sub-sections.

Bio-assay data. Phinney and Spray (35) made a detailed comparison of the bio-activities of *ent*-kaurene (Fig. 1), GA_{12}-aldehyde (Fig. 1 and Fig. 2) GA_{53}-aldehyde (Fig. 2), GA_{53} (Fig. 2), GA_{20} (Fig. 2 and Fig. 3) and GA_1 (Fig. 3) on *dwarf*-1, *dwarf*-2, *dwarf*-3 and *dwarf*-5. The results were consistent with the presumed pathway (Fig. 1 → Fig. 2 → Fig. 3) and suggested that the metabolic steps blocked in each of the mutants are as shown in Fig. 4.

160

$$\rightarrow \textit{ent-}\text{kaurene} \rightarrow GA_{12}\text{-aldehyde} \rightarrow GA_{53}\text{-aldehyde} \rightarrow GA_{53} \rightarrow GA_{20} \rightarrow GA_1$$

$$\qquad\quad \wedge \qquad\qquad\qquad\qquad\qquad\qquad \wedge \qquad\qquad\quad \wedge \qquad\qquad\qquad \wedge$$

dwarf-5   *dwarf-3*   *dwarf-2*   *dwarf-1*

FIG. 4. Metabolic steps at which various dwarf mutants block the synthesis of GA_1.

Thus for *dwarf-1*, only GA_1 had high activtity; GA_{20} showed less than 1% of the activity of GA_1 and the other compounds had zero activity. For *dwarf-2*, GA_{20} and GA_1 had the same high activity; GA_{53} had 10% of the activity of GA_{20} and GA_1; and the other compounds had zero activity. For *dwarf-3*, GA_{20} and GA_1 had equal high activity; GA_{53} and GA_{53}-aldehyde showed 10% of the activity of GA_{20} and GA_1; and GA_{12}-aldehyde and *ent-*kaurene showed zero activity. For *dwarf-5*, there was a gradation of activity from *ent-*kaurene (1%) through GA_{12}-aldehyde (5%), GA_{53}-aldehyde (10%), GA_{53} (10%), to GA_{20} (100%) and GA_1 (100%). The positions of the metabolic blocks, shown in Fig. 4, are also supported by bio-assay data (53) for GA_{19} which has similar activity to GA_1 on *dwarf-2* and *dwarf-5* but very low activity on *dwarf-1*. It should also be noted that GA_{29} and GA_8 (Fig. 3), the 2β-hydroxy metabolites of GA_{20} and GA_1, respectively, show little or no biological activity on *dwarf-1*, *dwarf-2* and *dwarf-5* (57).

These bio-assay data lead to the conclusion that GA_1 is the only native gibberellin that is active *per se* for stem elongation in maize. The bio-assay data are also consistent with the pathway (Fig. 1 → Fig. 2 → Fig. 3) in which the mutants are blocked at the positions shown in Fig. 4. Other GAs in the pathway are active only through their metabolism to GA_1.

Endogenous levels of GAs. The endogenous levels of GA_{53}, GA_{44}, GA_{19}, GA_{20}, GA_{29}, GA_1 and GA_8 have recently been quantified (Phinney et al., unpublished) in normal, *dwarf-1*, *dwarf-2*, *dwarf-3* and *dwarf-5* seedlings of maize. Using $[1,7,12,18-^{14}C_4]$-labelled GA_{53} and GA_{44}, $[17-^2H_2]GA_{19}$ and $[17-^3H_2{}^{13}C]$-labelled GA_{20}, GA_{29}, GA_1 and GA_8 as internal standards, the endogenous levels were determined from the isotopic dilution of ^{14}C, ^2H or ^{13}C by capillary GC-SICM. The results were consistent with the positions (Fig. 4) of the metabolic blocks in the mutants as determined by bio-assay. Thus the levels of each of the GAs were zero (<0.2ng per 100g fresh weight seedling) for *dwarf-5*, *dwarf-3* and *dwarf-2* in which the blocks steps occur before GA_{53}. For *dwarf-1* the levels of GA_{20} and GA_1 were respectively 554 and 0.3ng per fresh weight, compared to 40ng of GA_{20} and 12ng of GA_1 per 100g fresh weight of normal seedlings.

Metabolic studies. In these studies the native GAs, doubly labelled with a radio and stable isotope, were fed to normal and mutant seedlings and the metabolites were identified by GC-MS. The experimental protocol in which the metabolites are identified by full scan mass spectra and Kovats Retention Indices have been briefly reviewed by MacMillan (28).

To date the later steps in the pathway (Fig. 3) have been investigated in most detail by Spray et al. (52) from feeds of $[17-^3H_2, 17-^{13}C]GA_{20}$ to seedlings of normal, *dwarf-1* and *dwarf-5*. In normal seedlings $[^{13}C]GA_1$ and $[^{13}C]GA_{29}$-catabolite were identified, and in *dwarf-5* seedlings $[^{13}C]GA_1$ and $[^{13}C]GA_{29}$ were identified, as metabolites of $[^{13}C]GA_{20}$. However in *dwarf-1* seedlings $[^{13}C]GA_1$

161

was not detected although $[^{13}C]GA_{29}$ and $[^{13}C]GA_{29}$-catabolite were formed. These results show that *dwarf-5* is blocked before GA_{20}, that *dwarf-1* is blocked for the step from GA_{20} to GA_1 and that the conversion of GA_{20} to GA_1 occurs in normal seedlings. These conclusions are in complete accord with the bio-assay data and the endogenous levels of native GAs.

The steps in the GA-metabolic pathway before GA_{20} (Fig. 1 and Fig 2) have not yet been examined using doubly labelled substrates. However Hedden and Phinney (10) have provided evidence that the formation of *ent*-kaurene from copalyl pyropophosphate (Fig. 1) is controlled by the normal *D-5* allele. Using cell-free enzyme preparations from maize coleoptiles these authors found that combined amounts of *ent*-$[^{14}C]$kaurene and its double bond isomer, *ent*-$[^{14}C]$kaur-15-ene, formed from $[^{14}C]$ copalyl pyrophosphate were similar for both normal and *dwarf-5*. However, the ratio of *ent*-$[^{14}C]$ kaurene to its double bond isomer was *ca.* 8:1 for normal coleoptiles and *ca.* 1:8 for *dwarf-1* coleoptiles. In these experiments the *ent*-$[^{14}C]$ kaurene isomers were separated by thin layer chromatography, diluted with unlabelled standards and crystallised to constant specific radio-activity.

Other feeds with radio-labelled substrates in which the metabolites were identified by co-chromatography with authentic standards include the conversion of $[1,2-^3H_2]GA_{20}$ to $[^3H]GA_1$ and $[^3H]GA_8$ in normal maize seedlings (47) and the metabolism of $[1,2-^3H_2]GA_1$ to $[^3H]GA_8$ in normal and *dwarf-5* seedlings (7).

Conclusions. The evidence from the biological activity, the endogenous levels and the metabolic studies for the native GAs in seedlings of normal, *dwarf-1*, *dwarf-2*, *dwarf-3* and *dwarf-5* maize lead to the following conclusions:

(a) the early 13-hydroxylation pathway (Fig. 2 and Fig. 3) occurs in maize seedlings;

(b) the GA-responding mutants are blocked for the metabolic steps, shown in Fig. 4;

(c) GA_1 is the only native GA that is active *per se* for stem elongation in maize;

(d) the other native GAs which show activity are precursors of GA_1 and are active only because they are metabolised to GA_1.

The status of GA_1 as a natural plant hormone for stem elongation is discussed later in this chapter

PEA

Stem-length mutants. Eight genes (*Le, Na, Ls, Lh, La, Cry, Lk* and *Lm*) for stem elongation have been identified (ref. 43 and references therein). The *Le, Na, Ls* and *Lh* genes appear to operate in the GA-biosynthetic pathway since mutations at these loci give dwarf phenotypes which respond to applied GA_1 by stem elongation (39, 40, 44). A second recessive allele, *le^d*, has been characterised recently (48); it confers a more severe dwarfing phenotype than the *le*-allele.

Mutants involving the other four loci are not GA-responders. The *lk*-mutant contains similar amounts of GA-like bioactivity (44). The combination *la cry*^{s} gives a slender phenotype in the presence of either *le* or *na* (31, 41). The combination of *la* and *cry*^{c} gives crypto dwarfs in the presence of *le* but the crypto tall phenotype has not been reported. The *lm*-allele results in a microphenotype giving rise to microtall, microdwarf, microcryptodwarf and microslender depending on the alleles at the *Le*, *le* and *cry* loci, respectively.

Native GAs in pea shoots and their metabolic relationship. Guided by the earlier identification of GAs in developing seeds of pea, GA_{19}, GA_{20}, GA_{29}, GA_{29} -catabolite, GA_1, GA_8 and GA_8-catabolite have been identified in normal pea seedlings (see ref. 17 for leading references). These GAs are all members of the early 13-hydroxylation pathway, indicating the metabolic pathway (Fig. 1 $\rightarrow$ Fig. 2 $\rightarrow$ Fig. 3). This presumptive pathway is the same as that for maize seedlings. Thus the studies on the biochemical genetics of GAs in pea and maize are complementary and supplementary. The evidence for the pathway (Fig. 1 $\rightarrow$ Fig. 2 $\rightarrow$ Fig. 3) in pea seedlings, for the position of the metabolic blocks in this pathway for the GA-responding mutants and, hence, for the role of GAs in stem elongation is summarised in the following three subsections.

Bio-assay data. No specific qualitative comparisons have been reported for the bio-activites of the members of the pathway (Fig. 1 $\rightarrow$ Fig. 2 $\rightarrow$ Fig. 3) with the GA-responding mutants of pea. However, as noted earlier, the *le*, *na*, *ls* and *lh* mutants respond to GA_1 (39, 40, 44). In addition, Ingram and Reid (15) have reported the following information: (a) *ls* and *lk* mutants respond to GA_{12}-aldehyde and GA_{53}-aldehyde; (b) the *na*-mutants do not respond to *ent*-kaurene, *ent*-kaurenol, *ent*-kaurenal, *ent*-kaurenoic acid or *ent*-7α-hydroxykaurenoic acid; and (c) the *lh* and *ls* mutants respond to *ent*-kaurene and *ent*-kaurenoic acid. These results indicate that the metabolic blocks for GA-biosynthesis in the *lh* and *ls* mutants are before *ent*-kaurene (Fig. 1) and in the *na* mutant between *ent*-7α-hydroxykaurenoic acid and GA_{12}-aldehyde (Fig. 2).

Bio-assay data for the normal and *le* mutant indicate that the *le* gene controls the 3ß-hydroxylation of GA_{20} to GA_1 (Fig. 3). Thus normal (Le) responds as well to GA_{20} as GA_1 but genotypes homozygous for the *le*-allele respond only moderately to GA_{20} (18). It is also important to note that GA_{29} and GA_{29}-catabolite (Fig. 3), the branch metabolites of GA_{20}, and GA_8 and GA_8-catabolite (Fig. 3), the metabolites of GA_1, show very low activity in the dwarf cultivar Progress No. 9 which is homozygous for the *le* allele (50, 57).

Endogenous GA levels. No specific quantitative analyses has been reported for the endogenous GA levels in normal and GA-responding dwarf seedlings of pea. However estimates for the GAs shown in Fig. 3 can be derived (19) from the $^{13}C{:}^{12}C$ ratios of those GAs detected by GC-MS after the feeds of $[^3H_2, ^{13}C]GA_{20}$ (see following section). These ratios indicate the following approximate amounts per seedling. For the two *le* mutants used, 25 and 50ng GA_{20}, 12.5 and 5.7ng GA_{29}, and no GA_1 or GA_8. For the two normal lines used: 18ng GA_{20}, 4 and 3.5ng GA_1; 3 and 4ng GA_{29}; and 5 and 6ng GA_8. For the *na le* double mutant and the *na* mutant, there was no dilution of the ^{13}C-label in the $[^{13}C]$-labelled GA_1, GA_8, GA_{20} and GA_{29}, identified after feeding $[^{13}C]GA_{20}$, indicating the absence of endogenous GA_1, GA_8, GA_{20} and GA_{29}.

163

These results are consistent with the bio-assay data and support the conclusion that the *na*-gene blocks for a step in the GA-metabolic pathway before GA_{20} and that the *le*-gene blocks for the conversion of GA_{20} to GA_1.

Metabolic studies. As in the case of maize seedlings, detailed metabolic studies have so far concentrated on the later steps (Fig. 3) in the GA-pathway. The biochemical basis of the *Le/le*-gene difference has been examined by feeding $[^3H_2,{}^{13}C]GA_{20}$ to the young expanding tissue of 21-day old seedlings of *Na le, Na Le, na le* and *na Le*-genotypes and identifying the metabolites by full scan GC-MS (19). In all *Le* genotypes $[^3H_2,{}^{13}C]GA_{20}$ was metabolised to $[^3H_2,{}^{13}C]GA_1$, $[^3H_2,{}^{13}C]GA_8$ and $[^3H_2,{}^{13}C]GA_{29}$. In contrast, for all *le*-genotypes, no $[^{13}H_2,{}^{13}C]GA_1$ was formed; $[^3H_2,{}^{13}C]GA_{29}$ and in one line, $[^3H_{21}{}^{13}C]GA_{29}$-catabolite, were the only metabolites detected. These results indicate that the *le*-gene controls the 3ß-hydroxylaton of GA_{20} to GA_1 (Fig. 3.).

Recently $[^2H_1]GA_{12}$-aldehyde has been shown (15) to be metabolised to $[^2H_1]GA_{20}$ $[^2H_1]GA_{29}$, $[^2H_1]GA_1$ and $[^2H_1]GA_8$ in *na* genotypes. These metabolites, identified by full scan GC-MS, showed no isotopic dilution. These results are consistent with the bio-assay evidence for the block in the *na*-mutants being before GA_{12}-aldehyde.

Metabolic studies have also been described in which only radio-labelled substrates were used, the radio-labelled metabolites being identified by chromatography and radio-counting. For example Ingram and Reid (15) have fed *ent*-$[17-{}^3H_2]$kaurenoic acid to the young expanding tissue of normal and *na, ls* and *lh* mutants. $[^3H]$-labelled metabolites that co-chromatograph with GA_{20}, GA_{29}, GA_1 and GA_8 were observed from the normal and *ls* and *lh* mutants but not from the *na* mutant.

Conclusions. The evidence from the biological and metabolic studies with normal and GA-responding stem-length mutants is consistent with the biosynthesis of GAs in peas by the early 13-hydroxylation pathway. This evidence also leads to the conclusion that the *le*-gene controls the 3β-hydroxylation of GA_{20} to GA_1 (Fig. 3); that the *na*-gene controls a step between *ent*-7α-hydroxykaurenoic acid and GA_{12}-aldehyde (Fig. 2) and the *lh*- and *ls*-genes control steps before *ent*-kaurene (Fig. 1). The most significant conclusion, however, is that GA_1 is the only native gibberellin necessary *per se* for stem elongation in pea. The hormonal status of GA_1 is discussed in the following section.

HORMONAL STATUS OF GIBBERELLIN A₁ FOR STEM ELONGATION IN MAIZE AND PEA

With the proviso that there is no bio-active native GA after GA_1 in shoots of maize and pea, the biochemical genetic studies presented in the last two sections show that GA_1 is the only native GA that is qualitatively required for stem elongation in these two species. To establish that GA_1 acts as a hormone it is also necessary to establish that it is quantitatively required for stem elongation. The classical definition of a hormone would also require that GA_1 formed at a distance from, and translocated to, the active site. These two aspects are now considered.

Quantitative relationship between gibberellin A_1 and stem elongation. Evidence that GAs are quantitatively required for stem elongation is provided by the very existence of stem length mutants that respond linearly to the logarithm of the amount of applied GA. This evidence is reinforced by the isolation of double mutants, such as the *na lh* and *na ls* mutants of pea (43) with even shorter internodes than their single mutants. Evidence for the quantitative relationship between endogenous GA_1 and stem elongation is being obtained from a study of the nature of the mutation at the *Le* and *D*-1 loci in pea and maize, respectively. Ingram et al. (16) have published data which suggests that the *le*-gene in pea results in an altered enzyme in which enzyme activity for the 3ß-hydroxylation of GA_{20} to GA_1 (Fig. 3) is impaired but not lost. High levels of $[^3H,^{13}C]GA_{20}$ were fed to the double mutant, *na le*, and the metabolites were identified by GC-MS; small amounts of $[^3H,^{13}C]$-labelled GA_1, 3-epiGA$_1$ and GA_8 were identified in addition to the expected $[^3H,^{13}C]GA_{29}$ and $[^3H,^{13}C]GA_{29}$-catabolite. These results establish that the *le*-mutation is "leaky" and provide an explanation for the fact that GA_{20} shows some bio-activity when applied to the *le* mutant. When a range of amounts of $[^3H]GA_{20}$ were applied to seedlings of the double mutant, *na le* and the *na* mutant, a linear relationship was observed (16) between the logarithm of the amount of $[^3H]GA_8$ formed and internode extension. The same relationship was found for the three alleles *le*, *le^d* and *Le*. Unfortunately in these experiments a direct correlation between internode elongation and the amount of GA_1 was not possible since the $[^3H]GA_1$ could not be cleanly separated, and quantified, in the presence of relatively larger amounts of 3-epi-$[^3H]GA_1$. Nevertheless the results provide evidence that internode extension and 3ß-hydroxylation of GA_{20} are directly related. These experiments with the *le*-alleles of pea are currently being repeated, using $[^3H,^{13}C]GA_{20}$ and monoclonal antibodies (24) to purify and quantify the GA_1 formed, and similar experiments with d_1-alleles of maize are underway.

Tissue localisation of gibberellin A_1 formation. The occurrence of GA_1 in normal (Le) phenotypic seedlings of pea appears to be confined to the young expanding apical tissue (18, 40). Grafting studies with pea have shown that stocks of normal mature tissue promote internode extension in *na* scions (45) but not in *le* scions (25, 26, 45). The implication from these results is that GA_1 is biosynthesised only in expanding apical tissue from a precursor, probably GA_{20}, that is biosynthesised in mature tissue. Using approach grafts of maize, Katsumi et al. (23) observed, *inter alia*, the promotion of stem elongation in *dwarf*-1 and *dwarf*-5 members, grafted to normal plants. Their results indicate the transmissibility of members of the biosynthetic pathway, including GA_1.

Hormonal status of gibberellin A_1 for stem elongation. Present evidence from the biochemical genetics of GA-dependent stem elongation in maize and pea indicates that GA_1 can be regarded as a plant hormone for stem elongation. As a naturally occurring compound, active at very low concentrations that are quantitatively related to the growth-response, GA_1 meets the classical requirements of hormone. With respect to the criterion of "active at a distance", GA_1 may be similar to several animal steroids in being formed in target cells from a precursor outside the target cells and transported to them. For example, 5α-androstan-5β-ol-3-one acts on the nuclear chromatin of prostate cells; and is produced in these target cells from testosterone which is formed in the testes and transported in the prostate cells (1, 4).

As noted by Phinney et al. (38) the early 13-hydroxylation pathway (Fig. 2 $\rightarrow$ Fig. 3) probably occurs in the shoots of many plant species as evidenced by the natural occurrence of members of the pathway (17). Thus GA_1 may be the universal GA for stem elongation.

FUTURE DIRECTIONS FOR STUDIES IN MOLECULAR GENETICS OF GA$_1$-ACTION.

A *prima facie* case has been presented for GA_1 as a plant hormone, regulating stem elongation in pea and maize. Attention can now be directed from a semantic discussion of whether GAs should be regarded as hormones to such topics as: (a) The regulation of GA_1 synthesis; (b) the cellular and subcellular localisation of GA_1; and (c) the primary action of GA_1. In these areas of current ignorance, molecular genetics can play an important role.

Regulation of GA$_1$ levels. In principle, regulation of GA_1-levels can occur at any step in the metabolic pathway. Control may be exercised through enzyme levels (gene expression) and enzyme activity (rate-determining kinetics, allosteric inhibition). To investigate these possibilities, the enzyme of the pathway, and their genes, must be identified and characterised. Several research groups are now engaged in the isolation of the pure enzymes from shoot tissue to determine their kinetic parameters. Antibodies to these enzymes would provide a means to identify the mRNA for the enzymes and hence, through c-DNA, the study of gene expression. Enzymes for which allelic genes are known are obvious targets and a start has been made (Smith et al., unpublished) with antibodies to the enzymes of the latter part of the pathway (Fig. 3).

A more direct approach is to identify, isolate and clone the DNA sequence that is genetically linked to a particular phenotype. Such an approach is being adopted by Phinney et al. (38) in their investigations of dwarf mutants of maize which originate from the insertion of Robertson's Mutator (Mu-1) into the genome.

Cellular and sub-cellular localisation of GA$_1$ and of GA$_1$-synthesis. The cytological distribution of the native GAs within the shoots of maize and pea is of interest in its own right. It also has an important bearing on the study of regulation of GA_1 levels and on the identification of the primary site of action of GA_1. For example the sub-cellular localisation of GA_1, its formation from GA_{20} and its metabolism to GA_8 (Fig. 3) should yield valuable information on the site of GA_1-action. Immunohistochemical studies with the mutations of pea and of maize may provide answers to these questions. Such studies would provide an interesting comparison between a dicot and a monocot. Highly specific antibodies to the GAs and enzymes (Fig. 3) are required and progress has recently been made (24) with the preparation of monoclonal antibodies to the GAs in Fig. 3. Monoclonal antibodies to the 2 -hydroxylases for GA_{20} and GA_1 and to the 3ß-hydroxylase for GA_{20} are in the course of preparation (Smith et al., unpublished).

166

Primary action of GA_1 for stem elongation. If GA_1 acts as a classical hormone, it should form a protein receptor complex that initiates biochemical reactions leading to stem elongation. However there is no such recognised biological event that is associated with GA_1 action and stem elongation. There are unconfirmed suggestions in the literature that the effect of applied GA_3 is to provide more substrate(s) for cell wall formation. Thus Nanda and Dhindsa (32) have shown that GA_3-induced internode extension in soybean seedlings is inversely correlated with starch levels and directly correlated to hydrolytic activity. In dwarf and normal cultivars of pea Broughton and Carr (3) have reported that applied GA_3 stimulates α-amylase, ß-fructosidase and starch phosphorylase activities and causes cell elongation, cell division and secondary wall synthesis; these authors also reported that glucose derivatives mimicked the morphological effects of GA_3. On the other hand Katsumi (21) concluded that the effect of applied GA_3 on the internode extension of *dwarf*-5 maize and on α-amylase activity were independent of each other. The relationship between GA_1 levels, α-amylase activity and internode extension merits further study. So also does the observation that GA_3, applied to etiolated *dwarf*-5 seedlings of maize, causes transverse microtubule orientation together with axial cell elongation (22). However, in the absence of a biochemical probe for GA_1-induced stem elongation, it will be difficult to recognise a GA_1-receptor.

It should also be borne in mind that GA_1 may not act through a classical protein-receptor complex. Recent metabolic studies (14) with the slender phenotype (*cry^s la*) of pea show that introduction of the *le* gene blocks the conversion of GA_{20} to GA_1. Similarly introduction of the *na* gene blocks the metabolism of *ent*-kaurenoic acid. However the presence of the *le* and *na* genes does not confer a dwarf growth habit on the slender phenotype. These results indicate that internode extension of slender seedlings is independent of GA_1 levels. A simple explanation is that stem growth in pea is controlled by an effector/repressor system in which GA_1 is the effector and the product of the *Cry* and/or *La* gene is the repressor.

Acknowledgments--We thank the Agricultural Research Council for a grant to J.M. and the National Science Research Foundation for NSF Grant DMB 85-06998 to B.O.P. We also thank Dr. Clive R. Spray for his generous help in preparing the diagrams.

LITERATURE CITED

1. ANDERSON KM, LIAO S 1968 Selective retension of dihydrotestosterone by prostate nuclei. Nature (London) 219: 277-279
2. BRIAN PW, HEMMING HG 1955 The effect of gibberellic acid on shoot growth and pea seedlings. Physiol Plant 8: 669-681
3. BROUGHTON WJ, McCOMB AJ 1971 Changes in the pattern of enzyme development in gibberellin-treated pea internodes. Ann Bot 35: 213-228
4. BRUCHOVSKEY N, WILSON JD 1968 The conversion of testosterone to 5α-androstan-17ß-ol-3-one by rat prostate *in vivo* and *in vitro*. J Biol Chem 243: 2012-2021

5. COE EH, NEUFFER MG 1977 The genetics of corn. *In* GF SPRAGUE(ed) Corn and Corn Improvement. Agronomy 18: 111-123
6. CROZIER A, ed 1983 The Biochemistry and Physiology of Gibberellins Vol. 1 Praeger Publishers, New York.
7. DAVIES LJ, RAPPAPORT L 1975 Metabolism of tritiated gibberellins in *d*-5 dwarf maize. II. [^{3}H]Gibberellin A$_1$, [^{3}H]gibberellin A$_3$, and related compounds. Plant Physiol 56: 60-66
8. EMERSON RA, BEADLE GW, FRAZER AC 1935 A summary of linkage studies in maize. NY State Agr Coll Expt Sta Geneva NY Mem 180: 1-89
9. FINSHAM JRS 1958 The role of chromosomal loci in enzyme formation. Proc 10th Intern Cong Genet I pp 355-363
10. HEDDEN P, PHINNEY BO 1979 Comparison of *ent*-kaurene and *ent*-isokaurene synthesis in cell-free systems from etiolated shoots of normal and *dwarf*-5 maize seedlings. Phytochemistry 18: 1475-1479
11. HEDDEN P, PHINNEY BO, HEUPEL RC, FUJII D, COHEN H, GASKIN P, MACMILLAN J, GRAEBE JE 1982 Hormones in young tassels of *Zea mays*. Phytochemistry 21: 391-393
12. HEUPEL RC, PHINNEY BO, SPRAY CR, GASKIN P, MACMILLAN J, HEDDEN P, GRAEBE JE 1985 Native gibberellins and the metabolism of [^{14}C]gibberellin A$_{53}$ and of [17-^{13}C, 17-^{3}H$_2$]gibberellin A$_{20}$ in tassels of *Zea mays*. Phytochemistry 24: 47-53
13. HOROWITZ NH 1950 Biochemical genetics of *Neurospora*. Adv Genetics 3: 33-71
14. INGRAM TJ, REID JB 1987 Internode length in *Pisum*. Biochemical expression of the *le* and *na* mutants in the slender phetotype. J Plant Growth Reg 5: 235-243
15. INGRAM TJ, REID JB 1987 Internode length in Pisum. Gene *na* may block gibberellin biosynthesis between *ent*-7α-hydroxykaurenoic acid and gibberellin A$_{12}$-aldehyde. Plant Physiol 83: 1048-1053
16. INGRAM TJ, REID JB, MACMILLAN J 1986 The quantitative relationship between GA$_1$ and internode growth in *Pisum sativum* L. Planta 168: 414-420
17. INGRAM TJ, MACMILLAN J, SPONSEL VM 1985 Gibberellin distribution and metabolism: A comparison between seeds, shoots and roots. Acta Univers Agric, in press
18. INGRAM TJ, REID JB, POTTS WC, MURFET IC 1983 Internode length in *Pisum* IV. The effect of the *le* gene on gibberellin metabolism. Physiol Plant 59: 607-616
19. INGRAM TJ, REID JB, MURFET IC, GASKIN P, WILLIS CL, MACMILLAN J 1984 Internode length in *Pisum*. The *le*-gene controls the 3ß-hydroxylation of gibberellin A20 to gibberellin A1. Planta 160: 455-463
20. KAMIYA Y, GRAEBE JE 1983 The biosynthesis of all major gibberellins in a cell-free system from *Pisum sativum* L. Phytochemistry 22: 681-689

21. KATSUMI M 1970 Effect of gibberellin A_3 on the elongation, α-amylase activity, reducing sugar content and oxygen uptake of the leaf sheath of dwarf maize seedlings. Physiol Plant 23: 1077-1084

22. KATSUMI M, MITA T 1986 Gibberellin control of microtubule arrangement in the mesocotyl epidermal cells of the *d5* mutant of *Zea mays* L. Plant Cell Physiol. 27: 651-659.

23. KATSUMI M, FOARD DE, PHINNEY BO 1983 Evidence for the translocation of gibberellin A_3 and gibberellin-like substances in grafts between normal and dwarf-1 and dwarf-5 seedlings of *Zea mays* L. Plant Cell Physiol 24: 379-388

24. KNOX JP, BEALE MH, BUTCHER GW, MACMILLAN J 1987 Preparation and characterisation of monoclonal antibodies which recognise different gibberellin epitopes. Planta 170: 86-91

25. LOCKARD RG, GRUNWALD C 1970 Grafting and gibberellin effects on the growth of tall and dwarf peas. Plant Physiol 45: 160-162

26. McCOMB AJ, McCOMB JA 1970 Growth substances and the relationship between phenotype and genotype in *Pisum sativum*. Planta 91: 235-245

27. MACMILLAN J 1984 Analysis of plant hormones and metabolism of gibberellins. *In* A Crozier and JR Hillman, eds, The Biosynthesis and Metabolism of Plant Hormones. Cambridge University press, pp 1-16

28. MACMILLAN J 1985 Gibberellin metabolism: objectives and methodology. Biologia Plantarum (Prague) 27: 172-179

29. MACMILLAN J 1987 Gibberellin deficient mutants of maize and pea and the molecular action of gibberellins. *In* GV HOAD, JR LENTON, MB JACKSON, eds, Hormone Action in Plant Development - A Critical Appraisal, Butterworth, London, pp 73-87

30. MACMILLAN J, SUTER PJ 1958 The occurrence of gibberellin A_1 in higher plants: isolation from seed of runner bean. Naturwissenschaften 45: 46-47

31. MARX GA 1983 Plant stature as affected by the interaction of *na* and *le la cry^c*. Pisum Newslett 15: 235-245

32. NANDA KK, DHINDSA RS 1968 Effect of gibberellic acid on starch content of soyabean (*Glycine max* L.) and its correlation with extension growth. Plant Cell Physiol. 9: 423-432.

33. PHINNEY BO 1956 Growth response of single gene mutants of maize to gibberellic acid. Proc Nat Acad Sci (USA) 43: 398-404

34. PHINNEY BO 1961 Dwarfing genes in *Zea mays* and their relation to gibberellins. *In* RM Klein, ed, Plant Growth Regulation. Iowa State University Press, pp 489-501

35. PHINNEY BO, SPRAY CR 1982 Chemical genetics and the gibberellin pathway in *Zea mays* L. *In* PF WAREING (ed) Plant Growth Substance 1982, Academic Press, London, New York, pp 101-110

36. PHINNEY BO, WEST CA 1957 The growth response of single gene dwarf mutants of *Zea mays* to gibberellins and gibberellin-like substances. Proc Int Genet Symposium, pp 384-5

37. PHINNEY BO, WEST CA, RITZEL M, NEELY P 1957 Evidence for "gibberellin-like" substances from flowering plants. Proc Nat Acad Sci (USA) 43: 398-404

38. PHINNEY BO, FREELING M, ROBERTSON DS, SPRAY CR, SILVERTHORNE J 1986 Dwarf mutants of maize - the gibberellin pathway and its molecular future. *In* M. Bopp (ed) Plant Growth Substances, Springer-Verlag, Berlin-Heidelberg, pp 55-64
39. POTTS WC, REID JB 1983 Internode length in Pisum III. The effect and interaction of the *Na/na* and *Le/le* gene differences on endogenous gibberellin-like substances. Physiol Plant 57: 448-454
40. POTTS WC, REID JB, MURFET IC 1982 Internode length in *Pisum*. I. The effect of the *Le/le* gene difference on endogenous gibberellin-like substances. Physiol Plant 55: 323-328
41. POTTS WC, REID JB, MURFET IC 1985 Internode length in *Pisum*. Gibberellins and the slender phenotype. Physiol Plant 63: 357-364
42. RADLEY M 1956 Occurrence of substances similar to gibberellic acid in higher plants. Nature (London) 178: 1070-1071.
43. REID JB 1986 Internode length in *Pisum*. Three further loci, *lh*, *ls* and *lk*. Ann Bot 57: 577-592; and unpublished results
44. REID JB, POTTS WC 1986 Internode length in *Pisum*. Two further mutants, *lh* and *ls* with reduced gibberellin synthesis, and a gibberellin insensitive mutant, *lk*. Physiol Plant 66: 417-426
45. REID JB, MURFET IC, POTTS WC 1983 Internode length in *Pisum*. II. Additional information on the relationship and action of loci, *Le*, *La*, *Cry*, *Na* and *Lm*. J Exp Bot 134: 349-364
46. ROBERTSON DS 1978 Characterisation of a mutator system in maize. Mutation Res 51: 21-28
47. ROOD SB, KOSHIOKA M, DOUGLAS TJ, PHARIS RP 1982 Metabolism of tritiated gibberellin A_{20} in maize. Plant Physiol 70: 1614-1618
48. ROSS JJ, REID JB 1987 Internode length in *Pisum*. A new allele at the *Le* locus. Ann Bot 59: 107-109
49. SPONSEL VM 1980 Gibberellin metabolism in legume seeds. *In* J.R. Lenton, ed., Gibberellins-Chemistry, Physiology, and use. British Plant Growth Regulator Group, Wantage, Monograph 5, pp 49-62
50. SPONSEL VM 1983 The localisation, metabolism and biological activity of gibberellins in maturing and germinating seeds of *Pisum sativum* W. Progress No. 9. Planta 159: 454-468
51. SPONSEL VM, MAcMILLAN 1978 Metabolism of gibberellin A_{29} in seeds of *Pisum sativum* cv. Progress No. 9; use of [^{2}H] and [^{3}H]GAs and the identification of a new GA catabolite. Planta 144: 69-78
52. SPRAY CR, PHINNEY BO, GASKIN P, GILMOUR SJ, MACMILLAN J 1984. The dwarf-1 mutation controls the 3ß-hydroxylation of gibberellin A_{20} to gibberellin A_1. Planta 160: 464-468
53. TAMURA S, TAKAHASHI N, MUROFUSHI N, KATO J 1966. Growth promoting activities of bamboo gibberellin. Plant Cell Physiol 7: 677-681
54. WEST CA, PHINNEY BO 1956 Properties of gibberellin-like factors from extracts of higher plants. Plant Physiol 31(suppl.): XX

55. WEST CA, PHINNEY BO 1959 Gibberellins from flowering plants. I. Isolation and properties of a gibberellin from *Phaseolus vulgaris* L. J Amer Chem Soc 81: 2424-2427
56. WHITE OE 1917 Studies of inheritance and variation in peas. Proc Am Phil Soc 56: 487-588
57. YOKOTA T, MUROFUSHI N, TAKAHASHI N 1971. Biological activities of gibberellins and their glycosides in *Pharbilis nil*. Phytochemistry 10: 2943-2949

Physiology of Cell Expansion During Plant Growth, D.J. Cosgrove and D.P. Knievel Eds.,
Copyright ©1987, The American Society of Plant Physiologists

THE USE OF GIBBERELLIN AND PHOTOMORPHOGENETIC MUTANTS FOR GROWTH STUDIES

M. KOORNNEEF[1], P. ADAMSE[3], G.W.M. BARENDSE[4], C.M. KARSSEN[2] AND R.E. KENDRICK[3]

*Departments of Genetics (1), Plant Physiology (2)
and Laboratory of Plant Physiological Research (3),
Agricultural University, Wageningen,
Department of Botany (4), University of Nijmegen,
The Netherlands*

INTRODUCTION

Many single-gene mutants having modified plant height have been described in higher plants. Compared to their wild type they may be reduced (dwarfs) or, less often, increased in height. Mutants characterized at a biochemical level are extremely useful for physiological research, since the relevance of the factor which is changed (often deleted) is directly indicated by the difference between the mutant and its isogenic wild type. Since changes in one factor within a plant often result in other changes, these interactions also complicate the interpretation of physiological experiments when using a genetic approach. Remnant levels of the factor affected by the mutation--below detection limits, but physiologically significant--can lead to problems of interpretation. A severe limitation in the use of mutants is that mutations with respect to essential cell-division factors may be lethal. Furthermore, not all mutants with a reduced plant height are cell-division or cell-elongation mutants. For example, mutants defective in normal photosynthesis will differ in their cell growth with respect to wild type, but in an indirect way which does not help in understanding cell growth *per se*. Therefore, the isolation of the gene defective in a particular dwarf mutant will certainly not answer all questions concerning the regulation of cell growth. However, it may help to identify some of the factors involved. What is therefore extremely important are the criteria used for selecting potentially interesting mutants.

To date, mutants which are defective in factors known to influence plant height have been mainly used to carry out further analysis of the roles played by these factors in controlling growth. The use of such mutants can be compared to the use of chemical inhibitors. However, the absence of problems connected

172

with penetration and distribution of such substances within the plant and the presence of non-specific effects makes the use of mutants preferable. The genetic-physiological approach is illustrated here by research on the role played by gibberellins (GAs[1]) in fruit growth and by photoreceptors in hypocotyl growth using mutants.

THE USE OF GIBBERELLIN-DEFICIENT MUTANTS

Gibberellin-deficient mutants have been described, and have been characterized biochemically in plant species such as maize, rice, pea, *Arabidopsis* and tomato (12, see also chapter by MacMillan and Phinney). Mutants of all these species are dark-green dwarfs which can be restored to the wild-type phenotype by the application of GAs. Some of the tomato and *Arabidopsis* mutants also require GAs for seed germination (9) and these have helped in understanding the role played by GAs in this process (3,4).

It has been suggested that seed-produced GAs regulate fruit growth (10). In *Arabidopsis thaliana*, fruit (siliqua) length is positively correlated with the number of seeds per siliqua (Fig. 1). In GA- deficient mutants this relationship can also be observed, although the stimulative effect of an increase in seed number is reduced (Fig. 1).

The positive relationship between seed number and siliqua length exhibited by GA-deficient mutants indicates that factors other than GAs that are produced by the developing seeds are involved in siliqua growth. Whether the extra stimulation of siliqua growth observed in the wild type could be attributed entirely to GAs produced by the developing seeds was investigated with a genotype possessing GA-producing embryos on a GA-deficient mother plant. This independent *in vivo* manipulation of hormone levels in seeds and surrounding fruit tissues, virtually impossible to carry out by application of chemicals, is readily attainable with mutants. The genetic procedure using reciprocal crosses is shown in Fig. 2.

Examination of the time course of siliqua growth (Fig. 3) indicates that growth is determined by the capacity of both the seed and the fruit tissue to produce GAs. The effect of different seed numbers (cf. A and B in Fig. 3) and the effect of the embryo genotype on siliqua length can be observed at an early stage of embryo development (2 days after pollination). After 1 week, siliqua length shows no further increase and is similar to that of 3-week-old mature fruits. During the first week of embryo development, the *ent*-kaurene synthesizing capacity, a measure of GA synthesizing capacity of the siliqua plus embryo tissue is much higher than at later stages of development (2). This enzyme activity, as well as the endogenous levels of GA in the mutant, is extremely low (2), but despite this, these low levels may be responsible for the limited siliqua growth. A similar stimulative effect of GA producing embryos on fruit growth has been found in GA-deficient mutants of tomato (4).

---

[1]Abbreviations: GAs = Gibberellins, Pfr = far-red light-absorbing form of phytochrome, Ptotal = total phytochrome (Pr + Pfr), Pr = red light-absorbing form of phytochrome.

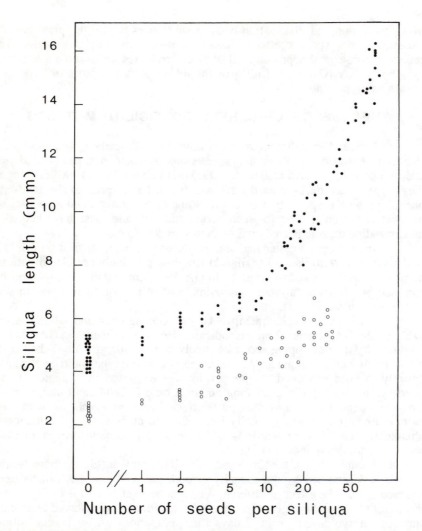

FIG 1. The effect of seed number and genotype (closed circles: wild type, open circles: GA-deficient mutant) on mature siliqua length. For experimental procedures see (2).

THE USE OF PHOTOMORPHOGENETIC MUTANTS

The amount or concentration of a photomorphogenetic photoreceptor pigment is difficult to manipulate chemically. The amount of physiologically active pigment can sometimes be manipulated by applying appropriate light treatments, however such treatments can have differential effects upon subsequent photosensitivity. Complications also arise with this physiological approach because of the coaction of several photoreceptor systems regulating the same process or because of multiple effects induced by a single photoreceptor. For the best characterized higher-plant photomorphogenetic photoreceptor, phytochrome,

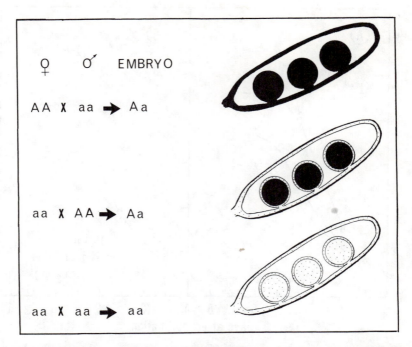

FIG. 2. Crosses between GA-deficient (*aa*) and wild-type (*AA*) plants that result in different combinations of GA-producing (wild type) and GA-non-producing (GA-deficient mutant) embryos and maternal tissues.

the possible presence of two phytochrome types (stable and labile) is an additional complication.

Mutants that lack a particular photomorphogenetic response may be defective in the photoreceptor, the transduction chain between photoreceptor and physiological response, or the response itself (6). Photomorphogenetic mutants can therefore be simplistically classified into three groups:

Photoreceptor mutants	*Transduction chain mutants*	*Response mutants*
		-----> R_1
hv		-----> R_2
----->	-----> -----> ----->	-----> R_3
		-----> R_4

True photoreceptor mutants will lack the photoreceptor or contain the modified photoreceptor which is non-functional. Spectrophotometric and immunological (8, 11) evidence has shown that the *aurea* mutant of tomato is the best characterized example of a higher-plant photomorphogenetic photoreceptor mutant, having a phytochrome content of less than 5 % of that of the wild type.

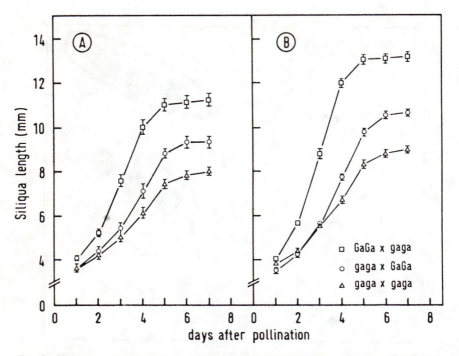

FIG. 3. Siliqua growth after reciprocal crosses and self-pollinations with wild type (*GaGa*) and gibberellin-deficient mutant (*gaga*): (A) mean number of seeds per siliqua for *GaGa* x *gaga*, *gaga* x *GaGa* and *gaga* x *gaga* were, respectively, 41.7 ± 2.7, 41.8 ± 1.9, 42.1 ± 1.9, and (B) 60.5 ± 15, 60.3 ± 0.9, and 59.1 ± 1.7 (from 2).

In addition, this mutant has no, or a strongly reduced, photoregulation of *cab* protein synthesis, chloroplast development, anthocyanin synthesis, seed germination and hypocotyl growth inhibition (8, 10). Similar mutants are the *hy-1* and *hy-2* mutants of *Arabidopsis* (7). The *Arabidopsis hy-4* mutants appear to be defective with respect to blue light-absorbing pigment(s). Mutants that contain a particular photoreceptor, but lack all photomorphogenetic responses associated with it, are, most probably, transduction chain mutants. Possible mutants which pass in this category are those at the *hy-3* and *hy-5* loci of *Arabidopsis* (7). A common characteristic of all mutants mentioned above is their increased hypocotyl growth relative to wild type when they are grown in white light.

      The presence of such a long-hypocotyl (*lh*) mutant of cucumber is especially interesting for research on hypocotyl growth, because its large size facilitates analysis of rapid growth changes in response to light treatments (1). The *lh* mutant shows the normal rapid reduction in growth rate upon exposure with blue light after incubation in darkness. In contrast, the growth rate of the mutant is hardly affected by red light compared to the wild-type response (Table I). It might appear to be a paradox that a mutant that retains inhibition of hypocotyl growth by blue light should ultimately grow long in white light containing a considerable blue component. The explanation of this probably lies in a temporal pattern of responsiveness to the different photoreceptor systems. In the experiment where dark-grown seedlings are transferred to continuous blue light of relatively low irradiance, growth is ultimately limited by photosynthesis and the hypocotyls remain relatively short. However, in high irradiance white light the blue light inhibition is only transient and photosynthesis sustains hypocotyl growth resulting in the long hypocotyls typical of the mutant phenotype.

Table I. *Provisional measurements of the hypocotyl growth rates of 4-5 day old dark-grown lh mutant and wild-type cucumber seedlings maintained in darkness (D) and upon transfer to continuous blue (BL) or red (RL) light (both $\approx 15\ \mu mol\ m^{-2}\ s^{-1}$ at 25 °C.*
Measured with a horizontal microscope under green safelight. The slower growth rate in darkness in the case of wild type probably reflects response to the repetitive exposures of green safelight whereas the mutant is insensitive.

Treatment	Wild type		*lh* mutant	
	Growth rate	% D	Growth rate	% D
	mm h^{-1}		*mm h^{-1}*	
D	1.30	-	1.65	-
BL	0.75	57	0.95	57
RL	0.80	60	1.40	87

The relative insensitivity of the *lh* mutant to red light is also obvious when seedlings are grown under continuous irradiation, presumably due to the retarded cessation of cell-elongation in the case of the mutant seedlings (Fig. 4). Apparently cell maturation that is normally induced by red light is severely reduced in the mutant (1).

In dark-grown seedlings of the *lh* mutant normal levels of phytochrome can be measured (1). However, when plants are grown in white light, phytochrome levels in Norflurazon treated plants and in non-green flower parts are reduced to approximately 50% of the wild-type levels. This observation, together with results from physiological experiments with both etiolated and de-etiolated seedlings (1), suggests that the phytochrome system in de-etiolated plants, which is probably predominantly represented by the light stable phytochrome type, may be defective. However, it can not be completely ruled out that the mutant is partially defective with respect to the phytochrome transduction chain.

The *lh* mutant shows normal responses to blue light which indicates the independent action of the blue light-absorbing photoreceptor on hypocotyl growth. This conclusion was also drawn from observations with the *Arabidopsis* mutants (6). These mutants may be extremely useful, especially for the study of hypocotyl growth, since they allow the construction of action spectra for the blue light-absorbing photoreceptor, without the intervention of

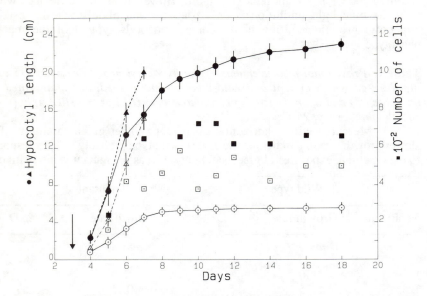

FIG. 4. Hypocotyl length ± S.E. and estimated number of epidermal cells per hypocotyl during growth in 14 μmol m⁻²s⁻¹ red light (circles) of cucumber wild type (open symbols) and *lh* mutant (closed symbols) at 25°C. Triangles represent the growth in darkness. The actual mean cell lengths of the hypocotyl epidermal cells over the time period 8-18 days in wild type and *lh* mutant were 11 ± 1 and 32 ± 1 μm, respectively. The arrow indicates the onset of continuous irradiation.

blue light absorbtion by phytochrome. It is hoped that photomorphogenetic photoreceptor and transduction chain mutants will allow an evaluation of the relative importance of different photoreceptor systems, including the relative contribution of light-stable and light-labile phytochrome in particular responses during development.

One of the long-standing questions of phytochrome research is: which of its two forms (Pr or Pfr) is physiologically active? While circumstantial evidence implicates the Pfr form as physiologically active, mutants lacking phytochrome provide the first direct evidence to support this hypothesis. Mutants possessing different levels of phytochrome, a situation until now only attainable by pre-irradiations which might selectively influence subsequent response sensitivity, would be additionally particularly useful to determine if a response is due to the concentration of Pfr or the Pfr/Ptotal ratio.

It is anticipated that mutants will play an invaluable role in understanding the molecular mechanism of photoreceptor action. Physiologically characterized mutants identified as transduction chain mutants can be studied at the molecular level, thus providing a glimpse into the black box between photoperception and ultimate physiological response.

Acknowledgments--We thank P.A.P.M. Verhoeven-Jaspers for technical assistance and A.M. Bijlsma for conducting preliminary experiments.

LITERATURE CITED

1. ADAMSE P, PAPM JASPERS, RE KENDRICK, M KOORNEEF 1987 Photomorphogenetic responses of a long-hypocotyl mutant of *Cucumis sativus* L. J Plant Physiol 127: 481-491
2. BARENDSE GWM, J KEPCZYNSKI, CM KARSSEN, M KOORNNEEF 1986 The role of endogenous gibberellins during fruit and seed development: Studies on gibberellin-deficient genotypes of *Arabidopsis thaliana* Physiol Plant 67: 315-319
3. GROOT SPC, CM KARSSEN 1987 Gibberellins regulate seed germination in tomato by endosperm weakening: a study with gibberellin-deficient mutants. Planta 171: 525-531
4. GROOT SPC, J BRUINSMA, CM KARSSEN 1987 The role of endogenous gibberellin in seed and fruit development of tomato: studies with a gibberellin-deficient mutant. Physiol Plant: in press
5. KARSSEN CM, E LACKA 1986 A revision of the hormone balance theory of seed dormancy: studies on gibberellin and/or abscisic acid-deficient mutants of *Arabidopsis thaliana. In* M Bopp, ed, Plant Growth substances 1985. Springer Verlag, Berlin Heidelberg, pp 315-323
6. KOORNNEEF M, RE KENDRICK 1986 A genetic approach to photomorphogenesis. *In* RE Kendrick, GHM Kronenberg, eds, Photomorphogenesis in Plants Martinus Nijhoff Publ., Dordrecht, pp 521-546
7. KOORNNEEF M, E ROLFF, CJP SPRUIT 1980 Genetic control of light inhibited hypocotyl elongation in *Arabidopsis thaliana* L. Heynh. Z Pflanzenphysiol 100: 147-160
8. KOORNNEEF M, JW CONE, RG DEKENS, EG O'HERNE-ROBERS, CJP SPRUIT, RE KENDRICK 1985 Photomorphogenic responses of long hypocotyl mutants of tomato. J Plant Physiol 120: 153-165
9. KOORNNEEF M, JW CONE, CM KARSSEN, RE KENDRICK, JH VAN DER VEEN, JAD ZEEVAART 1985 Plant hormone and photoreceptor mutants of *Arabidopsis* and tomato. *In* M Freeling, ed, Plant Genetics. Allan R Liss Inc, New York pp 103-114
10. NAYLOR AW 1984 Functions of hormones at the organ level of organization. *In* TK Scott, ed, Encyclopedia of Plant Physiology, New Series, vol 10, Springer Verlag Berlin, pp 172-218
11. PARKS BM, AM JONES, P ADAMSE, M KOORNNEEF, RE KENDRICK, PH QUAIL 1987 The aurea mutant of tomato is deficient in spectrophotometrically and immunochemically detectable phytochrome. Plant Mol Biol 9: 97-107
12. REID JB 1986 Gibberellin mutants. *In* AD Blonstein, PJ King, eds, A genetic approach to plant biochemistry. Springer Verlag Wien, New York, pp 1-34

LEAF AND ROOT EXPANSIVE GROWTH IN RESPONSE TO WATER DEFICITS

THEODORE C. HSIAO AND JIAHAI JING[1]

*Department of Land, Air, and Water Resources
University of California, Davis, CA 95616*

INTRODUCTION

Expansive growth of individual cells, in well defined patterns and sequences and as constrained by the neighboring cells, give rise to organs of diverse shapes and functions on the plant. The organ that has been studied the most in terms of expansive growth is the *Avena* coleoptile, with the leaf and root receiving considerably less attention. It is well known that when other factors are not limiting, expansive growth is very sensitive to water deficits (9). Leaf growth appears to be the most sensitive compared to root growth. This chapter reviews briefly the effects of water deficit on growth in general and that of leaves and roots in particular. Attention will be directed at the biophysical aspects of growth responses to water deficits.

EFFECTS OF WATER DEFICITS ON CELL EXPANSION

Since water accounts for most of the increase in volume during cell expansion, studies of expansive growth have often focused attention on water related parameters. Current view considers the biochemical loosening of the cell wall under the mechanical stress caused by turgor pressure as the initial step of the process (see chapters by Cosgrove and by Ray). This stress relaxation, which lowers cell water potential (ψ), is followed by water uptake and cell enlargement. For sustained growth, turgor must be maintained at an adequate level to allow continuous wall loosening. Since solutes are diluted by the water taken up for growth, the maintenance of turgor requires a continuous accumulation of solutes, else the turgor would fall with growth until it is too low to effect stress relaxation. Thus, expansive growth depends on the biochemical processes of wall loosening and formation, and solute transport or generation, as well as the physical parameters of turgor pressure and water transport. The interplay among these parameters and processes are embodied in the combined equation of Lockhart (12), which is reviewed in the chapter by Ray.

With the development of water deficits, expansive growth slows. At the first glance, one may be tempted to attribute this to a mere slowing down in

---

[1]Present address: Northwestern College of Agriculture, Yanling, Shaanxi, People's Republic of China.

water uptake that is needed to make up for the increased volume. The reality, however, is much more complex and not yet fully understood. For one, usually over 99% of the water taken up by a green plant in the natural environment is not actually used for growth but is transpired into the atmosphere. A number of studies have shown that expansive growth can be stopped or greatly inhibited by water deficits while transpiration continued unabated (1, 3). In many cases, leaves of deficient plants would be at a lower ψ, to maintain an adequate ψ gradient to sustain water uptake in spite of the reduction in ψ of the rooting medium.

The question then arises, if the growing region of the plant is at a lower ψ, would turgor of the cells be less and result in a slower or even no stress relaxation of the walls? For a number of years it was assumed that reduced turgor was the main reason for reduced growth in water deficient plants (9, 21). Of course, substantial reduction in turgor could occur when water deficit is severe enough. With less severe deficits, however, turgor may be maintained or at least kept relatively high by osmotic adjustment. In that case, either stress relaxation of the cell wall has to be extremely sensitive to turgor pressure, or the ability of the wall to expand at a given turgor has to be reduced, to explain substantial reductions in growth under mild or moderate water deficits. To sort out these possibilities, it is necessary to know how much osmotic adjustment takes place in the growing cells deficient in water, the quantitative relationship between growth rate and turgor pressure, and how water deficit affects this relationship.

Osmotic adjustment. Osmotic adjustment in this paper refers to changes in solute content (in osmoles) per cell volume, with the volume set at some reference state, e. g., at saturation or zero water potential. By accumulating more solutes per reference volume, solute potential and hence ψ will be lower at a given turgor pressure. Thus, turgor can be maintained or kept high in spite of low tissue and medium ψ. Solutes can also become more concentrated and ψ more negative in a cell if the cell simply loses water. In that case though, turgor will be reduced and there is no gain in solutes per reference volume. Therefore, there would not be true osmotic adjustment. For the expanding cell, the gain in solutes would have to be more than that required to keep up with dilution by growth, to effect osmotic adjustment.

Osmotic adjustment in reponse to water deficit has been shown to be more pronounced in the growing regions than in the mature zones. This was true for maize leaves (16) and nodal roots (18), as well as when wheat leaves of different ages were compared to each other and to the growing shoot apex (2). Within the growth zone, however, the youngest cells may not have more ability to adjust than the more mature cells. When the primary root of maize seedling was examined at 1 mm length intervals, osmotic adjustment was found to become more pronounced from the root apex to the basal portion of the growth zone, where the cells had reached essentially their final sizes (Sharp, Hsiao, and Silk, in preparation). Regardless, it is clear that the growing region overall can adjust osmotically to water deficit, by as much as more than 20 bars in the case of the shoot apex of wheat (2).

The importance of osmotic adjustment for expansive growth under water deficit has been well documented (7, 11, 14, 15). Here we give some data obtained with maize as an example. As water deficit developed in maize, leaf elongation slowed and the rates were correlated with water potential of the

181

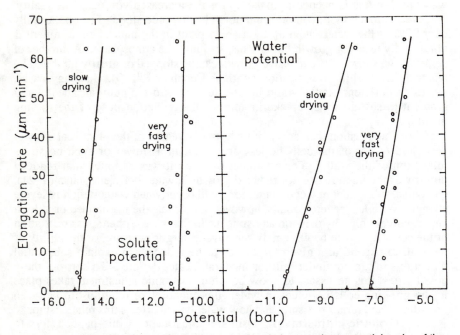

FIG. 1. Elongation rate of the fifth leaf of maize in relation to solute potential and ψ of the growth zone, as influenced by the speed of water deficit development. Plants were growing in a soil potting mixture in a controlled environment chamber at 29˚C. As water deficit developed, elongation rate decreased from the maximum to zero in 2 h for the very fast drying plants, and in 19 h for the slow drying plants. Basal (growth zone) segments (50 mm long) were excised from the leaves (approx. 300 mm long) and measured by the Shardakov method at 5˚C for ψ, and by isopiestic thermocouple psychrometry for solute potential. Growth was monitored with a linear variable position transducer and the rate represents the time interval of 10 to 15 min immediately prior to excision. Lines were fitted by linear regression. (Adapted from Jing & Hsiao, in preparation.)

growth zone (Fig. 1). For plants in a potting mixture drying slowly, leaf elongation was reduced as growth zone ψ was reduced below -8 bars and stopped as ψ reached -11 bars. In contrast, for plants subjected to much faster deficit development (very fast drying), elongation began to be affected as ψ dropped below -5 bars and ceased as ψ reached -7 bars (Fig. 1). The difference in behavior between the fast and slow drying plants was attributable largely to the difference in their osmotic adjustment. For any growth zone ψ and elongation rate, solute potential of the slow drying plants was lower than that of the very fast drying (Fig. 1). Slow drying apparently provided sufficient time for substantial osmotic adjustment, whereas very fast drying did not. By accumulating more solutes, the slow drying plants were able to grow at a lower tissue ψ.

Relationship between growth and turgor pressure. Quantitative relationships between growth rate and turgor pressure (P) can provide much information on the yielding characteristics of the cell wall. The relationship can often be described by one of the two equations of Lockhart (12):

$$\frac{1}{V} \frac{dV}{dt} = \emptyset \, (\psi_p - Y)$$

where V is cell volume, t is time, ø is volumetric extensibility of the cell, and Y is yield threshold turgor pressure, below which the cell will not expand. The model of Lockhart has been criticized for its inadequacies (20). To be sure, most of the criticisms are valid. However, some of the problems may lie with the users. There is a tendency to view ø and Y as constants whereas in fact they may change with condition and time (10), as will be exemplified below. Another problem is that the equation was derived originally for single cells or tissue of uniform growth, such as a small segment of a stem. For this reason, the relative growth rate (rate of change in volume per unit of total volume) was used instead of absolute growth rate, to eliminate or minimize the effect of cell or segment size. The equation, however, is often applied to the growth of the whole organ, where there is clearly a spatial gradient of growth rate (11). To overcome this problem, studies are carried out on organs whose overall growth rate is steady or nearly so. In that case, the size of the growth zone is probably invariant over a period of time, as is well known for roots under a constant environment. It is then possible to express growth rate in absolute terms and avoid the task of determining the exact size of and gradients in the growth zone. Much of the available data on organ growth and turgor pressure were obtained this way, as are all the data presented here.

Regardless of its limitations and in the absence of better alternatives, Lockhart's equation does provide a conceptual frame for the analysis of water deficit effects on growth. Much early data have shown a yield threshold turgor for growth. Particularly definitive are the results from experiments where turgor at which growth stopped was measured directly with pressure microprobes in cells (6, 8). More discussions on yield threshold turgor are found in the chapter by Cosgrove. As for extensibility, it is normally evaluated from plots of expansion rate vs. turgor as the slope of the plot. It has also been studied using the Instron or similar instruments, as discussed in the chapter by Cleland.

A number of earlier studies have shown that over a range of water status, plots of growth rate vs. turgor gave linear relationships with intercepts (the yield threshold) above zero (5, 7), consistent with the Lockhart model. More recent results were similar (14). In most cases, turgor pressure was calculated as the difference between solute potential and ψ measured in thermocouple psychrometers. In one way the results are surprising, for it is now known that ψ of the growing zone so measured can be erroneously low because of stress relaxation of the cell wall (see chapter by Cosgrove). When tissue is excised from the growing zone and placed in the psychrometer, the wall will continue to loosen as long as turgor is above the yield threshold. Since the tissue has been cut off from its water source, there will be no water uptake resulting from wall relaxation. Consequently turgor will drop, leading to a reduction in ψ, until the yield threshold is reached. The extent of this error will depend on the turgor and growth rate at sampling. The higher the initial turgor, the larger will be the drops in turgor and ψ during the measurement. The faster the growth, the more would be the wall relaxation. Yet, in spite of these

183

potential errors, the plots conform reasonably well to the equation of Lockhart.

In our study of maize leaf growth, we found that cold temperature apparently suppressed stress relaxation of the cell wall (Hsiao & Jing, in preparation). We measured ψ of the growing zone at 5°C by the Shardakov dye method (19), which has the additional advantage of requiring only a short equilibration time (less than 10 min) and hence minimizing the opportunity for stress relaxation. Growth rate was monitored with a position transducer and represents the 10 to 12 min period just prior to excision for ψ and solute potential measurement. The data we obtained show that growth rate was linearly related to the calculated turgor in the growth zone for plants subjected to a range of soil water status (Fig. 2). The relationship between growth and turgor had a

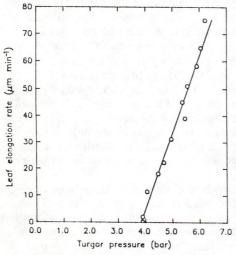

FIG. 2. Elongation rate of the fifth leaf of maize in relation to turgor in the growing zone. Growth and turgor were reduced by fast development of water deficit; growth rate slowed from the maximum to zero in 5 h. Turgor was calculated as the difference between ψ and solute potential, measured, respectively, by the Shardakov method at 5°C and isopiestic thermocouple psychrometry. The line was fitted by regression. (Adapted from 11)

very steep slope (indicating high extensibility), a drop in turgor of 2.5 bars from the maximum was sufficient to stop growth completely. Another key point is that the yield threshold (x-intercept) was very high. So growth ceased even though turgor was still nearly 4 bars. The steep slope and high yield threshold appears to be unusual. That could be partly due to the fact that growth rates were measured over periods of approximately 10 min, in contrast to over a number of hours in many other studies.

On the surface, the steep slope and high Y depicted in Fig. 2, assuming they hold true for leaves of other species as well, appear to account for the extreme sensitivity of leaf growth to water deficit. Any reduction in turgor, even if only a fraction of 1 bar, would lead to a substantial reduction in growth. It turned out, however, that the relationship shown may hold only for short moments if the water status of the plant is changing. The ephemeral nature of the relationship makes the causal linkage between growth inhibition and water deficits still more complicated, as will be elaborated on below.

Shifts in yield threshold and extensibility. In spite of the fact that reasonably linear plots can be obtained of growth vs. turgor pressure in different studies, there are ample data showing that Y and ϕ may change with time after the onset of water deficits. In an elegant study using micromanometers to monitor turgor in *Nitella* internodal cells, Green (8) first showed that Y began to diminish in a matter of minutes after a stepwise reduction of turgor and the associated reduction in growth. This had the effect of restoring growth to some extent while the turgor remained low. Similar adjustment in Y, but in the opposite direction, occurred when turgor was suddenly raised. Overall, Y appeared to adjust to stabilize the growth rate. For higher plants such as maize, the time course of leaf growth responses to stepwise shifts in medium ψ (Fig. 3) over short periods resembled closely

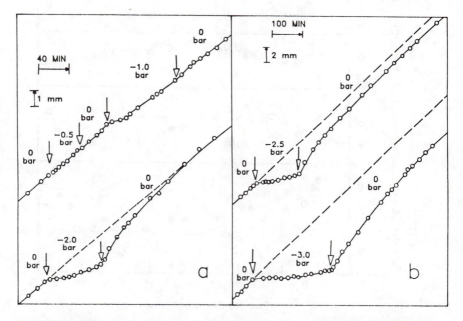

FIG. 3. Effects of step-wise changes in ψ of the root medium on length of growing maize leaf. The ψ of the root bathing solution, given in bars, was changed at the time indicated by the arrows. Dotted line indicates extrapolated leaf length. Purified Carbowax 6000 was the osmoticum. Note the difference in scales between a and b. (Adapted from 1)

those of *Nitella*. A sufficient drop in medium ψ first brought about a stoppage of growth, followed by resumption in a matter of minutes. When medium ψ was suddenly raised, there was a transitory rapid growth, which slowed down to a steady rate in a short while.

Water deficits over longer periods (in terms of days), on the other hand, often resulted in a situation where either growth rate was lower in the face of full turgor maintenance in the growth zone, or the reduction in growth was disproportionally large compared to the observed reduction in turgor (13, 14, 16). These studies were conducted under controlled conditions. In a field study on sorghum, we found a similar behavior in the natural environment over the daily course, that of same turgor but lower growth rate in the plants subjected to mild water deficit (Fig. 4). It may be expected that over a long period the whole growth system adjusts to a particular set of conditions so to have its own characteristic growth vs. turgor curves. That appears to be the basis for the

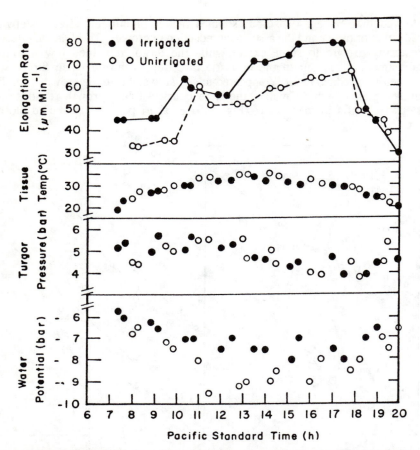

FIG. 4. Diurnal trends of leaf elongation rate, temperature, ψ and turgor of the growth zone of sorghum in the field at Davis, California, on Aug. 10, 1983. Water had been withheld for 12 days from the unirrigated plot. Turgor and ψ of the growth zone were determined as described under Fig. 2. (From 11)

different curves reported by Bunce (5) for soybean growing in different environments.

How long does it take for the tissue in the growth zone to adjust and change its growth rate vs. turgor characteristics? We addressed this question by following growth and turgor of maize leaves after a sudden drop in ψ of the root medium. The data (Fig. 5) show that significant osmotic adjustment took place in the growth zone within half of an hour after a reduction in medium ψ of 2.5 bars. That change might have accounted for some of the growth transients in Fig. 3. However, in spite of osmotic adjustment raising turgor to the level equal to that without water deficit in less than 1.5 h, growth rate recovered only to about a half of that without water deficit (Fig. 5). Thus, the data suggest that the wall yielding properties have shifted to cause slower growth in spite of the full osmotic adjustment, in agreement with the long term results. What was surprising was that the apparent change in Y or ø should have taken place so

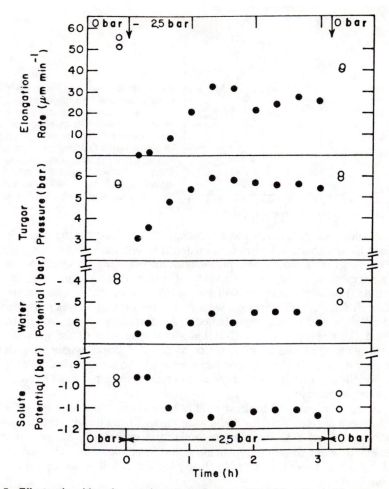

FIG. 5. Effects of sudden changes in root medium ψ on the elongation rate and ψ solute potential, and turgor of the growth zone of the fifth leaf of maize. The ψ of the rooting medium, given in bars, was changed at time zero and at 3 h 10 min, as indicated by the downward arrows (top of figure). Carbowax 6000 was the osmoticum. For any given sampling time, all of the parameters were measured on the same leaf. (From 11)

fast, in a matter of 1 h or so. In another experiment, growth and turgor were measured with or without exposing the plants to 4 h of prior water deficit. Plots of growth rates vs. turgor pressure (Fig. 6) showed quite clearly that the yield threshold was shifted to a higher turgor by the short period of prior water deficit. The slope of the growth rate vs. turgor plot, associated with ø, might have been reduced slightly by the prior stress, but the data (Fig. 6) are too variable to be sure. Other researchers have shown reduced slopes of such plots caused by different growing conditions (5) and by water deficiency (14). However, a change

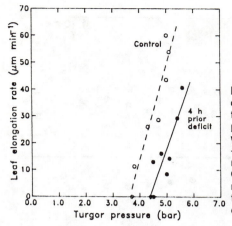

FIG. 6. Relationship between maize leaf elongation rate and turgor pressure of the growth zone as affected by 4 h of prior water deficit. Roots of control plants were exposed to medium of 0, -1.0, or -2.5 bars for 10 to 40 min to produce the range of elongation rate and turgor. Other plants were subjected to -2.5 bar medium for 4 h, then treated the same as the control. Carbowax 6000 was the osmoticum.

in slope does not necessarily mean a change in ∅, at least not the ∅ as originally defined by Lockhart. As discussed earlier, it is necessary to assume a constant size of the growth zone to interpret measurements made on whole organs or a large segment of an organ. In a study of maize roots, Sharp, Hsiao and Silk (submitted) showed that the growth zone was shortened during many hours under water deficiency. Similar results were found for sorghum leaves growing in the field (Walker and Hsiao, unpublished). A shortening of the growth zone by water deficits, even if the wall yielding characteristics of the growing cells remain unchanged, would give a plot of reduced slope, when the growth is measured for the whole organ or a large segment. So for experiments conducted over periods of many hours or days, these changes in slopes must be interpreted with extreme caution.

One interesting contrast with *Nitella* is clear at this point. Upon the development of water deficit, *Nitella* went through osmotic adjustment only very slowly but shifted Y to lower levels to maintain growth (8). Maize leaf, on the other hand, went through significant osmotic adjustment. In spite of that, growth became slower because Y actually shifted in a direction opposite that in *Nitella*.

CONTRASTING BEHAVIOR OF ROOTS

It is known that when water is deficient, root growth can be favored over that of shoot (18). This difference may not be attributed solely to the fact that roots are closer to the water source since Westgate and Boyer (22) showed that roots can maintain growth down to a lower tissue ψ than the shoot. Roots exhibit a strong ability to undergo osmotic adjustment under water stress (7; Sharp, Hsiao and Silk, in preparation). To delineate the roles of osmotic adjustment, turgor pressure, and wall yielding properties, a study was carried out on roots of very young maize seedlings using the pressure microprobe. Root growth rate was monitored continuously and altered by step-wise changes in external medium water potential. At the same time, turgor of its surface cells was measured with the pressure microprobe. An example of the results (Hsiao, in preparation) is given in Fig. 7. Turgor of cells in this root was between 6.5 and 7.3 bars when the root was in a medium of zero water potential (0.1 mM

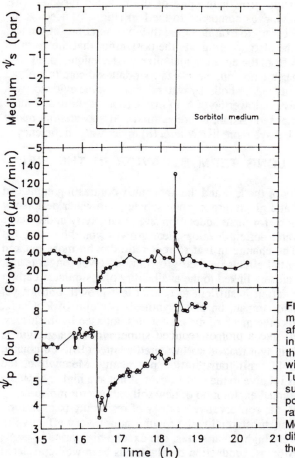

FIG. 7. Growth and turgor of maize seedling primary root as affected by step-wise changes in medium ψ. Growth rate of the whole root was monitored with a position transducer. Turgor was measured in surface cells of the root at a position of maximum growth rate (4 to 5 mm from the apex). Most turgor points represent different cells. Sorbitol was the osmoticum.

CaCl). When medium ψ was suddenly reduced by 4.2 bars, growth stopped momentarily, followed first by rapid recovery, then more slow recovery, to a rate nearly that before the reduction in ψ, all in a matter of 30 to 40 min. Turgor pressure, on the other hand, dropped by 3 bars or more as soon as ψ was reduced, and appeared to remain nearly unchanged during much of the rapid recovery of growth; a slow increase in turgor (about 0.7 bar/h), indicative of osmotic adjustment, was evident during the slow portion of growth recovery. However, turgor never reached the level attained before the reduction in ψ. In spite of the substantially lower turgor in the low ψ medium, growth recovered nearly to the rate prevailing prior to water deprevation. This behavior is in apparent contrast to the data on maize leaves, where turgor recovered fully but growth remained inhibited after a step-wise drop in medium ψ (Fig. 5).

The ability of roots to maintain substantial growth under water deficiency in spite of reduced turgor was confirmed by additional experiments with maize using the pressure microprobe (Hsiao, in preparation). It seems clear that after the onset of water deficit, the maize root had adjusted its wall yielding

189

properties in a way to sustain growth in spite of the reduced turgor. This apparently contrasting behavior, as compared to the leaf (Fig. 5), needs to be substantiated. It will be necessary to examine growth-turgor relations of the leaf with the pressure microprobe also, to eliminate the possibility that the contrast might have arisen somehow from the different measurement techniques used.

In terms of Lockhart's equation, the maize root data indicate that either ø was increased or Y was reduced, or both, by water deficit. Some evidence was obtained showing the successive lowering of Y in maize roots as more and more severe water deficit was imposed (Hsiao, in preparation). It appears that maize roots, in contrast to leaves, behave more like *Nitella* (8) under water deficiency.

IMPLICATIONS FOR LONG TERM BEHAVIOR IN THE FIELD

The reduced ability of the leaf and the apparently contrasting enhanced ability of the root to grow at a given turgor when subjected to water deficiency have far reaching implications for plant adaptation and productivity under water limiting conditions. As water deficit develops, leaf growth would be inhibited more than root growth. The change in leaf to root ratio can be more marked than the immediate differential effects on their growth rate. Over longer terms, expansive growth must be closely linked to the availability of assimilated carbon and energy. When water deficit is suffient to restrict leaf growth but not yet strong enough to have major impact on photosynthesis per unit of leaf area, roots would presumeably receive much of the assimilates not used by the leaves. That may explain why the absolute amount of root biomass can actually be increased under water deficit over that of well watered control (18). Of course, the reduced leaf area would limit transpiration per plant. Meanwhile, the preferential growth of roots relative to the leaves, even if not at a higher absolute rate, would ensure the continued exploration of new soil volume for moisture.

On the other hand, the conservative strategy of restricting transpiration surface area and maintaining substantial root growth, though very effective in keeping the plant at relatively high water status, can exact a high price in terms of reduced productivity. Biomass production by plants has been well correlated with radiation interception by their canopies (17). When foliage is sparse and limiting radiation interception and absorption, biomass production is often exponential with time. This can be attributed to the fact that leaf expansive growth is self amplifying with time at that stage (4). Any increment in leaf area due to expansive growth will lead to more radiation interception and hence more photosynthesis per plant. The availability of more assimilates will, in turn, lead to still more expansive growth. If water deficit inhibits leaf growth at this stage, even if only slightly, the effect can be amplified with time. This effect can be quantified by first order kinetics (4) and appears to explain the marked reduction in biomass production observed for what appear to be very minor or negligible reduction in leaf water potential (Nyabundi, Bolanos & Hsiao, in preparation).

CONCLUDING REMARKS

This chapter has addressed the problems and prospects of plant expansive growth as affected by water deficits, largely from our own perspectives. There is virtually no disagreement that leaf growth is very sentitive to reductions in water status. Also, the evidence for root growth being more resistant to water deficits than shoot, though not extensive, is quite convincing. Much more controversial is the role turgor plays in the inhibition of growth by water deficits. Our view is that turgor reduction is very likely the primary initial factor that causes the slowing down of growth when a plant is deprived of water. At the same time, either the turgor reduction, or the resultant deceleration of growth, may serve as a signal that is transduced, apparently in a short time, into modifications of the cell wall yielding properties. Whether the wall becomes more hardened or more plastic will determine how the growth rate vs. turgor relationship is altered. In leaves, the limited evidence indicate that cell wall hardens (Y increases) under water deficits, leading in time to slower growth in spite of turgor maintenance. In roots, the even more limited data indicate that cell wall becomes more plastic and extensible. Roots can then grow at reasonable rates although turgor is lower. Even these tentative conclusions must be tempered with the realization that changes in morphology, such as the shortening of the growth zone, most likely plays a critical role in long term growth under water limiting conditions.

There is no doubt that techniques and instrumentation are critical in resolving these controversial aspects of growth. The success of monitoring turgor in growing cells with the microprobe and of measuring tissue ψ while minimizing stress relaxation of the cell wall, and the realization of the significance of morphology and spatial variations, presage the resolution of these conflicting view points in the near future.

Acknowledgments--This work was supported by NSF grant DCB 8417504 to W. K. Silk and T. C. Hsiao, by Hatch fund of the California Agricultural Experimental Station, by a fellowship from the Chinese government to J. J., and by an Alexander von Humboldt Award for U. S. Senior Scientists to T. C. H. Prof. E.-D. Schulze kindly provided the facilities at the University of Bayreuth, West Germany, where the pressure microprobe work was carried out.

LITERATURE CITED

1. ACEVEDO E, TC HSIAO, DW HENDERSON 1971 Immediate and subsequent growth responses of maize leaves to changes in water status. Plant Physiol 48: 631-636
2. BARLOW EWR, RE MUNNS, CJ BRADY 1980 Drought responses of apical meristems. *In* Adaptations of Plants to Water and High Temperature Stress. NC Turner and PJ Kramer, Eds., John Wiley and Sons, New York. pp. 191-205
3. BOYER JS 1970 Leaf enlargement and metabolic rates in corn, soybean, and sunflower at various leaf water potentials. Plant Physiol 46: 233-35
4. BRADFORD KJ, TC HSIAO 1982 Physiological responses to moderate water stress. *In* Encyclopedia of Plant Physiology, N.S. Vol 12B Physiological Plant Ecology II. Water relations and carbon assimilation. OL Lange, PS Nobel, CB Osmond, H Ziegler, Eds, Springer Verlag, Berlin. pp. 263-324

5. BUNCE JA 1977 Leaf elongation in relation to leaf water potential in soybean. J Exp Bot 28: 156-161

6. COSGROVE DJ, E VAN VOLKENBURGH, RE CLELAND 1984 Stress relaxation of cell walls and the yield threshold for growth: demonstration and measurement by micro-pressure probe and psychrometer techniques. Planta 162: 46-54

7. GREACEN EL, JS OH 1972 Physics of root growth. Nature 235: 24-25

8. GREEN PB 1968 Growth physics in *Nitella*: a method for continuous *in vivo* analysis of extensibility based on a micro-manometer technique for turgor pressure. Plant Physiol 43: 1169-1184

9. HSIAO TC 1973 Plant responses to water stress. Ann Rev Plant Physiol 24: 519-570

10. HSIAO TC, E ACEVEDO, E FERERES, DW HENDERSON 1976 Water stress, growth and osmotic adjustment. Phil Trans R Soc London Ser B 273: 479-500

11. HSIAO TC, WK SILK, AND J JING 1985 Leaf growth and water deficits. *In* Control of Leaf Growth. NR Baker, WJ Davies and CK Ong, Eds, Society of Experimental Biology Seminar 27, Cambridge Univ Press, Cambridge. pp 239-66

12. LOCKHART JA 1965 Cell extension. *In* Plant Biochemistry. J Bonner and JE Varner, Eds. Academic Press, New York. pp 826-49

13. MATSUDA K, A RIAZI 1981 Stress-induced osmotic adjustment in growing regions of barley leaves. Plant Physiol 68: 571-76

14. MATTHEWS MA, E VAN VOLKENBURGH, JS BOYER 1984 Acclimation of leaf growth to low water potentials in sunflowers. Plant, Cell, and Env 7: 199-206

15. MEYER RF, JS BOYER 1972 Sensitivity of cell division and cell elongation to low water potentials in soybean hypocotyls. Planta 108: 77-87

16. MICHELENA VA, JS BOYER 1982 Complete turgor maintenance at low water potentials in the elongating region of maize leaves. Plant Physiol 69: 1145-49

17. MONTEITH JL 1977 Climate and the efficiency of crop production in Britain. Phil Trans R Soc London Ser B. 281: 277-94

18. SHARP RE, WJ DAVIES 1979 Solute regulation and growth by roots and shoots of water-stressed maize plants. Planta 146: 319-26

19. SLAVIK B 1974 Methods of Studying Plant Water Relations. Ecological Studies, Volume 9, Springer Verlag, Berlin

20. TOMOS AD 1985 The physical limitations of leaf cell expansion. *In* Control of Leaf Growth, Society for Experimental Biology, Seminar Series 27. NR Baker, WJ Baker, CK Ong, Eds. Cambridge University Press, Cambridge. pp 1-33

21. VAADIA Y, FC RANEY, RM HAGAN 1961 Plant water deficits and physiological processes. Ann Rev Plant Physiol 12: 265-92

22. WESTGATE ME, JS BOYER 1985 Osmotic adjustment and the inhibition of leaf, root, stem and silk growth at low water potentials in maize. Planta 164: 540-49

Physiology of Cell Expansion During Plant Growth, *D.J. Cosgrove and D.P. Knievel* Eds.,
Copyright ©1987, The American Society of Plant Physiologists

REGULATION OF DICOTYLEDONOUS LEAF GROWTH

E. VAN VOLKENBURGH

Botany Department
University of Washington
Seattle, WA 98195

INTRODUCTION

The expansion of dicotyledonous leaves is a determinate, light-requiring process which differs in many respects from the growth of other organs. Cell enlargement contributes significantly to the final size of broad leaves; in these tissues, as in most others, cells enlarge 10-50 times during development (18). However, this process in leaves has only been studied directly in a few cases which will be described below. Most of the information available in the literature on the physiology of leaf expansion describes effects on the growth of whole leaves, without separating cellular responses of division from enlargement (see reviews in 1, 8, 13, 16). The reason for this is that cell enlargement and division are not spacially separated in expanding dicot leaves.

After dicot leaves are initiated on apical meristems, the primordia undergo extensive cell divisions determining early their rudimentary leaf shape. Subsequent cell enlargement, frequently accompanied by additional cell divisions, completes the expansion of the leaf. Development proceeds as a wave from the tip to the base of the leaf, and once reaching the base, cell division normally ceases. Cell enlargement continues until mature leaf size is obtained. Often in dicotyledonous leaves, the processes of cell division and cell enlargement occur simultaneously throughout a broad region of the leaf surface. These leaves do not retain meristematic zones, whereas organs with indeterminate growth patterns continue to divide and elongate new cells.

Dicot leaf expansion requires exposure to light which stimulates both processes of cell division and enlargement. The stimulation of cell division has been shown to be a low fluence phytochrome response (5, 12). Cell enlargement on the other hand requires high fluence light (5, 30) and its regulation will be discussed further below. The possible role of plant growth substances in dicot leaf growth has been investigated (13). Unfortunately very little work has focused on the effects of these substances on cell enlargement alone.

Finally, a significant factor affecting the expansion of dicotyledonous leaves is their exposure to changeable environmental conditions. Unlike the meristematic zones in the base of monocot leaves, or the apical meristems of stems, both of which are protected by other tissues, most of the development of dicot leaves occurs after the young tissue emerges and is totally exposed to open air. The growth of cells in this tissue must occur in spite of frequent changes in leaf water status, temperature, and illumination. It is likely, therefore, that several different mechanisms operate to control the rate of leaf cell growth in dicots. One such mechanism is described below.

CELL ENLARGEMENT

A model system for the study of cell enlargement in dicotyledonous leaves has been developed using primary leaves of bean, *Phaseolus vulgaris* (25). These leaves normally complete cell divisions when they are less than 20% full size (6). The cell division and enlargement phases of development in these leaves can be separated in time by growing the plants at first in low fluence red light (less than 4 μmol m^{-2}s^{-1}), stimulating cell divisions until that phase is completed about 10 days after planting. Subsequent exposure of the leaves to high fluence white light (200 μmol m^{-2}s^{-1}) induces rapid cell enlargement. Induction of cell enlargement in this way can be obtained both with intact leaves and also with discs excised from leaves and grown in minimal nutrient solution (e.g., 10 mM KCl plus sucrose). This model system has been employed to identify the photoreceptor mediating cell enlargement, and the cellular parameters that change in response to light, allowing rapid growth to occur

The involvement of phytochrome in bean leaf cell growth was suggested by De Greef et al. (11), but the response they observed in red light was somewhat less than the total response elicited by white light, unless illumination of the apical hook occured as well. Earlier work by Dale and Murray (9) showed that leaf cell enlargement in beans could be stimulated by blue and red light, but again the response was not maximal. Similar results have been published by Göring (14), who attributes the growth response to stomatal behavior.

Recently we have carried out experiments in conjunction with M. Watanabe and M. Furuya at the Okazaki Large Spectrograph in Japan to construct an action spectrum for bean leaf cell enlargement (Van Volkenburgh and Cleland, in preparation). Preliminary experiments showed that for intact leaves as well as excised discs, the maximum growth response occured at fluence rates above 100 μmol m^{-2}s^{-1}, that the light exposure must exceed 8 hours, and reciprocity did not hold. Data obtained with the Large Spectrograph showed that cell enlargement was stimulated primarily by red (660 nm) and blue (460 nm) light. Continuous exposure to far-red (730 nm) light inhibited the red light growth response.

These results confirm that cell enlargement in bean leaves is a high irradiance response of phytochrome (9, 11, 30). That chlorophyll is not a required photoreceptor was shown by the fact that stimulation of cell growth by blue and red light occured even in leaves devoid of chlorophyll due to treatment with tentoxin, a compound which inhibits chlorophyll synthesis. Normally, light-stimulated disc growth can occur when KCl is present without sucrose (Table I). For light stimulated growth to occur in bleached tissue, sucrose must

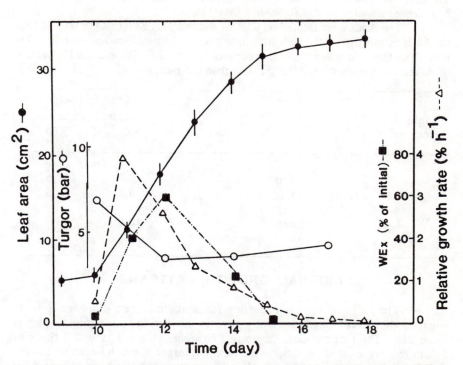

FIG. 1. Leaf area (o), relative growth rate (Δ), turgor pressure (o) and cell wall extensibility (o) of primary bean leaves exposed to white light (200 μmol m^{-2} s^{-1}) 10 days after planting in red light (4 μmol m^{-2} s^{-1}). Redrawn from (28). Leaf expansion during this period is due to cell enlargement and no cell division (25). Turgor was calculated from the difference between leaf disc water and osmotic potential measured psychrometrically. Wall (plastic) extensibility (WEx) was measured by the Instron technique and is expressed as the increase relative to the initial value.

be added. These results are similar to results obtained with discs incubated in the presence of DCMU. In combination, the results show that chlorophyll is not necessary for leaf cell growth and that the photoreceptor(s) involved in light-stimulated leaf cell enlargement is phytochrome (with perhaps an additional blue light receptor).

The physiological effect of the interaction of light with the photoreceptor, phytochrome, has been shown to be a stimulation of proton excretion from the leaf cells (26). Acidification of the cell walls increases their extensibility, and it is this effect on the cells which causes the cell growth rate to increase in the presence of light. The osmotic concentration of the leaves remains constant or declines slightly when growth is stimulated by light. As a result, leaf turgor does not increase, but may actually drop while growth rate accelerates (27) (see Fig. 1). Thus the limiting factor regulating leaf cell growth in the light is cell wall extensibility which is mediated by proton efflux and the capacity of the wall to be loosened when acidified.

Table I. *Growth of discs excised from green or tentoxin (Tx)-treated primary bean leaves sampled on day 10 in red light.*
The discs were floated on 0.01 M sucrose plus KCl (SK) or 0.02M KCl for 18 h in red light (RL) or white light (WL). Initial disc diameter was 7.0 mm. (The tentoxin was graciously provided by Dr. G.E. Templeton, University of Arkansas, Fayetteville, AR.) SE less than 0.3 mm, n = 10

Light Exposure		Disc Diameter	
		SK	KCl
		mm	
RL	green	8.7	8.6
	Tx	8.4	8.5
WL	green	10.0	11.0
	Tx	9.2	7.7

DIURNAL GROWTH PATTERNS

The fact that light is necessary for, and stimulates, dicot leaf growth suggests that leaves should normally grow in the day, but not at night. This is often true. The first trifoliate leaves of *Phaseolus vulgaris* grow 2-3 times as fast in the day as at night, despite the fact that turgor is much lower during the day (2 bars) than at night (4 bars)(10). This pattern has also been observed for sunflower leaves provided adequate water (19) and nitrogen (22). For this to occur, rapid daytime growth must be accompanied by increased wall extensibility sufficient to offset decreased daytime turgor. Such an adjustment in wall extensibility was observed by Bunce (4) for field-grown soybean leaves as compared to leaves growing in the greenhouse or growth chamber.

In many cases, however, leaves grow faster at night, as was documented for sunflower leaves by Boyer (2) who attributed the differences in growth rate to the higher turgor at night. Similar results have been obtained for leaves exposed to environmental stress such as water deficit (10) or nitrogen deficiency (22). Although in these cases it appears that the diurnal growth pattern has shifted (night > day), the growth rate of stressed leaves was comparable to the dark growth of unstressed leaves, and at no time did the stressed leaves grow as rapidly as the unstressed leaves. Deficit conditions, therefore, inhibit light-stimulated growth, in part by reducing turgor too far to be offset by increases in wall extensibility, but additionally by reducing wall extensibility (19, and see below, Table II).

Despite the fact that the maximum growth rate for some species can only be obtained in the light, it is apparent that these leaves have a mechanism enabling them to continue growing in the absence of light, and they may rely on this mechanism for most of their expansion. This appears to be true for sycamore leaves (*Acer pseudoplatanus* L.) which are shade-adapted and, therefore, unconditioned to evaporative demand in bright light and consequent turgor loss

196

Table II. *Comparison of hybrid poplar tree growth, leaf growth rate, wall properties, and epidermal cell size for plants growing in irrigated (WET) and nonirrigated (DRY) plots.*

Wall extensibility (WEx) was measured with the Instron technique. CAWL is expressed as the linear rate of extension of frozen-thawed leaf strips in pH 4.5 NaAcetate (0.01M)(see 28). Means followed by SE.

	WET	DRY
Tree height (m) (n=5)	4.2 (0.1)	3.6 (0.2)
Total leaf area (dm^2) (n=5)	174 (11.2)	140 (19.3)
Individual leaf area (dm^2) (n=100)	5.6 (0.3)	4.3 (0.40)
Leaf growth rate (cm^2 d^{-1}) (n=9)	34.0 (1.3)	28.0 (2.8)
WEx (%/100g) (n=15)	23.2 (2.1)	20.9 (4.2)
CAWL (μm/min) (n=10)	0.85 (.10)	0.23 (.05)
Epidermal cell diameter (mm) (n=25)	.027 (.001)	.028 (.001)
Epidermal cell number per leaf surface (x 10^6) (n=25)	81.9 (3.9)	63.8 (5.5)

(23). Taylor and Davies (23) have compared the diurnal growth patterns of sycamore and birch (*Betula pendula* Roth), a sun-adapted plant, and have found that two different patterns emerge. Birch leaves exhibit the typical bean leaf pattern: light-stimulated growth accompanied by increased wall extensibility and proton excretion in response to light. Sycamore leaves show none of these responses. They grow faster in the dark, with a marked increase in turgor, and no change in wall properties or proton excretion. The authors suggest that two different mechanisms might regulate leaf growth: light-stimulated proton efflux controlling wall properties, and perhaps the control of turgor by plant growth substances.

The involvement of plant growth substances in leaf cell enlargement is not clear, and few studies have addressed this process directly, as separated from cell division. Recent studies of Brock (3), using the bean leaf model system, have investigated the possibility that light might act to stimulate proton excretion via a second messenger system involving one of several growth-stimulating substances. He tested auxin (indole acetic acid, IAA), cytokinins (zeatin; benzyl adenine purine, BAP), and gibberellic acid (GA3) for their ability to stimulate leaf cell enlargement in the absence of white light. Both BAP and GA3 were able to do this, while IAA had no effect. However, BAP and GA3 also acted additively with light, suggesting that the cellular responses to hormones and light were different. This was supported by the observations that neither growth substance stimulated proton efflux, although they did cause increased cell wall extensibility and cell growth without significant increases in osmotic concentration.

How the growth of leaf cells is regulated in the dark remains an open question. There are several possibilities. One could be that the effect of light (e.g., acidification of the cell walls) is carried over into the dark period. Such a carryover effect could be augmented by the restoration of leaf turgor as the plant rehydrates in the dark. Or, an entirely separate mechanism may be activated in the dark by plant growth substances, such as utilization of stored carbohydrate for wall synthesis or turgor elevation.

LIMITATIONS TO LEAF GROWTH

When considering plant productivity and development of photosynthetic surface area, it becomes particularly interesting to consider what determines ultimate leaf size, and then how limitations to growth might be overcome. Anatomically, leaf size is limited by cell size and cell number. Physiologically the rates and duration of cell division and expansion are limiting. The following paragraphs briefly review what is known about cellular bases for leaf growth limitation.

The latter phase of leaf expansion is dominated by cell enlargement because cell division has ceased. Therefore, it is the cessation of cell enlargement which ultimately stops the expansion of the leaf. In bean leaves turgor is maintained as enlarging cells mature (28). In fact, turgor exceeds the yield threshold more in maturing cells than in rapidly growing younger cells (29). The growth rate of maturing leaf cells decreases because cell wall extensibility is reduced as the cells enlarge (28). The mechanism responsible for stiffening the walls has not been fully described. It is clear that leaf cells continue to excrete protons into the cell wall space (28). However, the capacity for the walls to extend in response to acidification (CAWL) is lost in maturing cells (28). These results indicate that the biochemistry of the walls changes developmentally such that wall extensibility is lost. The CAWL assay provides a useful test for studying possible changes in wall structure or enzymatic activity (26, 28).

Environmental stresses such as salinity and water deficit may also inhibit cell enlargement and reduce final cell and leaf size. Water deficit inhibits sunflower leaf expansion by reducing wall extensibility (WEx) and yield threshold, while turgor is maintained (19). Inhibition of *Populus* leaf growth by water deficit is also due to reduced wall extensibility and CAWL (Table II).

Interestingly, for *Populus* leaves, inhibition of leaf (and cell) growth rate by water deficit results in smaller leaves, but final cell size is not reduced (Table II). This implies that cells expand more slowly in deficit conditions, due to reduced wall extensibility, but reach a genetically determined final cell size. (The duration of leaf expansion was 30 days in both treatments.) Decreased leaf size in this case and others (16, 20, 24) is due to reduced cell number, not cell size. However, it is possible that the rate of cell enlargement directly affects the potential for cell divisions.

Considerable information exists in the literature about environmental influences on cell division in leaves (7). It appears that cell division under low irradiance levels is limited by the supply of photoassimilate (20). This idea is substantiated by CO_2 enrichment studies which show an enhancement of the exponential phase of growth which is dominated by cell division (17). Therefore, assimilate supply could be a factor limiting cell division under conditions of environmental stress. Inadequate nitrogen has also been reported to reduce cell number (24). In this case, final leaf size was limited by cell number; final cell volume of the nitrogen limited plants was very similar to the controls. The low nitrogen treatment greatly reduced the rate of cell enlargement and partially reduced the rate of cell division, indicating an interaction between these two processes.

Perhaps it is not necessary to consider separately the two processes of cell division and enlargement, and instead consider the growing leaf to be continuously increasing its area and volume, and that the partitioning of this volume has no controlling influence over the final size of the leaf (15, but see discussion in 21). In this view, the entire leaf can be considered as one large expanding cell, and the parameters controlling cell enlargement would totally describe the growth process. Alternatively, the rate of individual cell enlargement may directly determine the potential for that cell to divide (rapid enlargement would allow a cell to reach a critical size for division sooner, so more divisions could occur during a developmentally defined period). In this case also, the cellular mechanisms regulating the rate of cell enlargement would be critical in determining mature leaf size.

CONCLUSIONS

Cell enlargement in dicotyledonous leaves requires high fluence light and is mediated by the high irradiance response of phytochrome and perhaps also a blue light receptor. The process does not require chlorophyll, although an energy source must be provided, such as exogenously supplied sucrose when chlorophyll is absent. The cellular response to light is stimulation of proton efflux, resulting in cell wall acidification and increased wall extensibility. This response is sufficient to increase the growth rate of leaf cells and is usually accompanied by turgor loss as water uptake into the growing cells lags behind wall loosening. As long as the growing leaves are protected from water deficit, increased wall extensibility promoted by light offsets the loss of turgor and growth rate is stimulated in light.

In many cases, leaf growth rate in the day is slower than during the night because the daytime reduction in turgor precludes growth. As a result, most of the expansion of these leaves occurs in the dark, and some mechanism for storing the light response, or a different light-independent mechanism, permits leaf expansion to proceed in the dark.

The interaction between the rate of cell enlargement, and the rate of cell division may prove to be very interesting. If it is true that a developmental time window exists during which division can occur, and the rate of cell enlargement determines the frequency with which critical cell size for division is obtained during that window, then the cellular factors controlling cell expansion rate are important both for final cell size and number, and leaf size.

LITERATURE CITED

1. BAKER NR, WJ DAVIES, CK ONG, eds 1985 Control of Leaf Growth. Cambridge Univ Press, Cambridge
2. BOYER JS 1968 Relationship of water potential to growth of leaves. Plant Physiol 43: 1056-1062
3. BROCK TG 1985 The role of hormones in leaf cell enlargement. Ph.D. Dissertation. Univ. Washington, Seattle
4. BUNCE JA 1977 Leaf elongation in relation to leaf water potential in soybean. J Exp Bot 28: 158-163
5. BUTLER RD 1963 The effect of light intensity on stem and leaf growth in broad bean seedlings. J Exp Bot 14: 142-152
6. DALE JE 1964 Leaf growth in *Phaseolus vulgaris*. I. Growth of the first pair of leaves under constant conditions. Ann Bot 28: 579-589
7. DALE JE 1976 Cell division in leaves. *In* MM YEOMAN, ed, Cell division in higher plants. Acad Press, London, pp. 325-345
8. DALE JE, FL MILTHORPE 1982 General features of the production and growth of leaves. *In* JE DALE, FL MILTHORPE, eds, The growth and functioning of leaves. Cambridge Univ Press, Cambridge, pp. 152-175
9. DALE JE, D MURRAY 1968 Photomorphogenesis, photosynthesis, and early growth of primary leaves of *Phaseolus vulgaris*. Ann Bot 32: 767-780
10. DAVIES WJ, E VAN VOLKENBURGH 1983 The influence of water deficit on the factors controlling the daily pattern of growth of *Phaseolus* trifoliates. J Exp Bot 34: 987-999
11. DE GREEF JA, R CAUBERGS, JP VERBELEN, E MOEREELS 1978 Phytochrome-mediated inter-organ dependence and rapid transmission of the light stimulus. *In* HH Smith, ed, Light and Plant Development. Butterworths, London, pp. 295-315
12. DOWNS RJ 1955 Photoreversibility of leaf and hypocotyl elongation of dark-grown red kidney bean seedling. Plant Physiol 30: 468-475
13. GOODWIN PB 1978 Phytohormones and growth and development of organs of the vegetative plant. *In* DS LETHAM, PB GOODWIN, TJV HIGGINS, eds, Phytohormones and related compounds: A comprehensive treatise. Vol II. Phytohormones and the development of higher plants. Elsevier/North Holland, Amsterdam, pp. 31-174
14. GÖRING H, S KOSHUCHOWA, H MUNNICH, M DIETRICH 1984 Stomatal opening and cell enlargement in response to light and phytohormone treatments in primary leaves of red-light-grown seedlings of *Phaseolus vulgaris* L. Plant Cell Physiol 25: 683-690

15. HABER AH, DE FOARD 1963 Nonessentiality of concurrent cell divisions for degree of polarization of leaf growth. II. Evidence from untreated plants and from chemically induced changes of the degree of polarization. Amer J Bot 50: 937-944
16. HUMPHRIES EC, AW WHEELER 1963 The physiology of leaf growth. Ann Rev Plant Physiol 14: 385-410
17. KRIEDEMANN PE 1986 Stomatal and photosynthetic limitations to leaf growth. Aust J Plant Physiol 13: 15-31
18. MAKSYMOWYCH R 1973 Analysis of leaf development. Cambridge Univ Press, Cambridge
19. MATTHEWS MA, E VAN VOLKENBURGH, JS BOYER 1984 Adaptation of sunflower leaf growth to water deficit. Plant Cell Environ 7: 199-206
20. NEWTON P 1963 Studies on the expansion of the leaf surface. II. The influence of light intensity and daylength. J Exp Bot 14: 458-482
21. POETHIG RS 1984 Patterns and problems in angiosperm leaf morphogenesis. *In* GM MALACINSKI, SV BRYANT, eds, Pattern Formation, MacMillan Pub Co, New York, pp. 413-432
22. RADIN JW, JS BOYER 1982 Control of leaf expansion by nitrogen nutrition in sunflower plants. Plant Physiol 69: 771-775
23. TAYLOR G, WJ DAVIES 1985. Control of leaf growth of *Betula* and *Acer* by photoenvironment. New Phytol 101: 259-268
24. TERRY N, LJ WALDRON, A ULRICH 1971 Effects of moisture stress on the multiplication and expansion of cells in leaves of sugar beet. Planta 97: 281-289
25. VAN VOLKENBURGH E, RE CLELAND 1979 Separation of cell enlargement and division in bean leaves. Planta 146: 245-247
26. VAN VOLKENBURGH E, RE CLELAND 1980 Proton excretion and cell expansion in bean leaves. Planta 148: 273-278
27. VAN VOLKENBURGH E, RE CLELAND 1981 Control of light-induced bean leaf expansion: Role of osmotic potential, wall yield stress and hydraulic conductivity. Planta 153: 572-577
28. VAN VOLKENBURGH E, MG SCHMIDT, RE CLELAND 1985 Loss of capacity for acid-induced wall loosening as the principle cause of the cessation of cell enlargement in light-grown bean leaves. Planta 163: 500-505
29. VAN VOLKENBURGH E, RE CLELAND 1986 Wall yield threshold and effective turgor in growing bean leaves. Planta 167: 37-43
30. VINCE-PRUE E, DJ TUCKER 1982 Photomorphogenesis in leaves. *In* JE DALE, FL MILTHORPE, eds, The growth and functioning of leaves. Cambridge Univ Press, Cambridge, pp. 233-269

STIMULUS-RESPONSE COUPLING IN THE ACTION OF AUXIN AND GRAVITY ON ROOTS

MICHAEL L. EVANS AND KARL-HEINZ HASENSTEIN

*Department of Botany, Ohio State University,
Columbus, OH 43210*

INTRODUCTION

In spite of extensive research on hormone action and binding in plant cells, we know little about the nature of stimulus-response coupling in these cells. It is generally assumed that, in order for a hormone to elicit a response, it must first bind to a receptor in or on the plant cell. The hormone-receptor complex then causes the production or release of one or more secondary messengers which ultimately lead to the effects attributed to the hormone. In the case of animal cells there is a vast literature indicating secondary messenger functions for calcium and for certain small molecules including cAMP[1], DG, and IP$_3$ (1, 23, 24). Although cAMP appears to play no role as a secondary messenger in plant cells (12) there is evidence that calcium as well as certain products of phospholipid turnover may.

CALCIUM, INOSITOL TRISPHOSPHATES, AND DIACYLGLYCEROL AS SECONDARY MESSENGERS

Inositol phospholipid turnover. Figure 1 illustrates the role thought to be played by inositol phospholipid turnover in stimulus-response coupling. According to this model the agonist (e.g., hormone) first binds to a receptor on the membrane surface. The receptor-agonist complex in turn activates the enzyme phospholipase C which catalyzes the breakdown of a membranous pool of PtdIns(4, 5)P$_2$ releasing DG and IP$_3$. Both of these products are viewed as playing a secondary messenger role. IP$_3$, which is water soluble, is thought to move to nearby ER where it stimulates the release of stored calcium from the ER into the cytoplasm. Elevation of cytoplasmic free calcium is thought to lead to enhanced protein phosphorylation either by direct calcium stimulation of

---

[1]Abbreviations: ABA = abscisic acid, CaM = Calmodulin, cAMP = adenosine 3',5'-cyclic monophosphate, 2,3-D = 2,3-dichlorophenoxy acetic acid, 2,4-D = 2,4-dichlorophenoxyacetic acid, DG = diacylglycerol, EDTA = ethylenediamine tetraacetic acid, EGTA = ethylene glycol-bis (B-amino ethyl ether)-N,N,N',N'-tetraacetic acid, ER = endoplasmic reticulum, GA = gibberellic acid, IAA = indole-3-acetic acid, IP$_3$ = inositol 1,4,5-trisphosphate, PDE = phosphodiesterase, PtdIns(4,5)P$_2$ = phosphatidylinositol 4,5-bisphosphate.

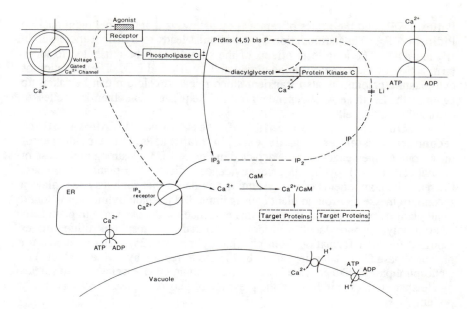

FIG 1. Schematic representation of the roles of DG, IP_3, and calcium as secondary messengers. The agonist binds to an external receptor leading to the stimulation of phospholipase C, an enzyme that catalyzes the hydrolysis of $PtdIns(4,5)P_2$ to IP_3 and DG. DG, in the presence of calcium, activates protein kinase C which in turn catalyzes the phosphorylation of target proteins. IP_3 released upon the hydrolysis of $PtdIns(4,5)P_2$ moves to the ER where it stimulates the release of stored calcium. The elevation of cytoplasmic Ca^{2+} participates in the activation of protein kinase C and also activates target proteins either directly or through CaM-mediated enzyme activation. DG and IP_3 are used in the resynthesis of $PtdIns(4,5)P_2$. The metabolism of IP_3 to inositol during this process is thought to be inhibited by lithium as indicated. Some of the factors thought to participate in the regulation of cytoplasmic Ca^{2+} levels are indicated in the diagram. These include a) Ca^{2+} influx across the plasmalemma through a voltage-gated Ca^{2+} channel, b) Ca^{2+} transport into the ER by a Ca^{2+}-ATPase, c) Ca^{2+} secretion from the cell across the plasmalemma by a Ca^{2+}-ATPase, and d) Ca^{2+} accumulation in the vacuole via a Ca^{2+}/H^+ antiport carrier. The participation of the vacuole is indicated particularly in reference to plant cells since, in plant cells, the vacuole is large and there is evidence for a Ca^{2+}/H^+ antiport mechanism moving Ca^{2+} into the vacuole. Release of Ca^{2+} from the vacuole may also contribute to transient elevation of cytoplasmic free calcium.

protein kinase or, as shown in the diagram, through calcium activation of the calcium-modulated protein, CaM. Once activated by calcium, CaM is thought to activate protein kinases leading to enzyme phosphorylation and hence altered (increased or decreased) enzyme activity.

The second product of $PtdIns(4,5)P_2$ breakdown, DG, also enhances protein phosphorylation (23). DG stimulates the membrane-bound enzyme protein kinase C which, in turn, catalyzes the phosphorylation of target proteins. The actions of the two secondary messengers, IP_3 and DG, are interdependent in a least one way. Protein kinase C activity requires calcium, and in the presence of DG, the affinity of protein kinase C for calcium increases. The intracellular levels of IP_3 and DG are determined by their rates of formation from

PtdIns(4,5)P$_2$ and their rates of removal by pathways that channel them back to PtdIns(4,5)P$_2$. Thus stimulation by the agonist leads to transient increases in IP$_3$ and DG. The long-term effects of the stimulus can be attributed to long-term action of the target proteins modulated by IP$_3$ and DG. The elevation of cytoplasmic calcium is also transient since released calcium can be taken up again by the ER in an ATP-dependent fashion, sequestered in other organelles, or extruded from the cell.

Do the components of the inositol phospholipid secondary messenger cascade exist in plant cells? The evidence for the secondary messenger cascade depicted in Figure 1 is very strong in the case of animal cells (1, 23). Although recent evidence indicates that plant cells possess the requisite components of the system, the evidence that such a system plays a secondary messenger role in plant cells is limited. Phosphatidylinositol is found in all plant membranes and there is some evidence for the existence in plant cells of both polyphosphorylated inositides (2, 22) and the phosphoinositide kinases required for their formation from phophatidylinositol (25). The evidence for phospholipase C in plants appears to be limited to reports by Irvine et al. (14) of a phospholipase C-like enzyme in celery (and certain other species) and to studies by Helsper et al. (11) indicating the presence of phospholipase C activity in lily pollen.

What about protein kinase C? Although there is considerable evidence for calcium- and calcium/CaM-stimulated protein kinase in plants (12), the evidence for a DG-regulated protein kinase C comparable to that found in animal cells is weak. Hetherington and Trewavas (13) reported a calcium-activated protein kinase in pea membranes, but its properties differed from those of animal protein kinase C. However, Schäfer et al. (26) reported protein kinase C activity in extracts from zucchini. The enzyme was soluble and it exhibited properties similar to those of animal protein kinase C. Protein kinase C activity was also reported in potato tubers by Ladyzhen et al. (15). The bulk of the activity was in the cytosolic fraction and, just as with animal protein kinase C, its activity was enhanced by Ca^{2+} and by phosphatidylserine. Treatment of the cells with 0.1 µM GA caused an increase in protein kinase C activity of the membrane fraction and a decrease in the activity of the cytosolic fraction. In animal cells, agonist-induced association of protein kinase C with the plasmalemma is thought to be an effective means of protein kinase C activation.

Regulation of cytoplasmic free calcium levels. Much of the evidence for secondary messenger action in both animal and plant cells focuses on agonist-induced elevation of cytoplasmic free calcium levels. Cytoplasmic calcium levels are normally maintained at submicromolar levels in spite of high ($\geq$ mM) external calcium levels and a membrane potential that favors calcium uptake. Low cytoplasmic calcium levels appear to be maintained by ATP-dependent Ca^{2+} secretion from the cell and/or by sequestering of Ca^{2+} in organelles (e.g., ER, mitochondria, vacuole) (Fig. 1 and ref. 12). Agonist binding is thought to lead either directly (e.g., agonist-induced increase in membrane permeability to calcium) or indirectly (e.g., via IP$_3$-induced release of Ca^{2+} from ER) to a transient increase in cytoplasmic calcium. Calcium, in turn, may directly or indirectly (by activation of CaM) enhance protein kinase activity resulting in selective enzyme activation leading to expression of the effects of the agonist.

Calmodulin occurs throughout the plant kingdom and there are reports of both Ca^{2+}- and Ca^{2+}/CaM-dependent protein kinases in plants (12). In addition there is evidence in plant cells for *a*) an ATP-dependent Ca^{2+} efflux ATPase at the plasmalemma (3, 12), *b*) ATP-dependent Ca^{2+} uptake by the ER and mitochondria (3, 12, 19), and *c*) a Ca^{2+}/H^+ antiport system that moves calcium into the vacuole (3). Thus plant cells have the requisite components for stimulus-response coupling based either on Ca^{2+}/CaM regulation of enzyme activity or on phospholipid turnover with IP_3 and DG acting as secondary messengers. However, the evidence for the involvement of either of these stimulus-response coupling mechanisms in plant cells is limited and largely circumstantial. The sections that follow will consider some of this evidence, especially as it relates to the participation of these secondary messenger systems in the response of plants to auxin and to gravitropic stimulation. For a more thorough review including discussions of other hormone responses and of stimulus-response coupling in the action of light and gravity on plants, see reference 12.

STIMULUS-RESPONSE COUPLING IN THE RESPONSE OF PLANT CELLS TO AUXIN

Auxin action and secondary messenger systems. Although there is evidence for components of the phospholipid secondary messenger system in plant cells, there is little evidence for its operation. Morré and coworkers (21) studied the effects of 2,4-D on phosphatidylinositol turnover in membranes isolated from soybean hypocotyl. They prelabeled the membranes by incubating them in [^{3}H] *myo*inositol. They then incubated the membranes with and without 2,4-D and measured phospholipid turnover as loss of ^{3}H from the phospholipid fraction (Table I). The active auxin, 2,4-D, enhanced phospholipid turnover and the concentration dependence of this effect paralleled that for the action of 2,4-D on cell elongation. The inactive auxin, 2,3-D, was

Table I. *Comparison of 2,4-D and the 2,4-D analogue, 2,3-D, on loss of [^{3}H] myoinositol from prelabeled membranes.*

Treatments were for 10 min at 25 °C with 1 μM 2,4-D or 2,3-D. Based on results from six experiments ±S.E.

Treatment	Radioactivity remaining in phospholipid	Change
	cpm/mg protein	%
No auxin	1100 ±146	
2,4-D	851 ±196	-23
2,3-D	1080 ±83	-2

(from 21)

without effect. Although these results are indicative of an auxin effect on membrane phospholipid, the authors pointed out that the data probably do not reflect agonist-induced phospholipid turnover of the sort associated with the classical phospholipase C-based PtdIns(4,5)P$_2$ catabolism occurring in animal cells. Phospholipid turnover in soybean membranes appeared to be catalyzed by phospholipase D, an enzyme that most likely acts as a transferase, replacing inositol with some other hydroxylated constituent and releasing free inositol (rather than IP$_3$) without the simultaneous release of DG.

Although there was no evidence of auxin-induced IP$_3$ release in the study by Morré et al., one can ask whether IP$_3$ has the potential for inducing Ca^{2+} release from endogenous stores in plant cells. There is one report that it does. Drøbak and Ferguson (5) showed that IP$_3$ induces the release of ^{45}Ca^{2+} (previously taken up by an ATP-dependent process) from zucchini microsomal vesicles (Fig. 2). The release was rapid (< 0.5 min) and inducible by IP$_3$ in the range of 10-20 μM. Based on their findings, Drøbak and Ferguson suggested that auxin may exert its effects through the IP$_3$ secondary messenger pathway. Lew et al. (19) also investigated Ca^{2+} uptake by microsomal vesicles from zucchini. They found ATP-dependent Ca^{2+} accumulation in the ER fraction and they noted that this uptake could be stimulated 2-fold by CaM. However, in contrast to the results of Drøbak and Ferguson, they found no effect of IP$_3$ on release of Ca^{2+} from the vesicles. They also reported that the accumulation or release of Ca^{2+} by ER vesicles was unaffected by 10 μM GA, IAA, or ABA. Their results indicate that plant hormones do not directly induce Ca^{2+} efflux from the ER and that IP3 is not a likely intermediary for hormone-induced Ca^{2+} release from ER stores.

Although auxin appears not to induce Ca^{2+} release from zucchini ER, there are other indications that auxin can alter cytoplasmic calcium levels and that the action of the hormone depends upon calcium and/or CaM. Auxin stimulates Ca^{2+} efflux from maize coleoptile and sunflower hypocotyl sections and also promotes Ca^{2+} movement into inside-out plasmalemma vesicles from soybean hypocotyl when the tissue is treated with hormone prior to isolation of the vesicles (6). The enhancement of calcium efflux from auxin-treated cells is consistent with the idea that auxin may induce the release of Ca^{2+} from sequestered sources within the cell. Ca^{2+} efflux from the cell may be viewed as part of a homeostasis mechanism returning cytoplasmic free calcium to pre-stimulus levels. Further indirect evidence for Ca^{2+} involvement in auxin action includes reports that CaM antagonists interfere with the stimulation of cell elongation by auxin in coleoptiles (6). Although such results are vulnerable to the usual uncertainties concerning the specificity of the inhibitors used, in at least one case the active CaM inhibitor, W7, was found to inhibit auxin action while its inactive structural analogue, W5, did not (6).

Gilroy et al. (9) studied the effects of the CaM antagonists trifluoperazine, W7, and tetracaine on the level of free calcium in protoplasts prepared from carrot cells maintained in suspension culture. All of the CaM inhibitors caused a large increase in cytosolic free calcium. In view of the evidence for CaM-regulated Ca^{2+} ATPases which pump Ca^{2+} out of the cytosol either across the plasmalemma or into the ER it seems possible to attribute the

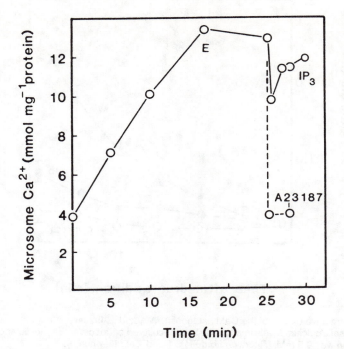

FIG. 2. Uptake and release of Ca^{2+} by zucchini hypocotyl microsomes. ATP (2 mM) was added at time zero (10 min after addition of $^{45}Ca^{2+}$). EGTA (final concentration 0.4 mM was added at E to reduce the external free Ca^{2+} concentration. IP_3 (final concentration 20 μM) and the Ca^{2+} ionophore A23187 (final concentration 0.1 mM) were added to separate samples at 25 min. (Redrawn from ref. 5).

rise in cytoplasmic calcium in the presence of CaM inhibitors to specific interference with such CaM-regulated Ca^{2+}-ATPase activity. The effectiveness of the compounds studied in causing a rise in cytosolic calcium was roughly parallel to their effectiveness as CaM inhibitors. The CaM inhibitor W7 was much more effective in causing rises in cytosolic calcium than the inactive structural analogue, W5, when both were applied at 10 μM (Fig. 3). Although this indicates that the inhibitor effects may result from specific inhibition of CaM-dependent Ca^{2+} ATPases, the authors point out that there are many other plant enzymes regulated by CaM and hence many potential indirect effects of the inhibitors that might ultimately lead to alterations in calcium distribution. Also, they noted that the inhibitory action of the compounds tested, including W5, was much greater at high concentrations, again raising the question of the specificity of the observed effects.

Auxin action in calcium-depleted cells. If the action of auxin depends upon calcium as a secondary messenger, one would expect auxin action to be impaired in calcium-depleted cells. We tested this idea by comparing the inhibitory effect of auxin on root elongation in maize seedlings raised under conditions expected to deplete endogenous pools of calcium (10). Grains were

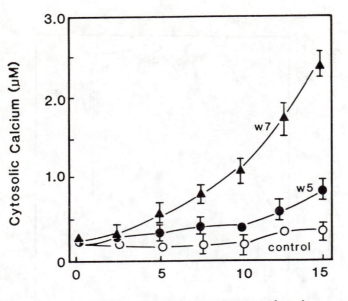

FIG. 3. Comparative effects of the CaM antagonist, W7, and its inactive structural analogue, W5, on cytosolic calcium levels in protoplasts from cultured carrot cells. Extracellular calcium concentration was 0.5 mM. (Redrawn from Figs. 1 and 2 of reference 9).

washed and soaked in distilled water and seedlings were allowed to develop on paper towels wetted with distilled water. The primary roots of these seedlings were further depleted of Ca^{2+} by treatment with 1 mM EGTA prior to measuring their auxin responsiveness. Figure 4 compares the auxin responsiveness of such calcium-depleted roots with that of control roots. The calcium-depleted roots showed little response to 10 μM IAA, whereas the growth of control roots was inhibited nearly 100% by this concentration or by lower (e.g., 1 μM) concentrations. Calcium-depleted roots became sensitive to auxin upon application of calcium (0.5 mM), and the combined Ca^{2+}-IAA inhibition of growth was readily distinguishable from the transient inhibition of growth induced by calcium alone (10). The results indicate that Ca^{2+} may be critical to the action of auxin, a relationship predicted by the Ca^{2+} secondary messenger model.

In its simplest form, the calcium secondary messenger model predicts that under the proper circumstances calcium should be able to mimick the effects of auxin. Of course this is based on the unlikely assumption that the broad spectrum of auxin effects can be attributed solely to auxin induced elevation of cytosolic free calcium. Nevertheless, we found a striking similarity in the action of calcium (1 mM) and IAA (0.1 μM) on the elongation of primary roots of maize seedlings. In both cases growth was strongly inhibited and in both casesthe inhibition was transient with growth recovering completely 60-75 min following the beginning of inhibition (10). There is, of course, no evidence that

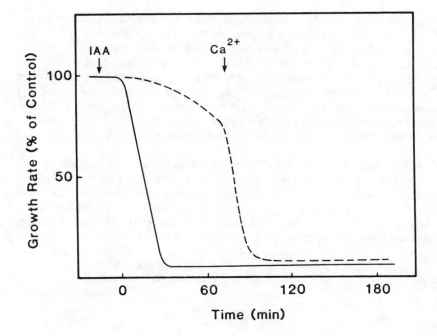

FIG. 4. Calcium dependence of auxin action on root elongation. Maize seedlings germinated and raised in 10 mM $CaCl_2$ (——) or germinated and raised in distilled water followed by exposure to 1 mM EGTA prior to experimentation (- - -). IAA (10 µM) was added to solution bathing roots of both types of seedlings at the first arrow. Calcium (0.5 mM $CaCl_2$) was added to the auxin-treated low calcium roots at the second arrow. (Redrawn from ref. 10).

calcium can substitute for auxin in its various effects including the stimulation of coleoptile and hypocotyl segment elongation and the induction of lateral root formation. The role of calcium in the multiple effects of auxin and the extent to which cytoplasmic calcium perturbations play a general role in auxin action remain to be determined.

STIMULUS-RESPONSE COUPLING IN ROOT GRAVITROPISM

Calcium as a secondary messenger in root gravitropism. In the case of gravitropism, the "agonist" is gravistimulation rather than a hormonal molecule which can bind to an external receptor. Nevertheless, there is indirect evidence for a secondary messenger role for calcium in both shoot and root gravitropism (12). In the case of roots several lines of evidence indicate that gravity-induced calcium movement across the root cap is necessary for the gravitropic response. When caps of primary roots of maize are treated with a calcium chelator such as EDTA, they continue to grow normally but become non-responsive to gravity. Gravitropic responsiveness can be restored by replacing the EDTA with Ca^{2+} but not with Mg^{2+} (17). Application of calcium-containing agar blocks to one side of the tips of vertically-oriented roots can cause the roots to bend toward the side of calcium application. Mg^{2+} and Mn^{2+} are ineffective (17). These observations indicate that gravity-induced calcium asymmetry might be important in the development of gravitropic curvature. Lee et al. (16) tested this idea further by measuring the transport of $^{45}Ca^{2+}$ across the tips of both vertically-oriented and horizontally-oriented (gravistimulated) roots of maize. $^{45}Ca^{2+}$ movement across the tips of vertically-oriented roots was symmetrical while in gravistimulated roots $^{45}Ca^{2+}$ transport was strongly polarized in the downward direction (Table II). The root cap

Table II. *Influence of gravity on $^{45}Ca^{2+}$ movement across roots of maize.*
Agar donor blocks containing 30,000 dpm $^{45}Ca^{2+}$ were applied to one side of the tips (0-1.5 mm region) of horizontally-oriented roots with a blank agar reciever block applied to the opposite side. Each number represents the average (with SD) in receiver blocks after 45 min of transport. Averages determined from five experiments with five roots in each experiment.

	CPM in Receiver Block	
	downward transport	*upward transport*
+ cap	2123 ± 416	612 ± 88
- cap	241 ± 33	192 ± 30

(from 16)

appeared to be especially important to gravity-induced polar Ca^{2+} movement. Decapped roots do not exhibit gravitropism, and calcium movement across the tips of horizontally-oriented decapped roots was weak and non-polar. Although these data are consistent with the proposal that calcium is important in the gravitropic response mechanism, the role played by calcium remains unknown. Gravity-induced polar Ca^{2+} movement can be inhibited by metabolic inhibitors (16) and, surprisingly, by auxin transport inhibitors (18). The latter observation is particularly intriguing since Dela Fuente (4) has shown that calcium is important in the polar transport of IAA. This raises the possibility that gravity-induced Ca^{2+} movement is linked to auxin movement in a manner which leads to asymmetric auxin distribution in the elongation zone, a factor considered by many to be the ultimate cause of the asymmetric growth leading to gravitropic curvature.

The role of calmodulin in root gravitropism. The importance of calcium to root gravitropism raises the question of whether or not CaM might be involved in the transport or action of Ca^{2+} in gravitropism. We examined this question by: *a*) testing for CaM and measuring its distribution in primary roots of maize, *b*) studying potential correlations between CaM activity and gravitropic sensitivity in roots of a cultivar of maize that exhibits light-dependent gravitropic responsiveness, and *c*) testing the effects of CaM inhibitors on calcium transport and gravitropism in primary roots of maize. We identified CaM in roots of light-grown maize and found that its activity was several fold higher in the apical mm (including the cap) than in 1 mm sections taken 2-4 mm behind the root tip (28).

The cultivar (Merit) of maize used for these studies is particularly interesting in that roots of dark-grown seedlings are not graviresponsive. The roots become graviresponsive soon after illumination and the light effect appears to be phytochrome-mediated (8). We compared CaM activity in the apical millimeter of roots of dark-grown (non-graviresponsive) seedlings with that in roots of light-grown seedlings. Calmodulin activity was about 4-fold higher in the tips of light-grown roots. In addition we found that CaM activity in the apical millimeter of dark-grown roots increased rapidly when the roots were illuminated. The activity increased to about the level found in tips of light-grown roots and this increase in CaM activity occurred prior to the gravitropic response in roots placed horizontally prior to illumination (Fig. 5).

These results provide indirect evidence that CaM might be important in the gravitropic response of roots. We tested this further by determining the effects of CaM inhibitors on root gravitropism. Application of low concentrations (e.g., 1 μM) of CaM inhibitors such as trifluoperazine, chlorpromazine, or calmidazolium to the root cap strongly retarded gravitropism without affecting growth (Stinemetz et al., unpublished). We also tested the effects of these inhibitors on gravity-induced polar calcium movement across the root tip. In all cases the inhibitors severely reduced polar calcium movement (Stinemetz et al., unpublished). The results indicate that CaM might be involved in gravity-induced polar calcium movement and that CaM inhibitors may retard gravitropism by slowing the development of calcium asymmetry across the tips of gravistimulated roots.

A model for calcium, calmodulin, and auxin interaction in root gravitropism. There is considerable evidence that calcium, CaM, and auxin interact in some fashion to mediate the response of roots to gravity. Some of the evidence for the participation of calcium and CaM has already been discussed. Although there is debate over which endogenous growth regulator (if any) mediates gravitropic curvature, the weight of the evidence favors auxin (20). We have proposed a model for auxin/calcium/CaM interaction in root gravitropism (7). According to this model, when amyloplasts sediment to the lowermost side of columella cells in the root cap of a horizontally-oriented root, they contact calcium-rich ER vesicles along that side, inducing the release of calcium as proposed by Sievers et al. (27). We propose that the elevation of cytoplasmic free calcium in the lowermost region of each cell activates two CaM-regulated pumps, a calcium pump which pumps calcium toward the lower side of the root cap and an auxin carrier which moves auxin downward, resulting in the diversion of both calcium and auxin toward the lower side of the root cap. To relate this to the overall pattern of auxin movement in the root, we may view the auxin transport pattern in the following way. Acropetal auxin transport through the apical portion of the root occurs primarily through the stele while basipetal movement occurs through the outer cell layers (29). We propose that the cells of the root cap determine the fate of IAA entering the central region of the cap through the acropetal transport system. In vertically-oriented roots a portion of the auxin entering the central region of the cap moves symmetrically toward the periphery of the cap while the remainder either continues to move

211

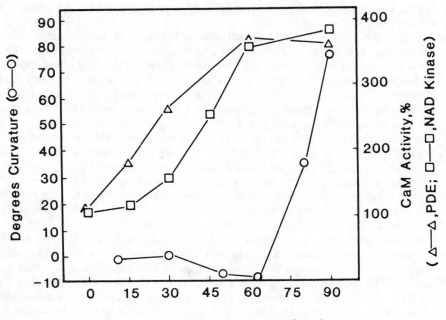

FIG. 5. Comparative time courses of photo-enhancement of extractable CaM activity and gravisensitivity in roots of dark-grown maize seedlings (cv Merit). CaM activity measured as in vitro stimulation of PDE or NAD kinase activity by crude extracts of root tips and shown as percent of activity found in extracts from dark-grown roots. CaM activity of 20 1-mm apical segments from dark-grown roots was equivalent to 0.68 units (PDE assay) or 0.64 units (NAD kinase assay) of bovine brain CaM per ml . (Redrawn from ref. 28).

through the cap into the surrounding medium or is metabolized. The auxin moving toward the periphery of the cap is loaded into a basipetally-moving stream and transported symmetrically toward the elongation zone through cells of the cortex and/or epidermis. In horizontally oriented roots auxin entering the cap is no longer symmetrically distributed toward the sides of the cap. Instead, as a result of the Ca^{2+}/CaM activation of Ca^{2+} and auxin pumps, both auxin and calcium are diverted toward the bottom of the cap. Since Ca^{2+} appears to enhance auxin transport (4), we propose that the accumulation of Ca^{2+} and auxin in the lower half of the root cap leads to enhanced loading of auxin into the basipetally-moving auxin transport stream along the lower side of the root. The increased auxin in the elongation zone on the lower side inhibits growth there, leading to downward curvature. Although this model is speculative it is consistent with existing data and it provides a framework for further investigation.

Note added in proof: IP$_3$-induced Ca^{2+} release has recently been reported for vacuolar membrane vesicles of oat roots (J Biol Chem 262: 3944-3946) and for microsomal vesicles of corn coleoptiles (J Biochem 101: 569-573).

Acknowledgments--The experiments on the effects of CaM antagonists on calcium transport and gravitropism were performed by Charles Stinemetz. Some of the research described herein was supported by NSF grants PCM 8305775 and DMB 8608673 and by NASA grant NAGW-297.

LITERATURE CITED

1. BERRIDGE MJ, RF IRVINE 1984 Inositol trisphosphate, a novel second messenger in cellular signal transduction. Nature 312: 315-321
2. BOSS WF, MO MASSEL 1985 Polyphosphoinositides are present in plant tissue culture cells. Biochem Biophys Res Commun 132: 1018-23
3. BUSH DR, H SZE 1986 Calcium transport in tonoplast and endoplasmic reticulum vesicles isolated from cultured carrot cells. Plant Physiol 80: 549-555
4. DELA FUENTE RK 1984 Role of calcium in the polar secretion of indoleacetic acid. Plant Physiol 76: 342-346
5. DRφBAK BK, IB FERGUSON 1985 Release of Ca^{2+} from plant hypocotyl microsomes by inositol-1,4,5- trisphosphate. Biochem Biophys Res Commun 130: 1241-1246
6. EVANS ML 1985 The action of auxin on plant cell elongation. Crit Rev Plant Sci 2: 317-365
7. EVANS ML, R MOORE, K-H HASENSTEIN 1986 How roots respond to gravity. Sci Amer 254: 112-119
8. FELDMAN LJ, WR BRIGGS 1987 Light-regulated gravitropism in seedling roots of maize. Plant Physiol 83: 241-243
9. GILROY S, WA HUGHES, AJ TREWAVAS 1987. Calmodulin antagonists increase free cytosolic calcium levels in plant protoplasts in vivo. FEBS Lett 212: 133-137
10. HASENSTEIN K-H, ML EVANS 1986 Calcium dependence of rapid auxin action in maize roots. Plant Physiol 81: 439-443
11. HELSPER HP, PF deGROOT, JF JACKSON, HF LINSKENS 1985 Phosphatidylinositol phosphodiesterase in lily pollen. Plant Physiol 77: Suppl 98
12. HEPLER PK, RO WAYNE 1985 Calcium and plant development. Annu Rev Plant Physiol 36: 397-439
13. HETHERINGTON AM, A TREWAVAS 1984 Activation of a pea membrane protein kinase by calcium ions. Planta 161: 409-417
14. IRVINE RF, AJ LETCHER, RMC DAWSON 1980 Phosphatidylinositol phosphodiesterase in higher plants. Biochem J 192: 279-283
15. LADYZHEN EP, NP KORABLEVA, TM MOROZOVA, OM SIDORKIN 1987 The activation of protein kinase C in plant cells under the action of gibberellic acid: Translocation of enzyme from cytoplasm into membranes. Akad Nauk SSSR 292: 763-765

16. LEE JS, TJ MULKEY, ML EVANS 1983 Gravity-induced polar transport of calcium across root tips of maize. Plant Physiol 73: 874-876

17. LEE JS, TJ MULKEY, ML EVANS 1983 Reversible loss of gravitropic sensitivity in maize roots after tip application of calcium chelators. Science 220: 1375-1376

18. LEE JS, TJ MULKEY, ML EVANS 1984 Inhibition of polar calcium movement and gravitropism in roots treated with auxin-transport inhibitors. Planta 160: 536-543

19. LEW RR, DP BRISKIN, RE WYSE 1986 Ca^{2+} uptake by endoplasmic reticulum from zucchini hypocotyls. Plant Physiol 82: 47-53

20. MOORE R, ML EVANS 1986 How roots perceive and respond to gravity. Amer J Bot 73: 574-587

21. MORRÉ DJ, B GRIPSHOVER, A MONROE, JT MORRÉ 1984 Phosphatidylinositol turnover in isolated soybean membranes stimulated by the synthetic growth hormone 2,4-dichlorophenoxyacetic acid. J Biol Chem 259: 15364-15368

22. MORSE MJ, RC CRAIN, RL SATTER 1986 Phosphatidylinositol turnover in *Samanea* pulvini: A mechanism of phototransduction. Plant Physiol 80: Suppl 92

23. NISHIZUKA Y 1986 Studies and perspectives of protein kinase C. Science 233: 305-312

24. RASMUSSEN H 1970 Cell communication, calcium ion, and cyclic adenosine monophosphate. Science 170: 404-412

25. SANDELIUS AS, M SOMMARIN 1986 Phosphorylation of phosphatidylinositols in isolated plant membranes. FEBS Lett 201: 282-286

26. SCHÄFER A, BYGRAVE F, MATZENAUER S, D MARME' 1985 Identification of a calcium- and phospholipid-dependent protein kinase in plant tissue. FEBS Lett 187: 25-28

27. SIEVERS A, HM BEHRENS, TJ BUCKHOUT, D GRADMANN 1984 Can a Ca^{2+} pump in the endoplasmic reticulum of the *Lepidium* root be the trigger for rapid changes in membrane potential after gravistimulation? Z Pflanzenphysiol 114: 195-200

28. STINEMETZ CL, KM KUZMANOFF, ML EVANS, HW JARRETT 1987 Correlations between calmodulin activity and gravitropic sensitivity in primary roots of maize. Plant Physiol 84: 1337-1342

29. TSURUMI S, Y OHWAKI 1978 Transport of ^{14}C-labeled indoleacetic acid in *Vicia* root segments. Plant Cell Physiol 19: 1195-1206

Physiology of Cell Expansion During Plant Growth, *D.J. Cosgrove and D.P. Knievel* Eds.,
Copyright ©1987, The American Society of Plant Physiologists

COOPERATION BETWEEN OUTER AND INNER TISSUES IN AUXIN-MEDIATED PLANT ORGAN GROWTH

U. KUTSCHERA

*Department of Plant Biology
Carnegie Institution of Washington
Stanford, CA 94305*

INTRODUCTION

The growth of plant organs such as coleoptiles or internodes is predominantly the consequence of cell elongation. The process of elongation growth has been extensively studied by use of excised plant sections after addition of the growth hormone auxin (indole-3-acetic acid, IAA)[1] (6). In maize coleoptiles and pea third internodes one can simulate *in situ* growth by use of excised sections incubated in distilled water containing IAA (9, 15). This simple model system of plant organ growth has been used to elucidate the interaction between outer and inner tissues during elongation growth. In the present contribution I first give a brief survey of some classical experiments. Then, I describe a series of investigations which led to a model of plant organ growth which takes into account tissue tension and the extensibility of the outer epidermal wall as decisive growth parameters.

TISSUE TENSION

The botanist W. Hofmeister was one of the first to discover that rapidly growing axial plant organs such as coleoptiles, internodes or petioles are composed of tissues having opposing tendencies to expand or contract on isolation (22). Hofmeister's experiments were repeated and extended by Sachs, who coined the term tissue tension ("Gewebespannung") in 1865 (21). The two key experiments demonstrating the existence of physical tensions in plant organs are described in Fig. 1. Splitting excised sections from a growing plant organ results in an immediate outward bending of the split halves. Peeling off the outer cell layer leads to a spontaneous contraction of the peel, whereas the inner tissue, placed on water, expands. The peeled contracting outer cell layer (OT) is composed of the epidermis and, depending on the plant organ, one or two

---

[1]Abbreviations: ABA = Abscisic acid, Fc = Fusicoccin, IAA = indole-3-acetic acid, [^{3}H]Ins = *myo*-[2-^{3}H(N)] Inositol, E_{e1}, E_{p1} = elastic and plastic extensibility, respectively, $E_{tot} = E_{e1} + E_{p1}$, IT = inner tissue, ITW = inner tissue walls, OT = outer tissue (epidermis), OEW = outer epidermal wall.

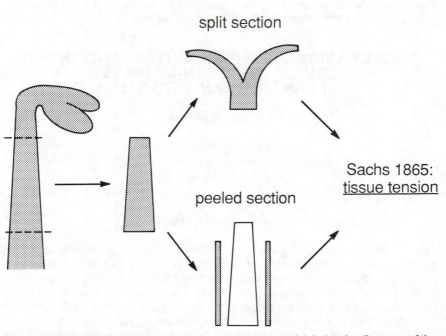

split section

peeled section

Sachs 1865:
tissue tension

FIG. 1. Scheme illustrating the two classical experiments which led to the discovery of the phenomenon of tissue tension. See text for explanation.

attached subepidermal cell layers (1, 13). This result shows that in the intact organ the inner tissue is maintained in a state of compression, while the outer cell layer is under physical tension (21, 22). Sachs concluded from his experiments that the inner cells "take up water with high power," "their walls are very extensible" and "they have not reached their potential maximal turgor pressure in the intact organ" (22). The existence of tissue tension has been confirmed on a variety of higher plants and was considered by Sachs in 1882 as a basic feature of growing plant organs. Moreover, Sachs suggested that a plant organ can only grow as long as tissue tension is established (22).

It has been postulated on theoretical grounds that the outer epidermal wall (OEW) of growing plant organs must bear a much higher physical stress than the inner tissue walls (ITW) (7). A comparison of the thickness of OEW versus ITW in IAA-sensitive plant organs reveals that OEW is, in fact, about 5 times as thick as the walls of the inner tissues (13, 19). Moreover, the sturdy rigid OEW shows a crossed polylamellate structure in which lamellae having a mainly longitudinal orientation of cellulose microfibrils alternate with lamellae of mainly transverse orientation (16, 20). The microfibrillar orientation of the ITW, on the other hand, seems to be predominantly transverse (16, 20). These structural differences in wall thickness and architecture of microfibrils may explain why the epidermis contracts and the inner tissue (IT) expands when the two are separated from each other (Fig. 1).

In spite of its obvious importance for the physics of plant organ growth, the existence of tissue tension has been largely ignored in the current literature concerning this subject (see 13). Moreover, the "plant cell wall" has been implicitly regarded as a uniform structure of equal thickness and architecture throughout the growing organ, a simplification which seems not to be justified (20).

AUXIN-INDUCED GROWTH OF ISOLATED PLANT SECTIONS

In 1931, i.e., only a few years after the "growth hormone" had been discovered, Heyn formulated the first theory on the mechanism by which auxin induces growth (8). Heyn discovered that auxin-treated sections from oat coleoptiles, subjected to a constant stress, show a higher plastic extensibility than nontreated segments. He therefore concluded that IAA induces growth by increasing the plastic extensibility of the cell walls (8). A constant stress extensiometer as used by Heyn in his original study is shown in Fig. 2a. A coleoptile section was fixed between two clamps and subjected to a constant stress by application of a weight of 10 g. The extension of the section was recorded using a horizontal microscope. The term "extensibility" is used here in its original sense as introduced by Heyn in 1931, i.e., it is the ability of a material (plant organ) to be extended (stretched) unidirectional under a constant stress (force per area).

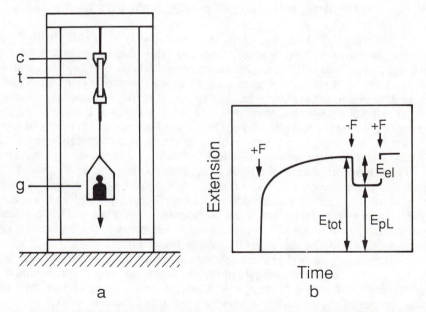

FIG. 2. a, Constant stress extensiometer as used by Heyn in his original study (Redrawn in modified form from 8). b, Kinetic of extension of a plant tissue subjected to a constant stress (force F, corresponding to a weight g). After removing the force F, E_{tot}, E_{el} and E_{pl} can be determined graphically as indicated. c = clamp, t = plant tissue, g = weight.

Fig. 2b shows a typical extension kinetic (viscoelastic deformation) of a plant tissue. The plastic extensibility (E_{pl}) is the irreversible part of the total extensibility (E_{tot}) which remains after removing the force F (corresponding to a weight g). The elastic extensibility (E_{el}) is the reversible difference between E_{tot} and E_{pl}. It should be stressed that E_{pl} and E_{el} are relative, operational quantities which can not give a direct measure of the yielding coefficient of the cell walls (see 4) of the growing plant organ (12).

In spite of numerous qualitative confirmations of Heyn's original finding on a variety of plant organs the temporal relationship between the IAA-induced increase in growth rate and increase in wall extensibility is unclear (3).

In 1937, Van Overbeek and Went discovered that sections from oat coleoptiles and pea internodes grow considerably less in response to IAA, after their epidermis has been removed (26). This finding has been subsequently confirmed for a variety of plant organs (see 13). More recently, Tanimoto and Masuda (24) have shown that the synthetic auxin 2,4D changes the "stress relaxation parameters" of the cell walls in pea internodes preferentially in the epidermis. It should be pointed out, however, that these changes are not correlated with the growth response of the intact segments (compare Figs. 2 and 4 in 24).

Nevertheless, largely on the basis of the above observations it has been postulated that the epidermis is the principal target of IAA-action and determines the growth of the whole plant organ (e.g., 1, 18, 24).

EXPERIMENTS WITH MAIZE COLEOPTILES

Effect of IAA on growth and *in-vitro* extensibility of intact sections. Two methods are available for determining of the extensibility of cell walls (see also chapters by Cleland, by Cosgrove, and by Ray). First, it is possible to stretch killed tissue (*in-vitro* extensibility) and second, living, turgid segments can be used to assess the mechanical properties of the cell walls (*in-vivo* extensibility) (11, 12). In order to kill the tissue and to remove the turgor pressure of the cells for the first method, the sections were frozen by spraying them briefly with a dermatological freezing spray (11). Subsequently the segment was incubated in distilled water, reextended to its *in-vivo* length by application of a force of 0.196 N (corresponding to a weight of 20g) and overstretched by an additional weight of 20 g (11). A constant load extensiometer, similar to that described in Fig. 2a but equipped with a linear-displacement transducer and a chart-recorder, was used (11). In these experiments, in contrast to Heyn's original investigations, the segments were incubated in distilled water during the time of stretching.

With this method a close temporal correlation between the IAA-induced increase in E_{pl} and the growth rate of the sections was found. E_{el} was virtually unaffected. After 3-4 h the growth rate slowed down, presumably the consequence of the reduction in osmotic pressure of the cells with dilution of osmotic solutes. It was concluded that E_{pl} represents a relative measure of the actual extensibility ("wall loosening") of the cell walls at the very moment when the tissue was killed. In addition, the abscisic acid (ABA)-induced inhibition of the IAA-promoted growth was accompanied by a reduction of E_{pl} (11).

Effect of IAA on growth and *in-vivo* extensibility of intact sections. As mentioned above, it is also possible to measure the mechanical properties of the cell walls by stretching fully turgid, metabolically active sections (12). This "constant stress method" has practical and theoretical advantages as compared to the *in-vitro* assay (12). A reinvestigation of the temporal relationships between IAA-induced growth and change in extensibility with this method revealed a close correlation between increase in growth rate and E_{pl}, whereas E_{el} was virtually unaffected. Moreover, cessation of growth caused by removal of IAA was closely correlated with a rapid decrease of E_{pl}. Hence both step up (+IAA) and step down (-IAA) growth modulations are accompanied by corresponding E_{pl}-changes. The ABA-induced growth inhibition was found to be accompanied by a reduction of E_{pl} as assayed by the *in-vivo* stretching technique as well (12). In summary, the results show that *in-vitro* and *in-vivo* assays could be used interchangeably to measure the extensibility of the cell walls in relative, operational units (11, 12).

Cooperation between epidermis and inner tissues during growth. In the maize coleoptile the isolated outer epidermis contracts spontaneously by about 18% on isolation, whereas the peeled IT (epidermis removed by >90%) expands by about 5% after incubation in water (13). The epidermal peel contracts in air, whereas IT expands only after transfer to water.

The contracted peels do not respond to IAA with an increase in length. Likewise the peeled tissue (IT) fails to respond to IAA with normal growth. Instead, a virtually IAA-independent, constant expansion of IT is detectable. The amounts of expansion of IT and contraction of the peeled epidermis on separation are virtually constant during IAA-induced growth of the intact section over a period of 6 h. This result shows that the IT is kept under constant physical compression by OEW, i.e., tissue tension is maintained during IAA-induced growth of the maize coleoptile (13).

The growth process of the coleoptile can therefore theoretically be induced by loosening of the expansion-limiting OEW (E_{pl}-increase). In order to test this hypothesis segments were peeled and their *in-vivo* extensibility determined after subsequent incubation in IAA. Whereas E_{el} was only slightly changed by the peeling process, E_{pl} of IT was found to be almost twice as high as that of the nonpeeled segment. This result shows that ITW do have a high E_{pl}, highly unlikely to be growth-limiting. IAA (and ABA) had no significant effect on E_{pl} of the peeled segments (IT). It was concluded that the IAA-induced growth process of the maize coleoptile (and the ABA-induced growth inhibition) can be attributed to a hormone-induced E_{pl}-change of OEW (12, 13).

Auxin-induced growth and proton secretion. The mechanism by which IAA loosens OEW is unknown. Since IAA induces a secretion of protons into the cell-wall solution it has been postulated that protons are acting as "second messenger" or "wall loosening factor" during the first phase of IAA-induced growth (2). A quantitative reinvestigation of the predictions of the "acid growth theory" using abraded sections (epidermis ~70% intact), however, revealed that IAA-induced growth and proton secretion are independent, not causally linked processes (9). The pH of about 5, which is presumably induced by IAA in the cell wall solution, was found to be insufficient to induce wall-loosening (E_{pl}-increase) and growth (9, 13). By contrast, the predictions of the "acid-growth theory" are fully fulfilled in the case of fusicoccin (Fc)-induced growth (10). This result shows that there are no basic experimental difficulties in proving the "acid-growth theory" if it is valid. It has been concluded that Fc (but not IAA) causes growth by acidification of the cell walls (OEW and ITW, see 9, 10, 13).

The IAA-sensitive proton pumps were found to be predominantly localized in the outer epidermis (13). It is therefore conceivable that the IAA-mediated proton secretion might have the function of maintaining tissue tension (and hence the high E_{pl}-value) in ITW. This hypothesis was investigated by comparing the sensitivity of abraded versus peeled sections with respect to acid-induced wall loosening (E_{pl}-increase) and growth. The results indicated that ITW alone likewise do not respond to a pH of 5; hence the IAA-mediated proton secretion can not have the proposed function in this tissue (13).

Electron-dense (osmiophilic) particles at OEW; auxin-binding protein in the outer epidermis. An electron microscopic investigation of the cell walls of maize coleoptiles revealed the appearance of electron-dense (osmiophilic) particles with a maximal diameter of about 3 μm at the outer epidermal cell wall/plasma membrane interface (13). These structures were not detectable at ITW and only as an exception occasionally at the outer epidermal wall of the inner epidermis. The electron-dense particles were found in sections which were fixed directly after cutting (*in situ*). They disappeared during incubation in water and reappeared rapidly after subsequent incubation of the sections in IAA. Acid buffer (pH 3.5) and Fc did not induce such structures. Although the chemical nature and functional significance of these particles have not been elucidated, it seems possible that they may be related to the induction of wall growth (13).

Löbler and Klämbt (17) recently discovered an auxin-binding protein which was found to be localized at the plasmalemma of the outer epidermal cells in maize coleoptiles. Since the IAA-induced growth of the segments could be inhibited by monospecific antibodies against this auxin-binding protein it has been referred to as an auxin receptor (17).

These findings demonstrate that the outer epidermis is characterized by IAA-specific structural and biochemical features which separate this cell layer from IT.

EXPERIMENTS WITH PEA INTERNODES

Effect of IAA on growth and *in-vivo* extensibility of intact sections. In order to elucidate the temporal relationship between the IAA-induced increase in growth rate and an increase in extensibility a constant load extensiometer, similar to that shown in Fig. 2a, was again used (15). The sections were fixed between the clamps, incubated in distilled water and stretched by applying a force of 0.098 N (corresponding to a weight of 10 g). In addition, by use of this device, it was also possible to stretch isolated epidermal strips and hence to determine the viscoelastic properties of the epidermal cell layer.

As in the maize coleoptile a close temporal correlation between the IAA- induced increase in growth rate and increase in *in-vivo* extensibility (E_{tot}) of the sections was measured (15). In contrast to the maize system, where only E_{pl} was significantly changed (11, 12), both E_{pl} and E_{el} were increased in the pea internode in the presence of IAA by stretching of intact segments (15).

Cooperation between epidermis and inner tissues during growth. In pea internodes it is not possible to remove the epidermis as an unicellular layer. The peeled OT is composed of the epidermis with one layer of subepidermal cells attached (1, 15). The isolated OT contracts spontaneously by about 13%, whereas the peeled IT (epidermis removed by >90%) expands in water (1,15). In contrast to the maize coleoptile, where IT first expands rapidly by about 5% and then subsequently by a slow, steady rate (13), the peeled cortical cylinders from pea internodes increase in length over a period of several hours with slowly decreasing rate (1, 15). IAA has only a small growth-promoting effect on the expanding IT (1, 15, 26).

The *in-vivo* extensibility (E_{pl}) of IT was found to be nearly twice as high as the corresponding E_{pl}-value of the intact segment. Therefore, in pea internodes, as with maize coleoptiles, the extensibility of ITW is again obviously not growth-limiting. IAA had no significant effect on E_{pl} of ITW over a period of 4 h (15). In order to determine whether IAA increases E_{pl} of the epidermis, intact sections were incubated in auxin and one epidermal strip removed after a later time. The peels were first reextended to their *in-vivo* length and then stretched by application of a force of 0.049 N (corresponding to a weight of 5 g). This experiment revealed a close temporal relationship between the IAA-induced growth of the intact segments and an E_{pl}-increase of the epidermis. E_{el} of the peels was not significantly changed by IAA. Finally, the isolated, contracted epidermis, incubated in IAA, responded with an E_{pl}-increase even in the absence of IT, whereas the peeled cortical cylinder did not show an E_{pl}-change under the same conditions (15).

The amount of expansion of the cortical cylinder and contraction of the epidermis, respectively, was not significantly changed over a 4 h period of IAA-treatment. This observation shows that tissue tension is maintained during IAA-induced growth of the intact internode segment (15). This conclusion can also be reached from experiments using the "split-section test" (26). As shown in Fig. 3, sections will bend outward into two halves on splitting even after a 4 h incubation in IAA. No significant difference in the amount of bending (bending angle) could be measured between the rapidly growing (+IAA) and virtually nongrowing segments (-IAA). It was concluded from the above experiments that the IAA-induced growth process of the pea internode can be attributed to an auxin-stimulated E_{pl}-increase of OEW (15).

Auxin-induced growth and proton secretion. Although no quantitative reinvestigation of the predictions of the "acid-growth theory," as performed by Kutschera and Schopfer (9, 10) for the maize coleoptile, has been reported for pea third internodes, there is evidence against a "second messenger" function of protons in the loosening process of OEW also in this organ.

The IAA-sensitive proton pumps in the epidermis (see 13) can only be activated in the presence of cations like K^+ and Ca^{2+} (25). Although these ions are essential for the release of protons from isolated sections they did not

221

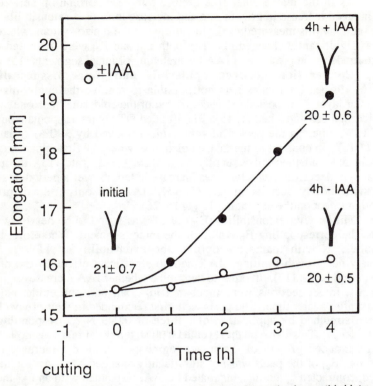

FIG. 3. Maintenance of tissue tension during IAA-induced growth of pea third internode sections. Fifteen mm-sections were incubated for 1h on distilled water (initial) and then incubated in distilled water lacking or containing IAA (10 µM) as described (15). The segments were trimmed to 15 mm, split, placed for 5 min in distilled water and photocopied. The bending angle was determined graphically. The numbers below the split sections represent the mean (±SE) bending angle of 30 measurements each.

influence the growth response. Thus IAA-induced proton secretion and growth can be experimentally separated. It was concluded that IAA- induced growth of pea stem sections is independent of the secretion of protons (25).

Rapid auxin-effects on dictyosome material and cell wall synthesis. Incubation of segments in IAA caused a rapid increase in the volume fraction of dictyosome material per unit area of cytoplasm in the epidermal cells compared to sections incubated in water (5). This increase was detected within 15 min of IAA-treatment and reached a maximum after about 30 min, followed by a decrease, presumably due to an increased utilization of the organelle (5). Unfortunately, as these measurements were restricted to the cells of the epidermis, it is not known whether similar effects also occur in the IT.

This finding suggests that IAA may rapidly stimulate the dictyosome-mediated deposition of matrix cell wall material in the growth limiting epidermal cell layer. This hypothesis has recently been investigated (14). In order to detect IAA-effects on the synthesis of matrix material *myo*-[2-^{3}H(N)] Inositol

([^{3}H]Ins) was used as a precursor. This tracer has been shown to be incorporated specifically into new matrix polysaccharides of growing cell walls, mostly as uronic acids and pentose residues, and only marginally into cellulose (23). IAA can rapidly enter the tissue via the cuticle to reach its reactive site (6), whereas the uptake of sugars is largely restricted to the cut surfaces (2, 14). Intact sections were therefore first preincubated in [^{3}H]Ins for 4 h to permit the uptake and internal distribution of the tracer, then IAA was added (14). Growth started after 15 min under these conditions. The sections were removed from the tracer and separated into epidermis and IT. In the epidermis, IAA- induced stimulation of [^{3}H]Ins incorporation started after a lag of 15 min. The amount of incorporation was 15% higher after 30 min and 24% higher after 2 h than in the water-control. In the IT, IAA-induced stimulation of [^{3}H]Ins-incorporation started only about 1h after adding of auxin. The uptake of tracer *per se* was not rapidly stimulated by IAA in either the epidermis or the IT (14).

In order to assess the possible relationships between IAA-induced growth, wall loosening and wall synthesis the ionophore monensin, which disrupts the dictyosomes, was used to inhibit wall synthesis (14). IAA- induced growth, wall-loosening (increase in E_{tot}), and matrix cell wall synthesis ([^{3}H]Ins incorporation) were all inhibited by monensin (20 µM) by $\geq$95%, although the oxygen uptake was unaffected. Hence all three IAA- responses appear dependent on normal Golgi-function and therefore may be mechanistically related processes (14).

CONCLUSIONS

In the present report I have summarized experiments performed with representative IAA-sensitive monocot-and dicot plant tissues (maize coleoptile, pea internode). The results support the following model of IAA- mediated plant organ growth (Fig. 4): IAA penetrates the cuticle (6) and loosens the sturdy OEW which, because of tissue tension, maintains IT in a state of constant compression. This loosening process, which can be measured in relative, operational units as an E_{pl}-increase of the epidermis (12, 15), permits organ growth, which is driven by the expansion of the compressed inner cells. The mechanism by which IAA loosens OEW is not yet clear. The reported IAA-induced appearance of electron dense particles at OEW in maize coleoptiles and the rapid IAA-induced stimulation of [^{3}H]Ins-incorporation in the epidermal walls of pea internodes suggest that a selective, rapid IAA-mediated stimulation of wall synthesis is possibly involved in this process (13, 14). It is conceivable that the increased incorporation of matrix polysaccharides into the existing wall (intussusception), which necessitates the breaking of load-bearing bonds, loosens OEW and hence permits organ growth, which is driven by water uptake of the extensible, compressed IT. Thus wall-loosening and wall synthesis are synchronous, mechanistically linked processes (14), a hypothesis which was put forward twenty years ago by Ray (19).

In order to get continuous growth a second, IAA-independent growth process must be postulated which keeps IT in a state of constant compression (maintenance of tissue tension). The mechanism by which this state is achieved is unknown. Several possibilities are conceivable. First, hydrolysis of stored

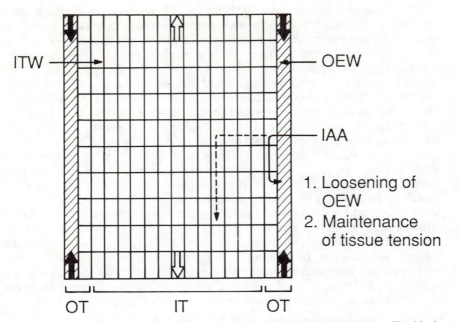

ITW — OEW

— IAA

1. Loosening of
   OEW
2. Maintenance
   of tissue tension

OT   IT   OT

FIG. 4. Model of IAA-mediated plant organ growth. See text for explanation. The black
and white arrows symbolize tissue tension exerted by OEW and ITW, respectively.

material could theoretically increase the osmotic potential of the inner cells and
hence maintain tissue tension. This hypothesis is not in agreement with
experimental results (11). Second, proton secretion, induced in the outer
epidermis, could increase E_{pl} of ITW and thus keep IT under compression. This
hypothesis has, at least in the maize coleoptile, been refuted (13). Finally,
matrix cell wall synthesis in IT could be involved in this process. This is
unlikely, since though IAA stimulates matrix wall synthesis in ITW (14), tissue
tension of these cells is independent of auxin (13, 15, Fig. 3). One should keep
in mind that IT is a complex structure composed of mesophyll (or cortical) cells,
vascular bundles and, in the case of the coleoptile, the inner epidermis. The
unknown process which keeps IT under constant compression possibly involves
an interaction of these inner cell layers about which nothing is known yet.

The model depicted in Fig. 4 states that it is not the actual turgor
pressure of the cells (4) but a hydrostatic pressure increment produced by tissue
tension that is the driving force of growth (13). The wall pressure of the
compressed inner cells in the intact organ is lower than it would be in the
absence of OEW (13), a conclusion which had already been reached by Sachs
more than one hundred years ago (22). The question as to whether tissue tension
is also involved in growth of other plant systems, i.e., whether the model shown
in Fig. 4 is of general significance, remains to be a challenge for future research.

Acknowledgments--This investigation was supported by the Alexander von Humboldt-Stiftung (Feodor Lynen Research Fellowship). CIW-DPB publication no. 979.

LITERATURE CITED

1. BRUMMEL DA, JL HALL 1980 The role of the epidermis in auxin-induced and fusicoccin-induced elongation of *Pisum sativum* stem segments. Planta 150: 371-379
2. CLELAND RE 1980 Auxin and H⁺-excretion: the state of our knowledge. *In* F Skoog, ed, Plant growth substances 1979, Springer, Berlin, pp 71-78
3. CLELAND RE 1984 The Instron technique as a measure of immediate-past wall extensibility. Planta 160: 514-520
4. COSGROVE DJ 1986 Biophysical control of plant cell growth. Annu Rev Plant Physiol 37: 377-405
5. CUNNINGHAME ME, JL HALL 1985 A quantitative stereological analysis of the effect of indoleacetic acid on the dictyosomes in pea stem epidermal cells. Protoplasma 125: 230-234
6. EVANS ML 1985 The action of auxin on plant cell elongation. CRC Critical Reviews in Plant Sciences 2: 317-365
7. GREEN PB 1980 Organogenesis-a biophysical view. Annu Rev Plant Physiol 31: 51-82
8. HEYN ANJ 1931 Der Mechanismus der Zellstreckung. Rec Trav Bot Neerl 28: 113-244
9. KUTSCHERA U, P SCHOPFER 1985 Evidence against the acid-growth theory of auxin action. Planta 163: 483-493
10. KUTSCHERA U, P SCHOPFER 1985 Evidence for the acid-growth theory of fusicoccin action. Planta 163: 494-499
11. KUTSCHERA U, P SCHOPFER 1986 Effect of auxin and abscisic acid on cell wall extensibility in maize coleoptiles. Planta 167: 527-535
12. KUTSCHERA U, P SCHOPFER 1986 *In-vivo* measurement of cell-wall extensibility in maize coleoptiles. Effects of auxin and abscisic acid. Planta 169: 437-442
13. KUTSCHERA U, R BERGFELD, P SCHOPFER 1987 Cooperation of epidermis and inner tissues in auxin-mediated growth of maize coleoptiles. Planta 170: 168-180
14. KUTSCHERA U, WR BRIGGS 1987 Rapid auxin-induced stimulation of cell wall synthesis in pea internodes. Proc Natl Acad Sci USA 84:2747-2751
15. KUTSCHERA U, WR BRIGGS 1987 Differential effect of auxin on *in-vivo* extensibility of cortical cylinder and epidermis in pea internodes. Plant Physiol 84:1361-1366
16. LANG JL, WR EISINGER, PB GREEN 1982 Effects of ethylene on the orientation of microtubules and cellulose microfibrils of pea epicotyl cells with polylamellate cell walls. Protoplasma 110: 5-14

17. LÖBLER M, D KLÄMBT 1985 Auxin-binding protein from coleoptile membranes of corn (*Zea mays* L.) II. Localization of a putative auxin receptor. J Biol Chem 260: 9854-9859
18. POPE DG 1982 Effect of peeling on IAA-induced growth in *Avena* coleoptiles. Ann Bot 49: 493-501
19. RAY PM 1967 Radioautographic study of cell wall deposition in growing plant cells. J Cell Biol 35: 659-674
20. ROELOFSEN PA 1965 Ultrastructure of the wall in growing cells and its relation to the direction of the growth. Adv Bot Res 2: 69-149
21. SACHS J 1865 Handbuch der Experimentalphysiologie der Pflanzen. Engelmann, Leipzig
22. SACHS J 1882 Vorlesungen über Pflanzenphysiologie. Engelmann, Leipzig
23. SASAKI K, IEP TAYLOR 1984 Specific labeling of cell wall polysaccharides with *myo*-[2-^{3}H]Inositol during germination and growth of *Phaseolus vulgaris* L. Plant Cell Physiol 25: 989-997
24. TANIMOTO E, Y MASUDA 1971 Role of the epidermis in auxin-induced elongation of light-grown pea stem segments. Plant Cell Physiol 12: 663- 673
25. TERRY ME, RL JONES 1981 Effect of salt on auxin-induced acidification and growth by pea internode sections. Plant Physiol 68: 59-64
26. VAN OVERBEEK J, FW WENT 1937 Mechanism and quantitative application of the pea test. Bot Gaz 99: 22-41

Physiology of Cell Expansion During Plant Growth, D.J. Cosgrove and D.P. Knievel Eds.,
Copyright ©1987, The American Society of Plant Physiologists

STUDIES ON INTERNODAL GROWTH USING DEEPWATER RICE

HANS KENDE

*MSU-DOE Plant Research Laboratory, Michigan State University,
East Lansing, MI 48824*

INTRODUCTION

Deepwater rice (*Oryza sativa* L.) is grown mainly in regions of Southeast Asia which are flooded each year during the monsoon season. Usually, ungerminated seeds are broadcast in dry river beds and flood plains. The seeds germinate after the first rains, and the plants become established under dryland conditions. Four to 20 weeks after germination, the flood waters begin to rise, and the partly submerged plants start to grow at fast rates. In Bangladesh, internodal growth of 25 cm a day and total plant heights of up to 7 m have been recorded (19). Survival of deepwater rice depends on its ability to keep part of its foliage above the rising waters. Total submergence of plants for more than one week results in heavy crop losses. The increase in water level coincides with the vegetative development of the plant. Flowering is under photoperiodic control and is timed such that panicle emergence occurs when the water level has stopped rising. Harvesting is done either from boats or from the ground after the flood waters have receded. Deepwater rice is of great agricultural importance because it is the only crop that can be grown in large, densely populated areas of Asia. It has been studied least among all rice types, and progress in the development of high-yielding varieties has been slow. A number of physio-morphological characteristics must be expressed to make an improved variety suitable for cultivation in deepwater areas. Most important among these traits is the ability of the plant to elongate rapidly when partially submerged. Therefore, understanding the growth physiology of deepwater rice is of immediate significance for ongoing efforts to introduce the deepwater trait into high-yielding varieties.

The magnitude of the growth response and the fact that it can be induced rapidly by a natural environmental signal, submergence, make deepwater rice a suitable plant for the study of internodal elongation. Accelerated growth of partially submerged plants can serve as reference to which growth elicited by hormonal and environmental factors can be compared. Internodal growth in whole plants can be reproduced in excised stem sections which contain the youngest, developing internode. Such stem sections are often easier to treat with gas mixtures and growth regulators than are whole plants, and a large number of them can be used per treatment. During the past five years, we have investigated internodal elongation of deepwater rice at the anatomical, physiological and biochemical levels and have reconstructed the chain of events that leads from submergence to accelerated growth.

THE ANATOMICAL BASIS FOR INTERNODAL GROWTH IN DEEPWATER RICE

Using light and scanning electron microscopy, we could distinguish three zones of internodal development in deepwater rice: the intercalary meristem at the base of the internode where cell division and cell elongation take place; a zone of cell elongation without concomitant cell division; and a zone of cell differentiation where neither cell division nor cell elongation occur (2). These zones can also be recognized by comparing the cell sizes along a growing internode (Fig. 1). The intercalary meristem contains small cells of constant

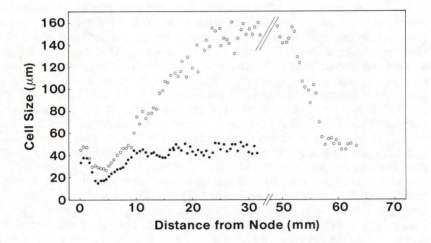

FIG. 1. The relationship between cell size and distance from the base of the uppermost, elongating internode of the stem of a rice plant. Points represent the average of measurements from three separate longitudinal sections through the same internode . (o) Air-grown; (o) submerged (from 2).

size and commences about 3-4 mm above the nodal septum. It is approximately 2 mm long in air-grown and 3 mm long in submerged plants. Cell-division activity has been localized in this region of the stem by autoradiographic determination of [3H]thymidine incorporation into cell nuclei (9). The average duration of the cell cycle within the intercalary meristem is 24 h in air-grown and 7 h in submerged plants (2). The length of cells increases linearly with distance above the intercalary meristem to a maximum of 40 μm in air-grown and 150 μm in submerged plants. This leads to an expansion of the cell-elongation zone from 5 mm to about 15 mm (Fig. 1). Above the zone of cell elongation, mean cell length remains constant in air-grown plants. In submerged plants, a transition zone with cell sizes decreasing from 140 to 40 µm separates growth that occurred prior to and during submergence (Fig. 1). Anatomical localization of growth processes is important for subsequent work on biochemical reactions and molecular controls that underlie the different phases of growth, e.g., cell division and cell elongation.

THE PHYSIOLOGY OF THE GROWTH RESPONSE IN DEEPWATER RICE

Ku et al. (7) were the first to show that applied ethylene promotes growth of rice coleoptiles. We reinvestigated this response using deepwater as well as traditional rice cultivars and found that the magnitude of the growth promotion by ethylene was about 20% (11). Musgrave et al. (10) extended this work using the semi-aquatic plant *Callitriche platycarpa* whose growth rate is greatly accelerated when the plant becomes submerged. It was shown that treatment of non-submerged plants with ethylene mimicked the effect of submergence on growth. Musgrave et al. (10) proposed that growth of submerged plants was enhanced because of accumulation of ethylene in the tissue. The role of ethylene in mediating the growth response of several semi-aquatic plants has been reviewed recently by Jackson (6).

Promotion of growth in deepwater rice by ethylene. In our initial experiments, we established the basic features of the response of deepwater rice to submergence and to ethylene under laboratory conditions (8). We chose the Bangladesh cultivar Habiganj Aman II for our work because it was reported to grow at particularly high rates when partially submerged and because this enhancement of growth could be observed with relatively young plants, i.e., within one month of sowing. The time course of growth and of ethylene accumulation in partially submerged plants is shown in Figure 2. Plants were

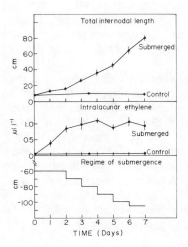

FIG. 2. Time course of the effect of partial submergence on internode elongation and endogenous ethylene concentration. The internodes of the main culm were measured. Each point was obtained with five separate plants. The bars denote SE (from 8).

229

immersed to a depth of 60 cm into a 300-liter plastic tank filled with deionized water. On subsequent days, the plants were lowered progressively deeper into the water until they reached the bottom of the tank. Under this submergence schedule, about one-third of the leaves remained always above water. During the seven days of this experiment, control plants in air grew very little if at all. The internodal length of the submerged plants increased by about 70 cm. The ethylene concentration in the internodal lacunae of submerged plants rose from 0.02 to around 1 μl l^{-1}.

When non-submerged plants were treated with ethylene (0.4 μl l^{-1}), total internodal length increased by about 30 cm; the internodes of control plants kept in air grew by about 5 cm during the same period of time (Fig. 3). The

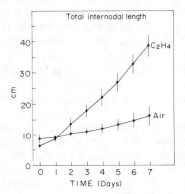

FIG. 3. Time course of the effect of ethylene at 0.4 μl l^{-1} on internode elongation of nonsubmerged deepwater rice plants. The internodes of the main culm were measured daily. Each point is the mean of the same five plants which were measured throughout the experiment. The bars denote SE (from 8).

role of endogenous ethylene in the submergence response was investigated using two inhibitors of ethylene biosynthesis, aminoethoxyvinylglycine (AVG) and aminooxyacetic acid (AOA). Plants treated with these compounds did not grow when partially submerged. This inhibition of growth could be reversed with 1-aminocyclopropane-1-carboxylic acid (ACC), the immediate precursor of ethylene (8). From these results, we concluded that the growth response in partially submerged deepwater rice is mediated, at least in part, by ethylene. However, since applied ethylene did not elicit as large a growth response as did submergence, the involvement of other growth-promoting factors besides ethylene had to be considered.

Since work with whole plants is quite cumbersome, especially if they have to be treated with gas mixtures of exact composition, we developed an experimental procedure using excised stem sections (12). Such sections, 20 cm in length, were excised from the main stem and tillers of six- to ten-week-old plants such that the second highest node of the culm was 2 cm above the basal cut, separated from the highest node by the youngest, ca. 2-7 cm long internode. The stem section above the highest node consisted of leaf sheaths which surrounded the youngest developing leaves. When such stem sections were submerged, internodal growth was greatly stimulated while leaf growth was inhibited (Table I). We determined the gaseous composition of the internodal lacuna and found that the O_2 content declined to as low as 2% during the dark

Table I. *Effects of O_2, CO_2, ethylene (C_2H_4) and submergence on the elongation of rice stem sections incubated under a 13-h photoperiod for 3 d.* Air and gas mixtures (all v/v) were passed through the incubation cylinders at 80 ml min⁻¹. Each value is the average of 11 sections ± SE (from 12).

Treatment	Length increase (mm)			C_2H_4 concentration in internode
	Internode	Leaf	Total section	
	------mm------			($\mu l\ l^{-1}$)
Air	9.5 ± 1.2	105.5 ± 9.9	115.0 ± 0.4	0.02
Air + 1 μl l⁻¹ C_2H_4	65.2 ± 7.8	62.9 ± 11.9	128.1 ± 7.8	0.9
21% O_2 + 6% CO_2 + 73% N_2	13.3 ± 2.8	118.3 ± 6.4	131.6 ± 6.8	0.02
21% O_2 + 6% CO_2 + 73% N_2 + 1 μl l⁻¹ C_2H_4	88.1 ± 1.1	90.6 ± 2.5	178.7 ± 21.4	1.0
3% O_2 + 0.03% CO_2 + 97% N_2	41.4 ± 5.1	76.2 ± 7.7	117.6 ± 7.2	0.01
3% O_2 + 0.03% CO_2 + 97% N_2 + 1 μl l⁻¹ C_2H_4	95.9 ± 1.7	35.1 ± 3.5	141.0 ± 12.9	0.8
3% O_2 + 6% CO_2 + 91% N_2	54.7 ± 4.7	111.9 ± 6.9	166.6 ± 10.4	0.02
3% O_2 + 6% CO_2 + 91% N_2 + 1 μl l⁻¹ C_2H_4	95.3 ± 9.6	52.8 ± 4.2	148.1 ± 8.1	1.1
Submerged	93.4 ± 0.1	61.8 ± 7.8	155.2 ± 14.3	1.0

period while the level of CO_2 and ethylene increased to ca. 6% and 1 μl l^{-1}, respectively. Stem sections incubated in a gas mixture containing 3% O_2, 6% CO_2 and 1 μl l^{-1} ethylene in N_2 exhibited the same growth response as did submerged sections. Internodal growth was increased tenfold while leaf growth was inhibited by 40-50% (Table I). Internodal growth was also promoted at low partial pressures of O_2 and by ethylene alone. An atmosphere containing increased levels of CO_2 did not, by itself, stimulate internodal growth. The growth response observed with isolated stem sections was similar in every respect to that obtained with whole plants. Promotion of internodal growth in stem sections was also based on enhanced cell division activity in the intercalary meristem and on increased cell elongation (9, 13).

Ethylene biosynthesis in submerged deepwater rice. The increased concentration of ethylene in submerged rice internodes (Fig. 2, Table I) may be due to reduced diffusion of ethylene from the tissue into the water or may be the combined result of enhanced ethylene biosynthesis and physical trapping of ethylene under water. Since it is not possible to compare ethylene evolution in submerged and non-submerged sections directly, we transferred, for ethylene determinations, internodes from submerged stem sections into test tubes containing a gaseous atmosphere of a composition similar to that found in internodal lacunae of submerged sections (3% O_2, 6% CO_2, 91% N_2). As control, internodes from sections that had been incubated in air were transferred into test tubes containing air. The rate of ethylene synthesis in internodal tissue from submerged sections was five times higher than that of internodal tissue from sections incubated in air (12). Promotion of ethylene biosynthesis by submergence could be mimicked by incubating stem sections in an atmosphere containing low levels of O_2. While ethylene formation in the internode was stimulated at low partial pressures of O_2, ethylene synthesis in the leaves was inhibited under the same conditions (Fig. 4).

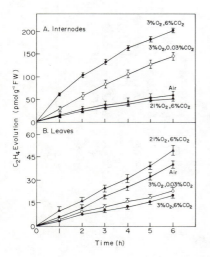

FIG. 4. Ethylene production by internodes (A) and leaves (B) excised from rice stem sections that had been incubated in a stream of various gas mixtures (shown as % by vol; remainder was N_2) under continuous light. Air and the three respective gas mixtures were passed through the incubation cylinders at 80 ml min^{-1}. For ethylene determinations, internodal and leaf tissues were isolated and incubated in 30- or 60-ml test tubes, respectively, containing 2 or 3 ml H_2O and the same gas mixtures with which the whole stem sections had been treated earlier. Each point is the average of three replicate test tubes, each containing internodes or leaf tissue from three sections. The bars denote SE (from 12).

The role of gibberellin in the growth response of deepwater rice. The discovery of gibberellins was based on the observation that the culture filtrate of the fungus *Gibberella fujikuroj* contained an active principle that induced growth of rice internodes. This active principle was, of course, later identified as a mixture of gibberellins. Therefore, the question arose whether or not gibberellins are involved in eliciting the growth response of submerged deepwater rice. Exploratory experiments showed that applied gibberellic acid (GA_3) indeed promoted internodal growth of air-grown deepwater-rice plants and of isolated stem sections. When plants were pretreated with an inhibitor of gibberellin biosynthesis, tetcyclacis, neither submergence nor ethylene enhanced growth (13). The growth response could be fully recovered by treating the plants with GA_3. How, then, is the action of ethylene related to that of gibberellin? Plants were treated with tetcyclacis to reduce endogenous gibberellin levels, and stem sections excised from these plants were used for a comparison of gibberellin action in the presence and absence of applied ethylene. Figure 5 shows that the responsiveness of stem sections to low concentrations

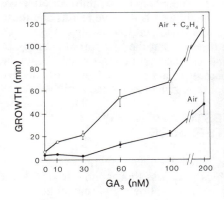

FIG. 5. Effect of low GA_3 concentrations on the growth of internodes of rice stem sections treated with 1 μM tetcyclacis (TCY) and incubated in a stream of air or air containing 1 μl l^{-1} ethylene. The sections, standing upright in 100-ml glass beakers containing 40 ml of 1 μM TCY solution with different GA_3 concentrations, were placed in 2.5-l plastic cylinders through which air or air with ethylene was passed at 80 ml min^{-1} and were incubated in continuous light for 3 d. Each point is the average of 14 sections ±SE (from 13).

of GA_3 is substantially enhanced by ethylene. Ethylene shifted the threshold of the response to lower GA_3 concentrations and increased the magnitude of the response. On the anatomical level, the response to gibberellin was indistinguishable from that to submergence and ethylene (13). The above experiments have shown that ethylene does not promote growth in the absence of gibberellin. Can gibberellin act in the absence of ethylene? Ideally, this questions should be examined using plants whose ethylene-biosynthetic pathway is blocked, e.g., with aminoethoxyvinylglycine (AVG). Unfortunately, the effect of AVG in rice is not fully reversible with ACC in long-term growth experiments (12). Instead of AVG, we used 2,5-norbornadiene (NBD), an inhibitor of ethylene action (17). NBD inhibits reversibly three ethylene-induced responses in deepwater rice: formation of adventitious roots, stimulation of internodal growth and reduction of leaf growth (1). In the presence of NBD, applied GA_3 promoted growth of deepwater rice internodes fully (1). In

233

summary, we conclude that gibberellin can stimulate growth independently of ethylene, but ethylene can promote growth only in the presence of gibberellin. Our experiments indicate that ethylene promotes internodal growth in deepwater rice by increasing the plant's responsiveness to gibberellin.

The time course of growth promotion. An angular transducer was used to measure the short-term growth response in stem sections of deepwater rice (16). Following submergence of stem sections, it takes 180 to 220 min for the rate of growth to increase. The lag time for growth induced by a gas mixture of 3% O_2, 6% CO_2, 91% N_2 and 1 µl l^{-1} ethylene was 60 min and that for gibberellin-induced growth between 40 and 60 min. When stem sections were incubated in a flow-through chamber through which a gas mixture containing 3% O_2, 6% CO_2 and 91% N_2 was passed, the growth rate accelerated slowly and only after a considerable lag (ca. 5 to 24 h). However, with time, growth reached the same rate as measured in sections treated with 3% O_2, 6% CO_2, 91% N_2 and 1 µl l^{-1} ethylene. The delay in the growth response was probably a consequence of the experimental conditions. To maintain the desired gas composition in the incubation chambers, the respective gas mixtures had to be circulated through these chambers at a fast rate. Under these conditions, newly synthesized ethylene was continuously removed from the plant (see ethylene levels, Table I). It may, therefore, take some time until an effective ethylene concentration is built up inside the tissue. The above results and those of Table I are important because they demonstrate that low levels of O_2 induce growth just as submergence does. We propose that the low partial pressure of O_2 in submerged plants is the signal for accelerated growth. Ethylene biosynthesis is enhanced under these conditions, and increased levels of ethylene promote growth by magnifying the response of rice to endogenous gibberellin.

BIOCHEMICAL INVESTIGATIONS WITH DEEPWATER RICE

The key biochemical reactions which regulate growth are not known. Molecular biology has enabled researchers to bypass this gap in our knowledge and to study genes whose expression is activated following hormonal treatment. We have chosen an alternative approach to investigating regulatory mechanisms that operate during growth. We have identified a number of biochemical events that are related to different phases of the growth response and have localized these reactions in distinct regions of the rice internode. Next, we shall study the regulation of these biochemical responses at the molecular level.

ACC synthase. Feeding experiments have indicated that ethylene synthesis in deepwater rice is limited by the activity of ACC synthase (8), as is the case in many other plant tissues (20). Because of this, we decided to determine the activity of ACC synthase in internodes prior to and during submergence (4). For reasons not understood, ACC-synthase activity in homogenates of vegetative tissues is often much lower than expected or not measurable at all. This is also the case with ACC synthase from deepwater-rice internodes (8). We have, therefore, employed an *in vivo* assay to determine the activity of this enzyme. This assay is based on the accumulation of ACC in tissue kept under N_2, i.e., under conditions where conversion of ACC to

234

ethylene is inhibited. Submergence of whole plants or stem sections enhanced the activity of ACC synthase in the internode. This stimulation of ACC-synthase activity was especially pronounced in the region of the internode which contained the intercalary meristem and part of the elongation zone above it. Enhancement of ACC-synthase activity was evident after 2 h of submergence and showed a peak after 4 h. Reduced levels of atmospheric O_2, which promote ethylene synthesis and growth in internodes of deepwater rice, also enhanced the activity of ACC synthase (Fig. 6).

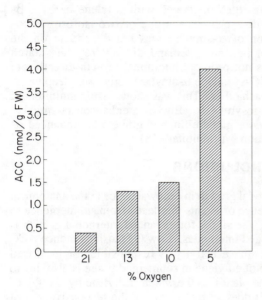

FIG. 6. Effect of low partial O_2 pressures on in vivo ACC-synthase activity in rice stems. Rice stem sections were incubated in gas mixtures containing 21, 13, 10, or 5% O_2 and 0.03% CO_2 (all by volume) in N_2 for 4 h. *In vivo* ACC-synthase activity was assayed by incubating stem sections in N_2 for 8 h and measuring ACC accumulation in the intercalary meristem (modified from 4).

Enzymes of polyamine biosynthesis. We chose polyamine biosynthetic enzymes as biochemical markers for cell-division activity in rice internodes (3). Upon submergence of the plant, the activity of S-adenosylmethionine (SAM) decarboxylase increased within 8 h six- to eightfold in the intercalary meristem. In other parts of the internode, SAM-decarboxylase activity rose much less or not at all. Investigating SAM decarboxylase was of particular interest because it and ACC synthase both utilize SAM as substrate. It has been suggested that ethylene and polyamine formation impose competitive demands on this precursor and that allocation of SAM to either pathway may constitute a point of regulation of ethylene and polyamine biosynthesis (5, 15). Since both pathways are activated by submergence, they do not appear to inhibit each other's functioning in deepwater rice. The activity of SAM decarboxylase was also enhanced by ethylene and GA_3. Submergence and hormonal treatments also increased arginine-decarboxylase activity in the intercalary meristem; the activity of ornithine decarboxylase was not enhanced under similar conditions.

Amylolytic activities in deepwater rice internodes. Rapid internodal growth in deepwater rice requires energy, substrates and osmotica.

These demands may be met, in part, by increased translocation of photosynthates into the growing internode and by breakdown of reserve starch (14). In air-grown plants, starch comprises about 10% of the dry weight of internodes. Three days of submergence led to the mobilization of 65% of the starch from those regions of the internodes which had been formed prior to submergence. Disappearance of starch was accompanied by a 70-fold enhancement of amylolytic activity. Isoelectrofocusing, product analysis, thermal denaturation studies and affinity chromatography were used to identify the amylolytic enzymes in deepwater-rice internodes (18). In air-grown plants, the amylolytic activity is predominantly β-amylase. When plants were submerged or treated with ethylene or GA_3, β-amylolytic activity decreased or disappeared, and one or two isoforms of α-amylase appeared. The induction of α-amylase was not dependent on the presence of the growing region, i.e., sink demand did not regulate starch breakdown. While ethylene does not promote internodal growth directly but through gibberellin, ethylene and GA_3 induce α-amylase activity in deepwater-rice internodes independently of each other. This was shown using inhibitors of ethylene action and gibberellin biosynthesis. Ethylene enhanced α-amylase activity in the absence of endogenous gibberellin, and gibberellin enhanced α-amylase activity when ethylene action was inhibited (18).

CONCLUSIONS

We have described internodal growth in deepwater rice at the anatomical level and have established a sequence of events that leads from submergence to the growth response of the plant. The signal for enhanced internodal growth is the reduced level of O_2 in submerged internodes. Low O_2 tensions induce ACC synthase, and enhanced ACC-synthase activity is responsible for increased ethylene synthesis. Accumulation of ethylene in submerged tissue is also due to slow diffusion of ethylene from the plant into the water. Ethylene by itself does not promote internodal growth. It does so by increasing the responsiveness of the tissue to endogenous gibberellin. An additional effect of ethylene and submergence on gibberellin biosynthesis has not been excluded. Gibberellin appears to be the immediate factor regulating growth of deepwater-rice internodes. It should be noted that acceleration of growth in submerged deepwater rice is the result of an increase in the level of a hormone, ethylene, and an increase in the sensitivity to a hormone, gibberellin. The growth response is based on enhanced cell-division activity in the intercalary meristem and on an increase in the final length of internodal cells.

We have identified several growth-related enzymes which are markers for different phases of the growth response. The development of their activities is temporally and spatially separated. The first enzyme found to be induced following submergence is ACC synthase (Fig. 7). Its activity is enhanced within 2 h in the intercalary meristem of submerged plants. The activities of two enzymes involved in polyamine biosynthesis, SAM decarboxylase and arginine decarboxylase, are stimulated within 4 h of submergence (Fig. 7). Both enzymes are localized predominantly in the intercalary meristem, and their activities appear to be related to the accelerated rate of cell division in this meristem. Enhancement of α-amylase activity by submergence has a 12-h lag

236

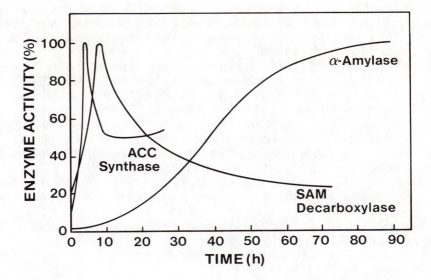

FIG. 7. Time course of development of ACC-synthase, SAM-decarboxylase and amylolytic activities in submerged deepwater-rice internodes. ACC-synthase and SAM-decarboxylase activities were measured in the intercalary meristem, amylolytic activity in the internodal region formed prior to submergence (data from 3, 4, 14).

and occurs mainly in the older part of the internode (Fig. 7). α-Amylase activity is responsible, at least in part, for the mobilization of reserves in support of the growth response. Future work will deal with the regulation of these and other growth-related enzyme activities by hormonal and environmental factors.

Acknowledgments--Research in the author's laboratory was supported by the National Science Foundation and by the U.S. Department of Energy.

LITERATURE CITED

1. BLEECKER AB, S ROSE-JOHN, H KENDE 1987 An evaluation of 2,5-norbornadiene as a reversible inhibitor of ethylene action in deepwater rice. Plant Physiol 84: 395-398
2. BLEECKER AB, JL SCHUETTE, H KENDE 1986 Anatomical analysis of growth and developmental patterns in the internode of deepwater rice. Planta 169: 490-497
3. COHEN E, H KENDE 1986 The effect of submergence, ethylene and gibberellin on polyamines and their biosynthetic enzymes in deepwater-rice internodes. Planta 169: 498-504
4. COHEN E, H KENDE 1987 *In vivo* 1-aminocyclopropane-1-carboxylate synthase activity in internodes of deepwater rice. Plant Physiol 84: 282-286

5. EVEN-CHEN Z, AK MATTOO, R GOREN 1982 Inhibition of ethylene biosynthesis by aminoethoxyvinylglycine and by polyamines shunts label from 3,4-[^{14}C]methionine into spermidine in aged orange peel discs. Plant Physiol 69: 385-388

6. JACKSON MB 1985 Ethylene and responses of plants to soil waterlogging and submergence. Annu Rev Plant Physiol 36: 145-174

7. KU HS, H SUGE, L RAPPAPORT, HK PRATT 1970 Stimulation of rice coleoptile growth by ethylene. Planta 90: 333-339

8. MÉTRAUX J-P, H KENDE 1983 The role of ethylene in the growth response of submerged deep water rice. Plant Physiol 72: 441-446

9. MÉTRAUX J-P, H KENDE 1984 The cellular basis of the elongation response in submerged deep-water rice. Planta 160: 73-77

10. MUSGRAVE A, MB JACKSON, E LING 1972 *Callitriche* stem elongation is controlled by ethylene and gibberellin. Nature New Biol 238: 93-96

11. RASKIN I, H KENDE 1983 Regulation of growth in rice seedlings. J Plant Growth Regul 2: 193-203

12. RASKIN I, H KENDE 1984 Regulation of growth in stem sections of deep-water rice. Planta 160: 66-72

13. RASKIN I, H KENDE 1984 Role of gibberellin in the growth response of submerged deep water rice. Plant Physiol 76: 947-950

14. RASKIN I, H KENDE 1984 Effect of submergence on translocation, starch content and amylolytic activity in deep-water rice. Planta 162: 556-559

15. ROBERTS DR, MA WALKER, JE THOMPSON, EB DUMBROFF 1984 The effects of inhibitors of polyamine and ethylene biosynthesis on senescence, ethylene production and polyamine levels in cut carnation flowers. Plant Cell Physiol 25: 315-322

16. ROSE-JOHN S, H KENDE 1985 Short-term growth response of deep-water rice to submergence and ethylene. Plant Science 38: 129-134

17. SISLER EC, SF YANG 1984 Anti-ethylene effects of *cis*-2-butene and cyclic olefins. Phytochemistry 23: 2765-2768

18. SMITH MA, JV JACOBSEN, H KENDE 1987 Amylase activity and growth in deepwater rice. Planta (in press)

19. VERGARA BS, B JACKSON, SK DE DATTA 1976 Deep water rice and its response to deep water stress. *In* Climate and Rice. International Rice Research Institute, Los Baños, Philippines, pp 301-319

20. YANG SF, NE HOFFMAN 1984 Ethylene biosynthesis and its regulation in higher plants. Annu Rev Plant Physiol 35: 155-189

Physiology of Cell Expansion During Plant Growth, *D.J. Cosgrove and D.P. Knievel* Eds.,
Copyright ©1987, The American Society of Plant Physiologists

APOPLASTIC SOLUTES AND LOWERED WATER POTENTIAL IN ELONGATING SUGARCANE LEAVES

F. C. MEINZER AND P. H. MOORE

Hawaiian Sugar Planters' Association, P.O. Box 1057, Aiea, HI 96701, and USDA, ARS, Aiea, HI 96701

INTRODUCTION

Values of apoplast osmotic potential reported for leaf and stem tissue range from -0.03 to -0.3 MPa in a number of species (1, 3, 6, 7). However, at present it is not clear how much of this order of magnitude range is due to differences in methodological approach and interpretation. Knowledge of the magnitude of apoplast osmotic potential for expanding plant tissue is particularly crucial in view of two distinct hypotheses proposed to explain the means by which water supply can limit growth. The discrepancy between the two hypotehses is centered around the origin of the lowered or "growth-induced" water potential observed in expanding plant tissues (2, 8). Lowered water potential in elongating tissue has been attributed to hydraulic limitations on water uptake by some workers (2, 7) and to the presence of apoplastic solutes by others (4, 5). Here we report on patterns of apoplast solute concentration and water potential in elongating sugarcane leaves.

MATERIALS AND METHODS

Leaf spindle and mature leaf tissue from greenhouse-grown sugarcane (*Saccharum* spp. hybrid) was used for all measurements. The sugarcane leaf spindle is a cylinder of several concentric whorls of tightly packed elongating leaves positioned above the shoot apical meristem.

Combined pressure chamber and psychrometric methods were used to determine water potential and its components in the symplast and apoplast. Measurements of apoplast osmotic potential were obtained with one direct and two indirect methods. In the direct methods the osmotic potential of small samples of apoplast solution extracted from leaf spindle segments by centrifugation was measured with an osmometer. In a second method total water potential of living spindle segments was measured psychrometrically. For this method it was hypothesized that total water potential would equal apoplast osmotic potential under conditions of water potential equilibrium and zero tension in the apoplastic space. In a third method exudate was collected from the exposed ends of spindle segments sealed in a pressure chamber. Bulk tissue osmotic potential was determined on sap expressed from frozen and thawed tissue. Symplast osmotic potential was determined from pressure-volume curves.

RESULTS AND DISCUSSION

The osmotic potential of apoplast solution collected directly by centrifugation of noninfiltrated tissue segments ranged from -0.25 MPa in mature tissue to -0.35 MPa in tissue just outside the elongation zone. The presence of these solutes in the apoplast manifested itself as a tissue water potential equal to the apoplast osmotic potential. Since the tissue was not elongating, the measurements were not influenced by growth-induced water uptake and no significant tension was detected with the pressure chamber. Further evidence for a significant apoplast solute concentration was obtained from pressure exudation experiments and comparison of methods for estimating tissue apoplast water fraction.

For elongating leaf tissue the centrifugation method could not be used to obtain direct measurements of apoplast solute concentration. However, several other observations suggested that the apoplast water potential of -0.35 to -0.45 MPa in elongating leaf tissue had a significant osmotic component and a small, but significant tension component.

Results of experiments in which exudate was collected from pressurized tissue segments of different ages suggested that a tissue age-dependent dynamic equilibrium existed between intra- and extracellular solutes.

LITERATURE CITED

1. BERNSTEIN L 1979 Method for determining solutes in the cell walls of leaves. Plant Physiol 47: 361-365
2. CAVALIERI AJ, JS BOYER 1982 Water potentials induced by growth in soybean hypocotyls. Plant Physiol 69: 492-496
3. COSGROVE DJ, RE CLELAND 1983a Solutes in the free space of growing stem tissues. Plant Physiol 72: 326-331
4. COSGROVE DJ, RE CLELAND 1983b Osmotic properties of pea internodes in relation to auxin action. Plant Physiol 72: 332-338
5. COSGROVE DJ, E VAN VOLKENBURGH, RE CLELAND 1984 Stress relaxation of cell walls and the yield threshold for growth. Planta 162: 46-54
6. JACOBSON SL 1971 A method for extraction of extracellular fluid: Use in development of a physiological saline for Venus'-flytrap. Can J Bot 49: 121-127
7. NONAMI H, JS BOYER 1987 Origin of growth-induced water potential. Plant Physiol 83: 596-601
8. WESTGATE ME, JS BOYER 1984 Transpiration- and growth-induced water potentials in maize. Plant Physiol 74: 882-889

Physiology of Cell Expansion During Plant Growth, *D.J. Cosgrove and D.P. Knievel* Eds.,

DOES LEAF TURGOR DETERMINE LEAF EXPANSION RATES IN DRY OR SALINE SOILS?

RANA MUNNS

CSIRO, Division of Plant Industry, Canberra, A.C.T. 2601, Australia

INTRODUCTION

Rapid changes in turgor cause rapid changes in rates of cell expansion, but not necessarily long-lasting changes. Use of a root pressurization technique to increase leaf turgor of intact plants showed that rates of leaf area expansion of plants in saline soils were initially increased by increased turgor, but that after a day, rates were similar to non-pressurized plants in saline soils, and were much lower than controls in salt-free soils (3). This technique was used more recently on plants in drying soils; root pressurization (i.e., increased shoot turgor) did not prevent stomates from closing as the soil dried out (1). Here I show it did not prevent leaf expansion from decreasing as the soil dried out.

MATERIALS AND METHODS

Barley (*Hordeum vulgare* L., cv Himalaya) was grown in soil in pots which could fit inside pressure bombs. Some pots were watered regularly, some were allowed to dry out, and some were pressurized as the soil dried out. The pressurization technique involved maintaining the xylem hydrostatic pressure at zero (i.e., atmospheric), so the leaf water potential was always near zero, and cell turgor and volume did not decrease as the soil dried out. The pressures needed to maintain near-zero are shown in Fig. 1 (dotted line). Note: the pressurization technique does not affect the root water status; neither root turgor nor soil suction is changed (for theory see ref. 2).

RESULTS AND DISCUSSION

Root pressurization did not prevent the rate of leaf expansion from decreasing as the soil dried out (Fig. 1), i.e., high leaf turgor did not prevent expansion rates from decreasing. Root pressurization caused an increase in leaf expansion on the first 1-2 days, when the soil was still wet, a phenomenon noted previously (3). Root pressurization also caused a shift in expansion rates from night to daytime; this was most pronounced on the first 1-2 days, but continued as the soil dried out (Fig. 1). This shift towards daytime growth suggests that the cell walls are more extensible during the day than the night. After the first few days, the plants apparently adapted to the increased turgor, and

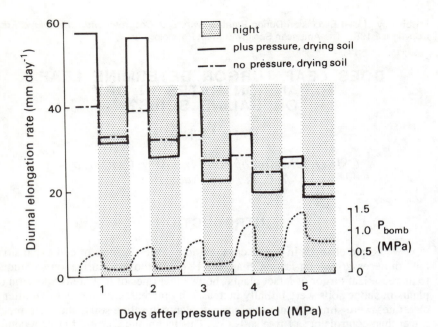

FIG. 1. Effect of root pressurization, to maintain a high leaf water status, on leaf elongation rates of barley plants in drying soil. The extent of soil drying is reflected in the bomb pressure (P_{bomb}) applied to the roots to keep leaf water potential near zero. Plants which were watered regularly had the same elongation ratio on days 1-5 as shown for days 1-2.

total daily expansion rates were the same for plants with and without pressurization, i.e., any increase in daytime expansion was balanced by a decrease in night-time expansion. This occurred in the drying soils (Fig. 1) and also occurs in saline and well-watered soils (data not shown). This suggests that some process controls daily leaf expansion rates independently of turgor and wall extensibility.

These results show that leaf turgor does not determine leaf expansion, and suggest that a message from the roots is determining rates of leaf expansion in dry and saline soils.

LITERATURE CITED

1. GOLLAN T, JB PASSIOURA, R MUNNS 1986 Soil water status affects the stomatal conductance of fully turgid wheat and sunflower leaves. Aust J Plant Physiol 13: 459-464
2. PASSIOURA JB, R MUNNS 1984 Hydraulic resistance of plants. II Effects of rooting medium, and time of day, in barley and lupin. Aust J Plant Physiol 11:341-350
3. TERMAAT A, JB PASSIOURA, R MUNNS 1985 Shoot turgor does not limit shoot growth of NaCl-affected wheat and barley. Plant Physiol 77: 869-872

Physiology of Cell Expansion During Plant Growth, *D.J. Cosgrove and D.P. Knievel* Eds.,
Copyright ©1987, The American Society of Plant Physiologists

LEAF WALL YIELD THRESHOLD AS DETERMINED BY VAPOR PRESSURE PSYCHROMETERY FOR FIELD-GROWN SOYBEANS

H. C. RANDALL AND T. R. SINCLAIR

*USDA-ARS, Agronomy Physiology Laboratory
Gainesville, FL 32611*

INTRODUCTION

The irreversible extension of plant cell walls requires that cell hydrostatic pressure (P) exceed a minimum pressure termed the cell wall yield threshold (Y). Consequently, Y may be an important regulator of growth. Measurement of Y has generally been conducted under controlled environment conditions (2, 3, 4), but because the growth environment significantly influences Y (1), this variable needs to be further evaluated under a range of field conditions. However, measurement of Y in field-grown plant tissue may be difficult because the evaporative conditions can result in short periods of water deficit. A necessary condition to measure Y in vapor pressure psychrometers is that P > Y (2). For growing soybean (*Glycine max.* (L) Merr.) leaves, values of Y were determined at night when P was expected to exceed Y. Specifically, the influence of drought on Y in field-grown soybean leaves was evaluated.

MATERIALS AND METHODS

Measurement technique of Y. Water and osmotic potentials were determined on growing and nongrowing leaf samples, and were measured in vapor pressure psychrometers (2). Equilibration of leaf tissue samples in psychrometer chambers for up to 10 h did not result in significantly different potentials compared with those obtained after 5 h. 'Apparent' P was calculated as the difference between the total and osmotic potentials. It has been shown in growing plant tissues that in the absence of water uptake, P declines to the value of Y (2). Maximum P from growing leaves during the night were interpreted as Y. All turgors estimated in mature, nongrowing tissue were interpreted as true P.

Response of Y to drought. Apparent P was measured hourly throughout the night when P was expected to exceed Y in growing and nongrowing leaves of irrigated and droughted plants. Rapidly growing leaves and mature, nongrowing leaves were sampled to determine P and Y during two droughted and two well-watered periods.

A second experiment was done to measure Y intensively during and subsequent to release from drought. On growing leaves, Y was measured every 4

243

h during the last day of an 8-day drought and the following day when irrigation was applied.

To identify changes in Y at different leaf areas, P was measured on leaves located at the 6th through 15th nodes of field-grown, droughted and well-watered soybeans.

RESULTS AND DISCUSSION

In mature, nongrowing leaves P rose throughout the night in well-watered plants and attained a maximum P of between 0.59 and 0.89 MPa. Rapidly growing leaves from the same plants were measured and found to have an apparent P or Y of between 0.26 and 0.38 MPa. Mature leaves from droughted-stressed plants regained P throughout the night but reached maximum values less than those measured in well-watered leaves. Rapidly growing leaves from these drought-stressed plants, however, had Y of approximately 0.39 MPa. Consequently, the value of Y was found to have increased in the rapidly growing leaves in response to drought.

This conclusion was supported by the second experiment where apparent P on growing leaves from drought-stressed and control (irrigated) plants was measured on the last day of an eight-day drought and upon release from drought. At the end of the drought phase Y in growing, drought-stressed leaves was 0.5 MPa, while in control leaves these values were between 0.03 and 0.23 MPa. Upon release from drought, Y decreased in growing leaves of drought-stressed plants to values similar to those of control leaves (0.25 MPa).

Measurement of apparent P for a range of leaf areas indicated that Y increased approximately linearly with an increase in the area of both drought-stressed and irrigated leaves. Despite higher P in mature leaves from irrigated plants relative to those from drought-stressed plants, Y measured in leaves recently emerged form the meristem were 0.5 MPa in drought-stressed plants and 0.3 MPa in irrigated plants.

LITERATURE CITED

1. BUNCE JA 1977 Leaf elongation in relation to leaf water potential in soybean. J Exp Bot 28: 156-161
2. COSGROVE DJ, E VAN VOLKENBURGH, R CLELAND 1984 Stress relation of cell walls and the yield threshold for growth. Planta 162: 46-54
3. DAVIES WJ, E VAN VOLKENBURGH 1983 The influence of water deficit on the factors controlling the daily pattern of growth Phaseolus trifoliates. J Exp Bot 34: 987-999
4. VAN VOLKENBURGH E, RE CLELAND 1986 Wall yield threshold and effective turgor in growing bean leaves. Planta 153: 572-577

Physiology of Cell Expansion During Plant Growth, *D.J. Cosgrove and D.P. Knievel* Eds.,
Copyright ©1987, The American Society of Plant Physiologists

SENSITIVITY OF VESSEL DIAMETER TO CLIMATIC CONDITIONS IN BUR OAK (*QUERCUS MACROCARPA*)

DEBORAH WOODCOCK

*Geography Department, Bucknell University
Lewisburg, PA 17837*

INTRODUCTION

A range of anatomical variables in the wood of a ring-porous oak (*Quercus macrocarpa*) are sensitive to year-to-year variance in the climate system. Although the motivation for this research is to determine the feasibility of using anatomical characters in climate reconstruction (1), the physiology of the response is also of interest. Factor analysis aided in the selection of variables suitable for climate reconstruction, and the variables selected relate to diameter of either the spring or summerwood vessels. The response of these diameter-related variables to temperature is presented here.

MATERIALS AND METHODS

Ten individuals of bur oak were cored by means of standard increment borers. The cores were sectioned lengthwise and thin-sectioned using a sliding microtome. Measurements were obtained both using a light microscope with micrometer and by a digitizing technique in which an image is projected onto a digitizing tablet connected on-line to a computer. The data comprised a 25-year record of tree growth.

RESULTS AND DISCUSSION

Preliminary analyses included a variety of factors cited as having functional significance in wood. Most of these variables were found to display climate sensitivity. Intercorrelations among the variables were evaluated by means of factor analysis (Table I), and a tentative functional interpretation of the rotated factors is as follows.

Among the variables identified with Factor 1 is the sum of the diameters to the fourth power, which according to Zimmermann (2) describes conductivity, or volumetric flow of water through the tree. In flow of this type, the large vessels contribute disproportionately, and it can be seen that the diameter of the largest springwood vessel is a related variable. Also identified with this factor is the total conductive area, which Carlquist (3) cites as having significance in some plant groups. Factor 1 thus appears to be significant with respect to water conductance.

Table I. *Factor analysis of wood-anatomical characters and ring width in bur oak. Factor loadings less than .250 not shown.*

	Factor 1	Factor 2	Factor 3
Spring wood: diameter of the largest vessel	.779	--	-.375
Spring wood: total conductive area	.959	-.262	--
Spring wood: sum of the diameters to the fourth power	.733	--	--
Spring wood: average vessel diameter	--	.521	.251
Summer wood: vessel density	--	-.872	--
Ring width	--	.986	--
Spring wood: average vessel diameter	.540	--	-.739
Spring wood: vessel density	--	--	.934
VP	2.421	2.108	1.739

Factor 2 is identified with vessel diameter and density of the summer wood, quantities that are inversely related and which Carlquist (4) suggests determine the vulnerability of the wood to water stress. The rationale here is that smaller, denser vessels make a plant less vulnerable to water stress since small vessels are less susceptible to breakage of the water column in times of water stress and greater density of vessels provides additional backup in case of loss of conductive elements. Note also that ring width is grouped with the summerwood variables, varying positively with vessel diameter and negatively with vessel density.

The third factor is identified with diameter and density of the springwood vessels and, again, according to Carlquist's interpretation, would correspond to vulnerability of the springwood to water stress.

Selection of variables suitable for climate reconstruction was made partially on the basis of the factor analysis and partially from practical considerations. Diameter of the largest springwood vessel and average diameter of the spring and summerwood vessels are representative of the three factors as seen in the factor analysis and also have the advantage that their measurement is both relatively easy and unambiguous (measurements involving determination of an area, as, for example, density or total conductive area, are problematic owing to the heterogeneity of the tissue). Thus it is possible to represent the major sources of variance among the anatomical variables, which include width of the growth increment, by means of diameter-related measures.

Both temperature and precipitation affect the diameter-related variables. The effect of precipitation, however, although sufficiently significant to permit its reconstruction from the variables, is somewhat less consistent from month to month and thus provides more difficulties in making physiological interpretations.

Temperature during the winter half-year appears to have a particularly pronounced effect on growth (Fig. 1). The springwood vessels, some of which

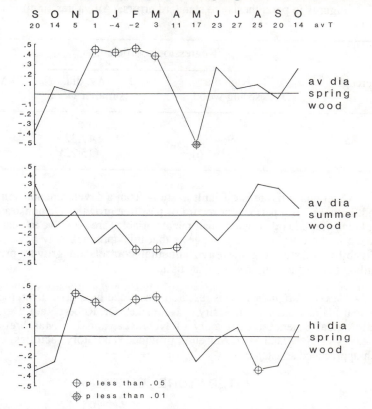

FIG. 1. Growth response for average springwood vessel diameter, average summerwood vessel diameter, and diameter of the largest springwood vessel. Values plotted are correlations to monthly temperature. The sample includes 25 years of data from 10 trees. Average monthly temperature (top) is given for comparison.

attain diameters of approximately 0.5 mm, reach their largest average size in years with warm winters. Even though ring-porous trees begin cambial activity earlier in the year than diffuse-porous species, the sensitivity shown here occurs during months when temperatures are below or just above freezing, conditions under which the trees would be presumed to be dormant. High May temperatures, on the other hand, are associated with small springwood vessels, perhaps because of the adverse effect of high temperatures and water stress at the time these large cells are expanding. The summer wood, although also affected by wintertime temperatures, exhibits an inverse response, warm winters being associated with smaller summerwood vessels. As an example of the use of these data in climate reconstruction, a multiple-regression equation for heating degree days is presented in Table II.

Table II. *Regression analysis for heating degree days during the winter half-year, 1960-84.*
Figures in parenthesis are standard errors. All values significant at the .01 level.

| y intercept | Regression coefficients | | Adj. r^2 |
	High dia spring wood	Av. dia. summer wood	
13332.83	-25944.32 (5539.96)	47.32 (15.22)	65%

Secondary tissue is difficult to study from a developmental standpoint. Wood, however, is a permanent record of past tree growth. Demonstration that vessel diameter and other wood-anatomical variables are sensitive to year-to-year variance in the climate system suggests that this approach may have potential applicability in studying the environmental controls on growth processes, including cell expansion, in secondary tissue.

Acknowledgments--This research was aided by a grant from the Center for Great Plains Studies, University of Nebraska. Dr. Robert Kaul, Department of Biological Sciences, University of Nebraska, and Dr. Margaret Bolick, University of Nebraska State Museum, provided laboratory facilities and other much-appreciated assistance.

LITERATURE CITED

1. WOODCOCK D 1987 Use of wood-anatomical variables of bur oak (*Quercus macrocarpa*) in reconstruction of climate. Ph.D. Dissertation, University of Nebraska
2. ZIMMERMANN MH 1983 Xylem structure and the ascent of sap. Springer-Verlag, New York
3. CARLQUIST S 1977 Wood anatomy of tremandraceae: Phylogenetic and ecological implications. Amer J Bot 64: 704-713
4. CARLQUIST S 1975 Ecological strategies of xylem evolution. University of California, Berkeley

Physiology of Cell Expansion During Plant Growth, *D.J. Cosgrove and D.P. Knievel* Eds.,
Copyright ©1987, The American Society of Plant Physiologists

EXTRUSION PORES: LOCALIZED DEFORMATION OF COTYLEDONARY CELL WALLS DURING IMBIBITION

STEPHEN C. SPAETH

*Grain Legume Genetics and Physiology Research Unit
USDA, ARS, 215 Johnson Hall, WSU
Pullman, WA 99164*

INTRODUCTION

Legume cotyledons leak intracellular substances during imbibition (1, 5). Such losses make seedlings less vigorous (2) and more susceptible to soil-borne pathogens (4). Mechanisms for leakage of protoplasm and protein bodies from hydrating tissues include pressure-driven extrusion from multicellular blisters (5), and cellular rupture (1) accompanied by nearly complete discharge of cell contents (Spaeth, unpublished). The presence of protoplasm and protein bodies located on surfaces of bean and pea cotyledons at some distance from blisters and single cell ruptures (Spaeth, unpublished) indicates that protoplasm and protein bodies may also be released from cells by a process analogous to pressure-driven extrusion from multicellular blisters.

Cell walls, like the tissue bounding blister cavities, may restrict volumetric expansion of hydrating protein bodies and protoplasm. Expansion of cell contents during hydration may locally deform cell walls and force protein bodies from cells through extrusion pores. But, cellular analogs of extrusion orifices in multicellular blisters have not been identified. The objective of this research was to test the hypothesis that cellular presssure-driven extrusion contributes to losses of protein bodies. If the extrusion hypothesis were correct, cell walls must, during imbibition, contain pores sufficiently large to permit protein bodies to be extruded through walls.

MATERIALS AND METHODS

Bean (*Phaseolus vulgaris* L. cv Royal Red Mexican) cotyledons were soaked and prepared for scanning electron microscopy as described previously (5). Briefly, cotyledons were soaked in distilled water for 1 h, fixed in 3% glutaraldehyde, post-fixed in 2% osmium tetroxide, dehydrated in ethanol, freeze fractured, critical-point dried, sputter coated, and observed at 20 Kv.

RESULTS AND DISCUSSION

When partially hydrated cotyledons were freeze fractured, they often separated through intercellular spaces, and between the interior surfaces of cell walls and exterior surfaces of fixed protoplasm (Fig. 1). Pores 0.5 to 2 μm in diameter passed through cell walls. Protoplasm protruded into pores in cell walls. Spherical bodies ranging from 0.5 to 2 μm in diameter were held on surfaces of the intercellular spaces by thin films of material. Protrusions were sufficiently large to contain protein bodies.

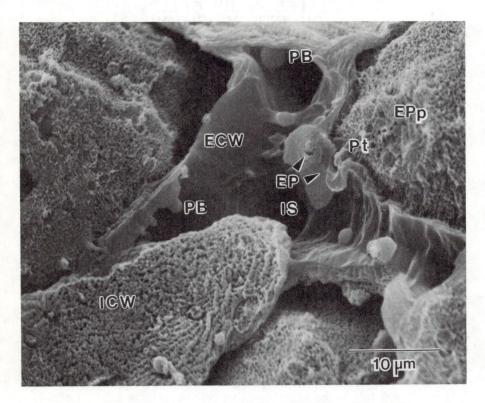

FIG. 1. Freeze fracture through cotyledonary cells and a junction of intercellular spaces (IS). Cells fractured between interior surfaces of cell walls (ICW) and exterior surfaces of fixed protoplasm (EP$_p$). Extrusion pores (EP) about 1 μm in diameter passed through cell walls. Protrusions (Pt) from exterior surfaces of protoplasm extended into pores. Protein bodies (PB) adhered to exterior surfaces of cell walls (ECW) lining intercellular spaces.

    The spherical bodies found in intercellular spaces were almost certainly protein bodies because no other constituents of cotyledonary tissues fall within this range of sizes (3, 6, 7). Protrusions were similar to those which have been described for other bean cultivars (3, 7), however, the mechanism for their formation was not identified.

    Cotyledonary cells apparently extruded protein bodies and protoplasm through cell walls (Spaeth, unpublished) by a process analogous to that observed for multicellular blisters (5). The cellular analogs of extrusion orifices in multicellular blisters were pores about 1 μm in diameter which passed through cell walls. Pressure of hydration expansion apparently deformed cell walls locally and forced protein bodies and protoplasm through walls (Fig. 1).

Deformation of cell walls during imbibition may alter leakage of intracellular substances from seeds by modulating cellular pressure-driven extrusion.

Acknowledgments--Electron Microscopy Center of Washington State Univ. and Joe S. Hughes provided technical assistance.

LITERATURE CITED

1. DUKE SH, G KAKEFUDA 1981 Role of the testa in preventing cellular rupture during imbibition of legume seeds. Plant Physiol 67: 449-456
2. DUKE SH, G KAKEFUDA, CA HENSON, NL LOEFFLER, NM VAN HULLE 1986 Role of the testa epidermis in the leakage of intracellular substances from imbibing soybean seeds and its implications for seedling survival. Physiol Plantarum 68: 625-631
3. HUGHES JS, BG SWANSON 1985 Microstructural changes in maturing seeds of the common bean (*Phaseolus vulgaris* L.). Food Microstructure 4: 183-189
4. SCHROTH MN, RJ COOK 1964 Seed exudation and its influence on pre-emergence damping-off of bean. Phytopathology 54: 670-673
5. SPAETH SC 1987 Pressure-driven extrusion of intracellular substances from bean and pea cotyledons during imbibition. Plant Physiol (In press)
6. VARNER JE, G SCHIDLOVSKY 1963 Intracellular distribution of proteins in pea cotyledons. Plant Physiol 38: 139-144
7. VARRIANNO-MARSTON E, GM JACKSON 1981 Hard-to-cook phenomenon in beans: Structural changes during storage and imbibition. J Food Sci 46(5): 1379-1385

Physiology of Cell Expansion During Plant Growth, *D.J. Cosgrove and D.P. Knievel* Eds., Copyright ©1987, The American Society of Plant Physiologists

SOLUTE SUPPLY AND STEM ELONGATION: EFFECT OF COTYLEDON REMOVAL ON GROWTH, SOLUTE AND TURGOR PRESSURES

JUDY GOUGLER SCHMALSTIG AND DANIEL J. COSGROVE

Biology Department, The Pennsylvania State University, University Park, PA 16802

INTRODUCTION

Solute uptake during cell expansion maintains osmotic pressure and provides substrates for respiration and synthesis of macromolecules such as cell wall material, membranes, etc. A reduction in solute influx may inhibit elongation by: 1) reduction in turgor pressure in response to reduced osmotic pressure; 2) lack of substrates for respiration and/or macromolecule synthesis; 3) direct effect on cell wall properties or hydraulic conductivity before lack of substrates becomes critical. Our experimental approach was to reduce the solute influx by removal of the cotyledons, and to measure the effect on elongation, osmotic pressure and turgor pressure.

MATERIALS AND METHODS

Pea (*Pisum sativum* L. var. Alaska) seedlings were grown in the dark in vermiculite soaked with half-strength Hoagland's solution. Seedlings were used after 4 days and experiments were performed under dim, green safe lights.

Elongation rate of the apical 1.5 cm of the pea epicotyl was measured by linear position transducer. Osmolality (Osm kg^{-1}) of expressed cell sap from the apical 1 cm of epicotyl was measured with a Wescor vapor pressure osmometer and was converted to bars (24.78 bars per Osm kg^{-1}). Turgor pressures of 6 to 10 individual cortical cells per plant were measured with a computer-assisted pressure probe (1).

Stems were securely fastened to lessen mechanical disturbance and cotyledons were excised with razor blade fragments. Dry weight (DW) and fresh weight (FW) changes were measured by cutting, weighing and drying a length of the growing region initially marked at 0.5 cm from three sets of ten plants each: 1) at the time of marking (0 h); 2) 5 h later with cotyledons excised at time zero; 3) at 5 h with cotyledons intact.

RESULTS AND DISCUSSION

Cotyledon excision reduced solute influx as evidenced by reduced osmotic pressure of expressed cell sap (Fig. 1) and no net increase in DW over 5

252

h (Table I). During the first 1.5 h after excision, the elongation rate decreased by about 20 % and osmotic pressure dropped by 0.5 bars from 8.75 to 8.25 bars (Fig 1). Between 2.5 and 4.25 h after excision, elongation rate was reduced by 60 % and osmotic pressure dropped by a total of 1.4 bars. Osmotic pressure of intact seedlings did not change over a 5 h period.

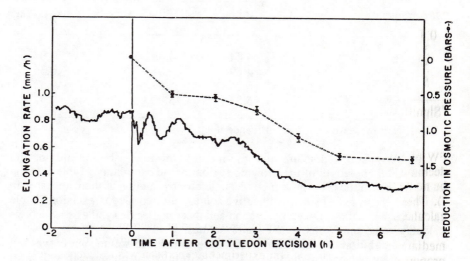

FIG. 1. Elongation rate (solid line) and reduction in osmotic pressure (--o--) with time after removal of cotyledons at time zero (vertical line). Elongation rate is average of 15 stems. Reduction in osmotic pressure from initial zero h is mean of 5 to 8 experiments with 12 plants per time point per experiment. Vertical bars show standard error. Reduction in osmotic pressure from 0 to 1 h is significant at 5 % significance level by ANOVA.

Table I. *Increases in fresh and dry weights and μOsmoles in 0.5 cm length of elongating region over 5 h, +/- cotyledons.*
        FW and DW are means of 8 experiments each with 10 plants per treatment.

	0 h	5 h	
		- cot.	+ cot.
DW (mg)	1.22[a]	1.24[b]	1.50[ab]
FW (mg)	11.4[c]	13.9[c]	16.7[c]
Osmolality (mOsm /kg)	350	292	350
μOsmoles	3.99	4.06	5.85

[a,b,c] Significant differences at 5 % significance level between like groupings.

        In contrast to osmotic pressure, turgor pressure did not decrease during the 5 h period after cotyledon removal. There was a significant increase in turgor pressure of 0.5 bar after 5 h in both controls and experimental plants (Table II).

Table II. *Change in turgor pressure (bars) with time +/- cotyledons.*
Pressures are means of 6 to 7 plants measured at time 0 and final time as indicated.

Time (h) of Final Measurement	Turgor Pressure (bars)		
	Initial	Final	Final-Initial
0.5	4.84	4.88	0.04
3	4.69	5.01	0.33
5 -cot.	5.06	5.58	0.52[*]
5 +cot.	5.33	5.80	0.47[*]

[*]Significant increase at 5 % significance level.

We estimated solute flux into the growing region by multiplying the FW of a 0.5 cm length of the growing region by the osmolality (Table I). The decrease in osmotic pressure can be accounted for by dilution as there appeared to be no net increase of osmotically-active solutes after cotyledon excision (Table I). These calculations ignore respiration and macromolecule synthesis, which we calculate to be small relative to osmotic solutes.

Inhibition of elongation in response to reduced solute supply was not mediated by a turgor reduction. The possibility that changes in cell wall properties may be responsible for the inhibition of elongation is being investigated.

LITERATURE CITED

1. COSGROVE DJ, DM DURACHKO 1986 Automated pressure probe for measurement of water transport properties of higher plant cells. Rev Sci Instrum 57: 2614-2619

Physiology of Cell Expansion During Plant Growth, *D.J. Cosgrove and D.P. Knievel* Eds.,
Copyright ©1987, The American Society of Plant Physiologists

THE BIOPHYSICAL MECHANISM OF REDUCED GROWTH IN *PISUM SATIVUM* SEEDLINGS UNDER SALINE CONDITIONS

CYNDE MARGRITZ AND DANIEL J. COSGROVE

Biology Department, The Pennsylvania State University, University Park, PA 16802

INTRODUCTION

Salinity is a leading cause of water stress which inhibits cell expansion during plant growth. This study was undertaken to shed light on the biophysical cause of saline-induced growth reductions in a salt-sensitive species. A need existed for direct data pertaining to the early primary effects of salt stress and the sequence of physiological events that occur at the onset of stress. Empirical values for the six growth-governing biophysical parameters were established through direct measurements on the growing region of etiolated pea epicotyl.

MATERIALS AND METHODS

Four-day-old *Pisum sativum* L. var Alaska seedlings were grown for 20 hours in vermiculite soaked with distilled water (controls) or 605 mmol/kg NaCl. Vapor pressure osmometry was used to measure the osmolality. A pressure microprobe was used to make turgor pressure measurements. The "stress relaxation" technique was used to measure yield threshold and the cell wall yielding coefficient ("wall extensibility"). Hydraulic conductance was measured in intact seedlings by monitoring elongation or shrinkage during the application of pressure or vacuum. Sodium content was analyzed using atomic absorption spectrophotometry.

RESULTS AND DISCUSSION

Growth rate was decreased by 50-60% and the growing region was reduced in length from 2 cm (from apex down the epicotyl) to 1.2-1.3 cm by salt treatment (Fig. 1). The internal osmolality increased by 30% (90-100 mmol/kg) with salt treatment (Fig. 2). The internal sodium content of the growing region increased by 82% (.33 mmol Na^+/g dry weight=control, 1.81 mmol Na^+/g dry weight=salt). There are two indications of a reduction in the rate of solute import into the growing region: 1) internal osmolality did not reach 631 mmol/kg which would have been obtained had the rate of solute import continued at the pre-stress rate of 27.8 mmol/kg/hr, 2) if freely distributed and osmotically active, sodium ions alone could increase the osmolality of the growing region to 459 mmol/kg. It is clear that the plant may have some

255

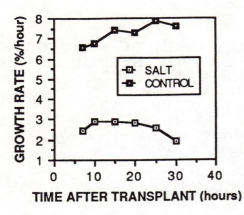

FIG. 1. Growth rate of pea seedlings over time. Seedlings were grown under nearly 100 percent relative humidity and growth was measured as the change in stem length over time. Average standard deviation was 0.7%/hr for n=40.

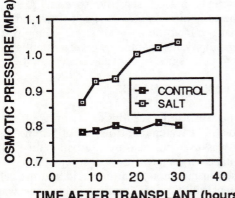

FIG. 2. Internal osmotic pressure changes over time. Average standard deviation was 0.36-0.44 MPa for n=40. Osmotic pressure was measured in the growing region.

osmoregulatory mechanism regulating solute import during salt-stress induced growth reduction.

Turgor pressure (P=.58-.59 MPa), yield threshold (Y=.25 MPa), hydraulic conductance (half-time for shrinking/swelling=16-17 sec.), and the volumetric elastic modulus (ε=4.8-4.9 MPa) were unaffected by salt treatment. Since the seedlings had no net loss or gain of water, redistribution of internal water reserves may have functioned to maintain turgor. The turgor values presented here are the first direct measurements of turgor pressure in growing tissue of higher plants under saline conditions, unlike estimated or indirect turgor values reported by others (1, 2).

The growth reduction can be attributed completely to a fifty percent reduction in the cell wall yielding coefficient (control=.63 $MPa^{-1}hr^{-1}$, salt=.32 $MPa^{-1}hr^{-1}$) as calculated from the rate of stress relaxation (Fig. 3). Changes in the cell wall yielding coefficient may be closely linked to cell wall synthesis and metabolism. Alternatively a physical or biochemical mechanism independent of cell wall synthesis may be operating to decrease wall "extensibility."

256

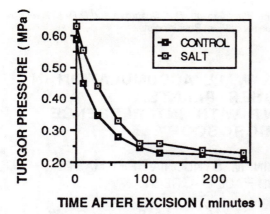

FIG. 3. Pressure changes during stress relaxation of pea seedlings. Pressures were obtained in 10-15 stem cortical cells per time point in growing segments 4.5 mm in length. This graph is typical of four others showing the same trend.

LITERATURE CITED

1. MASON HS, K MATSUDA 1985 Polyribosome metabolism, growth and water status in the growing tissues of osmotically stressed plant seedlings. Plant Physiol 64: 95-104
2. MEYER RF, JS BOYER 1972 Sensitivity of cell division and cell elongation to low water potential in the elongatin region of maize leaves. Plant Physiol 69:1145-1149

Physiology of Cell Expansion During Plant Growth, *D.J. Cosgrove and D.P. Knievel* Eds., Copyright ©1987, The American Society of Plant Physiologists

RATES OF CELL WALL ACCUMULATION IN HIGHER PLANTS: MEASUREMENT WITH INTERFERENCE MICROSCOPY

TOBIAS I. BASKIN, M. SYNDONIA BRET-HARTE AND PAUL B. GREEN

Biological Sciences, Stanford University, Stanford CA 94305

INTRODUCTION

We report a new method for measuring the rate of accumulation of mass in the outer cell wall of the epidermis. The method uses interference microscopy to quantify the amount of mass present in an excised strip of wall. Green (2) pioneered the use of interference microscopy to study changes in wall mass during growth in *Nitella* internodes. We have developed the method to study growth in higher plants. Populations are sampled at several times from an area on the plant where the growth rate is known, and measurements of the amount of mass present at each time are converted into rates of accumulation of mass by means of a continuity equation (3). Such rates represent the sum of rates for synthetic deposition of wall material and for its hydrolytic removal.

Efforts to understand how growth is controlled have focused on wall hydrolysis; however, incorporation of new material can also cause wall loosening (4). The presented method has certain advantages over previously used gravimetric or radio-isotopic methods: the rate of wall accumulation is measured for a localized region (e.g., the outer epidermal wall of a single cell), such rates can be obtained for intact plants, and a constant can be found that converts the mass units of the measurement to grams.

METHOD

We used the method to study IAA-induced, segment elongation in third internodes of *Pisum sativum*, and first positive phototropic bending in coleoptiles of *Zea mays* (both species grown under continuous red light). At the appropriate times, plants were harvested by placing the shoot (or segment) on ice.

Segments or marked zones were photographed through a dissecting microscope, and growth rates were measured from the photographs. Then, a strip of tissue, in which areas consisted of only the outer epidermal wall, was cut from within the marked zone with a microscalpel. The strip was washed in

258

water and then spread flat on a slide and dried. The contraction of the epidermis upon cutting was also measured.

Samples, mounted in air, were examined through an interference microscope (Perval Interphako, Aus Jena), with monochromatic light of 546 nm, such that a double image of the sample appeared in a field of fringes (Fig. 1). The sample's optical path difference (OPD) with respect to the surrounding medium was determined from enlarged negatives (1). A value was obtained for each peel from the mean OPD of 8 to 30 cells on that peel. The variability in OPD among cells was less than that among peels.

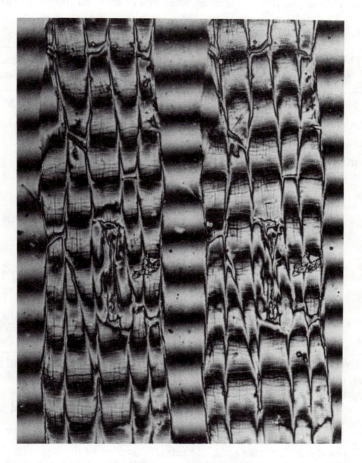

FIG. 1. Interference micrograph of the excised outer epidermal wall of a maize coleoptile. 330 x.

The refractive index and density of the sample relate OPD to mass per unit area (1). During cell elongation over the short term, wall density and refractive index are virtually constant; therefore wall OPD can be assumed directly proportional to the mass per area of the wall. The rate of accumulation of mass per area for a growing zone can be calculated, given the relative growth rate of the zone and the amount of mass per area found there in the wall over time (3).

RESULTS & DISCUSSION

For pea 3rd internodes, a low rate of wall accumulation occurred for auxin-treated segments, as compared either to the intact internode or to segments incubated in water only (Table I). This may be a consequence of the lack of exogenous carbohydrate. The negative sign of the accumulation rate found for treatment with auxin, if real, indicates a net loss of material from the wall.

Table I. *Rates of wall accumulation in Pisum sativum and Zea mays.*

TREATMENT:		OPD	Growth Rate	Accumulation Rate
		nm ± SEM	*(% h⁻¹)*	*(nm h⁻¹)*
Pea, Intact:	initial	394 ± 7.0	5.5	14.7
	4 h	367 ± 8.7		
Segments[a] - IAA:	initial	358 ± 10.8	1.6	10.9
	4 h	379 ± 11.0		
+ IAA:	initial	358 ± 10.8	6.1	-1.1
	4 h	276 ± 9.2		
Maize[b] Control:	initial	237 ± 6.6	2.3	3.7
	6 h	233 ± 6.6		
Shaded:	initial	219 ± 6.0	3.4	8.6
	6 h	226 ± 4.4		

[a]3rd internode segments (15 mm) incubated for 1 h in distilled water and then for 4h with or without 10^{-5} M IAA.
[b]Coleoptiles were marked with a 4.5 mm zone, given blue light (30 s pulse, 3 μmol m⁻²) or not and then clinostatted (3 rpm) for 6 h. Growth and bending rates on the clinostat were constant over the 6 h.

For the maize coleoptile, the refractive index of the wall was found to equal 1.54; hence, (estimating wall density at 1.4 g cm⁻³) absolute rates of wall accumulation were around 1 μg cm⁻² h⁻¹. For phototropism, the rate of wall accumulation on the shaded side of the coleoptile was stimulated along with the growth rate, as compared to unirradiated controls (Table I). Since the stimulation of growth on the shaded side of the coleoptile is likely caused by an increased amount of auxin reaching the tissue, we hypothesize that the increased wall accumulation rate results from an auxin-induced increase in the rate of delivery of wall precursors, although an inhibition of wall removal cannot be excluded.

LITERATURE CITED

1. DAVIS HG   1958   The determination of mass and concentration by microscope interferometry. *In* JF Danielli, ed, General Cytochemical Methods, Vol 1. Academic Press, New York, pp 55-161
2. GREEN PB 1958 Structural characteristics of developing *Nitella* internodal cell walls. J Biophys Biochem Cytol 4: 505-515
3. SILK WK, RC WALKER, J LABAVITCH 1984 Uronide deposition rates in the primary root of *Zea mays*. Plant Physiol 74: 721-726
4. RAY PM 1969 The action of auxin on cell enlargement in plants. Devel Bio (Suppl) 3: 172-205

Physiology of Cell Expansion During Plant Growth, *D.J. Cosgrove and D.P. Knievel* Eds.,
Copyright ©1987, The American Society of Plant Physiologists

XYLOGLUCAN FUCOSYLTRANSFERASE ACTIVITY IS LOCALIZED IN GOLGI DICTYOSOMES

A. CAMIRAND, D. A. BRUMMELL AND G. A. MACLACHLAN

*Biology Department, McGill University
Montreal, Quebec, Canada*

INTRODUCTION

Xyloglucan (XG) is a hemicellulosic polysaccharide widely distributed in growing plant tissues and which appears to play a crucial role in the regulation of cell wall rigidity, especially in dicotyledons. Treatment of stem segments with auxin is known to stimulate cell expansion, and this effect has been correlated with both a solubilisation (5) and a decrease in average molecular weight of cell wall XG (4, 6). Turnover of XG is thus a strong candidate for a biochemical event responsible for cell wall loosening. In pea stems, XG consists of a series of heptasaccharides ($Glc_4 Xyl_3$) and nonasaccharides ($Glc_4 Xyl_3 Gal Fuc$) found mostly in alternating sequence (2).

Studies on XG synthesis *in vitro* suggest that the addition of fucose is the last step in the synthesis of the nonasaccharide (1). Both glucosyl- and xylosyltransferase activities have been localized on Golgi membranes (3, 7). However, the addition of terminal fucose residues could occur in the Golgi dictyosomes or elsewhere in the secretory pathway, such as in secretory vesicles or at the plasma membrane. The present study investigates the subcellular localization of XG fucosyltransferase in pea stem cells.

MATERIALS AND METHODS

Microsomal membranes were prepared from growing regions of etiolated 1-week old *Pisum sativum* var. Alaska seedlings (1, 2). The membranes were fractionated at 100,000 x g for 1 h (4°C) on a linear 20-35% Renografin gradient to separate lighter Golgi secretory vesicles from heavier dictyosome membranes (8). Twelve fractions were obtained, the membranes centrifuged down, resuspended and incubated 30 min at 22°C with either 1.0 µM GDP[^{14}C]fucose, or 0.5 µM UDP[^{14}C]xylose, in the presence of unlabeled UDP glucose (2 mM), UDP xylose (20 µM) and UDP galactose (20 µM). Ethanol-insoluble, alkali-soluble radioactive products were analysed for radioactivity, and the label shown to be associated with XG by cellulase digestion followed by BioGel P-2 column chromatography (1). For *in vivo* experiments, stem slices were pulsed with 40 µCi L-[^{3}H]fucose 30 min 22°C, then chased with 50 mM unlabeled fucose for various periods. Microsomal membranes were isolated from the tissues, separated on a Renografin gradient as above, and radioactivity estimated for each of the twelve fractions as well as for the insoluble cell wall material. The following marker enzymes were assayed for all *in vitro* and *in vivo* gradient fractions: latent UDPase for Golgi, vanadate-sensitive ATPase for plasma membrane, NADH cyt c reductase for ER, and cyt c oxidase for mitochondria.

RESULTS AND DISCUSSION

When aliquots of membranes from individual Renografin gradient fractions were incubated in the presence of either UDP[^{14}C]xylose or GDP[^{14}C]fucose, the incorporation peaks of both radioactive xylose and fucose into ethanol-insoluble, alkali-soluble products co-migrated with the heavier latent UDPase marker activity peak (fractions 6 and 7), i.e., with the dictyosome (D) population (Fig. 1). No significant fucosyl or xylosyltransferase activity was detected in the lighter latent UDPase region, i.e., the secretory vesicle (SV) population (fraction 2). All radioactivity was digested by cellulase to compounds possessing typical XG subunit gel chromatography patterns (not shown). The localization of the XG fucosyltransferase with the xylosyltransferase in dictyosomes, and its complete absence from SVs suggest that XG structure is completed within the Golgi stacks. These results were confirmed by *in vivo* pulse-chase experiments which indicated that fucosylated XG chains move sequentially from dictyosomes to SVs and then to the cell wall (Fig. 2). It is concluded that the complete xyloglucan molecule, including fucose, is formed in Golgi dictyosomes and that the secretory vesicles perform a transport rather than a synthetic function for this polysaccharide.

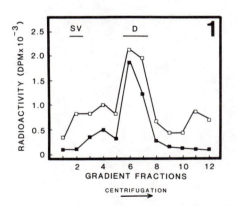

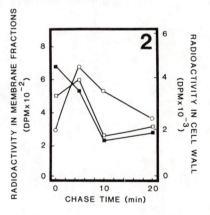

FIG. 1. Distribution of glycosyltransferase activities on Renografin gradients: label incorporated into ethanol-insoluble, alkali-soluble material from UDP[^{14}C]xylose (□) and GDP[^{14}C]fucose (■). Bars indicate location of secretory vesicles (SV) and dictyosome-enriched (D) fractions.

FIG. 2. Kinetics of secretion of fucosylated xyloglucan in pea stem slices. After an initial 30-min pulse with [^{3}H] fucose, the accumulated radioactivity was chased with 50 mM unlabeled fucose. Label was assayed in XG nonasaccharides isolated from dictyosomes (■), secretory vesicles (□), and in cell wall fractions (o).

LITERATURE CITED

1. CAMIRAND A, G MACLACHLAN 1986 Biosynthesis of the fucose-containing xyloglucan nonasaccharide by pea microsomal membranes. Plant Physiol 82: 379-383
2. HAYASHI T, G MACLACHLAN 1984 Pea xyloglucan and cellulose. I Macromolecular organization. Plant Physiol 75: 596-604
3. HAYASHI T, K MATSUDA 1981 Biosynthesis of xyloglucan in suspension-cultured soybean cells. Evidence that the enzyme system of xyloglucan does not contain β-1,4-glucan 4-β-D-glucosyltransferase activity (EC 2.4.1.12). Plant Cell Physiol 22: 1571-1584
4. HAYASHI T, YS WONG, G MACLACHLAN 1984 Pea xyloglucan and cellulose. II. Hydrolysis by pea endo-1,4-β-glucanases. Plant Physiol 75: 605-610
5. LABAVITCH JM, PM RAY 1974 Turnover of plant cell wall polysaccharides in elongating pea stem segments. Plant Physiol 53: 669-673
6. NISHITANI K, Y MASUDA 1983 Acid pH-induced structural changes in cell wall xyloglucans in *Vigna angularis* epicotyl segments. Plant Sci Lett 28: 87-94
7. RAY PM 1980 Cooperative action of β-glucan synthetase and UDP-xylose xylosyl transferase of Golgi membranes in the synthesis of xyloglucan-like polysaccharide. Biochim Biophys Acta 629: 431-444
8. TAIZ L, M MURRY, DG ROBINSON 1983 Identification of secretory vesicles in homogenates of pea stem segments. Planta 158: 534-539

Physiology of Cell Expansion During Plant Growth, *D.J. Cosgrove and D.P. Knievel* Eds.,
Copyright ©1987, The American Society of Plant Physiologists

PROPERTIES OF CELL WALL-ASSOCIATED ENZYMES IN THE BOUND AND FREE STATE

G. NAGAHASHI, S.-I. TU, AND P. M. BARNETT

USDA, ARS, Eastern Regional Research Center,
600 East Mermaid Lane,
Philadelphia, PA 19118

INTRODUCTION

Determination of enzymatic properties of cell wall-associated enzymes are usually performed on salt-solubilized or free enzymes in solution (2). Since limited studies have been reported on cell wall-bound enzymes *in situ* (1, 2, 3), we have investigated the *in situ* properties of two cell wall-associated enzymes. In one case (β-N-acetylglucosaminidase), the properties of a bound versus free enzyme were directly compared. In the second case (acid phosphatase), the effects of cations on the *in situ* enzyme activity was determined.

MATERIALS AND METHODS

Cell walls were isolated from Russet potato tubers and highly purified as described previously (4). Purified cell walls were either extracted with 3 M LiCl for 15 hr at 4°C or assayed without treatment. Salt extracted enzymes were concentrated in an Amicon ultrafiltration apparatus (PM 10 membrane) and dialyzed for 2 days with water containing 1 mM 2-mercaptoethanol. β–N-acetyl-glucosaminidase activity (β-GlcNAcase) and acid phosphatase activity (APase) were determined as described (5) except the APase assay was performed in 50 mM Na-acetate buffer at pH 4.5. When cell wall enzymes were assayed *in situ*, 0.1 to 0.5 mg dry weight cell wall were used per assay at 38°C and enzymatic reactions were stopped with 1 M Na_2CO_3. The assay tubes were centrifuged to pellet the cell walls and samples were read at 405 nm.

RESULTS AND DISCUSSION

Properties of bound versus free β–GlcNAcase. Approximately 30% of the total cellular β-GlcNAcase activity was associated with the cell wall. The cell wall-bound activity was almost completely solubilized (over 95%) with 3 M LiCl and therefore provided an ideal enzyme to study *in situ* (bound) versus the solubilized (free) condition. The bound enzyme had maximum activity at pH 5.0 while the free enzyme had maximum activity at pH 4.7. Most striking was the fact that the bound enzyme had little or no activity at pH 3.5 while the free enzyme at this pH had one half the maximum

265

activity observed. When kinetic studies were performed at pH 5.0, the bound enzyme had a Km value of 570 μM and the free enzyme had a Km value of 280 μM. These results indicated there are wall effects which influence enzyme activity.

The effects of cations on cell wall-associated acid phosphatase. Cell walls with low Ca^{2+} content were isolated from sycamore cell suspensions and shown to contain an APase with pronounced negative cooperative kinetics (1). In the presence of 1 mM $CaCl_2$, the negative cooperativity was drastically reduced and approached Michaelis-Menten kinetics. By increasing the ionic strength of the bulk phase, simple Michaelis-Menten kinetics was observed (1). Negative cooperativity was thought to be a result of the electrostatic repulsion of the negatively charged substrate from the polyanionic cell wall. Increased ionic strength suppressed the repulsion. Their theory on the ionic control of cell wall-associated enzymes relies on the existence of an electrostatic potential difference between the inside and outside of the cell wall which is regulated by Ca^{2+} (1). This theory has been recently extended to explain the control of cell wall expansion (5).

We observed negative cooperativity with potato tuber cell wall-associated APase; however, 1 mM $CaCl_2$ had no effect on this kinetic property. On the other hand, 1 mM $MgCl_2$ eliminated the negative cooperativity and generated Michaelis-Menten kinetics. This result indicated that the theory of enzymatic regulation by the electrostatic potential may not be sufficient to account for the specificity of Mg^{2+} over Ca^{2+} on the cell wall-APase of potato tubers. Further studies will be necessary to extend or modify the ideas presented by Ricard et al. (1, 5).

LITERATURE CITED

1. CRASNIER M, A-M MOUSTACAS, J RICARD 1985 Electrostatic effects and calcium ion concentration as modulators of acid phosphatase bound to plant cell walls. Eur J Biochem 151: 187-190
2. GOLDBERG R 1985 Cell-wall isolation, general growth aspects. *In* Modern Methods of Plant Analysis Vol. 1. Cell components. HF Linskens, JF Jackson, eds., Springer-Verlag, New York, pp 1-30
3. NAGAHASHI G, PM BARNETT, S-I TU, J BROUILLETTE 1986 Purification of primary cell walls from corn roots: Inhibition of cell wall-associated enzymes with indolizidine alkaloids. *In* Regulation of carbon and nitrogen reduction and utilization in maize. JC Shannon, DP Knievel, CB Boyer, eds, Waverly Press, Baltimore, pp 289-293
4. NAGAHASHI G, TS SEIBLES 1986 Purification of plant cell walls: Isoelectric focusing of $CaCl_2$ extracted enzymes. Protoplasma 134: 102-110
5. RICARD J, G NOAT 1986 Electrostatic effects and the dynamics of enzyme reactions at the surface of plant cells. I. A theory of the ionic control of a complex multi-enzyme system. Eur J Biochem 155: 183-190

Physiology of Cell Expansion During Plant Growth, *D.J. Cosgrove and D.P. Knievel* Eds.,
Copyright ©1987, The American Society of Plant Physiologists

AUTOLYSIS OF POTATO TUBER CELL WALLS

KEN SASAKI, GERALD NAGAHASHI, PAULINE BARNETT
AND LANDIS DONER

U.S.D.A. ARS Eastern Regional Research Center
600 East Mermaid Lane,
Philadelphia, PA 19118

INTRODUCTION

Cell wall autolysis has been studied in growing tissues of some monocots and dicots (2, 5, 6, 7, 9). Enzymes appear to be responsible for the cell wall autolysis in corn coleoptiles (3, 4) and pea epicotyls (1). Cell wall autolysis has also been studied in ripening tomato fruits (10). Studies of cell wall autolysis will not only help us to understand the physiology of cell walls, but also provide a tool for probing the chemical structure of cell walls.

We studied the autolysis of potato tuber cell walls to determine if the cell wall autolysis occurs in non-growing storage tissues. The optimum conditions for the autolysis were also investigated.

MATERIALS AND METHODS

Potato tubers (cv Russet) were purchased locally and the cell walls were prepared in HEPES-MES buffer (100 mM, pH 7.8) with or without 10 mM Na_2-EDTA, sodium acetate buffer (10 mM, pH 4.5), potassium phosphate buffer (10 mM, pH 6.7) or water using the Parr nitrogen bomb as the major cell disruption step. Cell walls were subsequently purified as described (8). The fresh cell wall preparations (approximately 10 mg dry weight) were incubated at 35°C in 2 ml of water, sodium acetate buffer (50 mM, pH 4.5 and 5), sodium citrate buffer (50 mM, pH 2.5 to 7) or potassium phosphate buffer (50 mM, pH 6 and 7). After centrifugation, the supernatant solutions were assayed for the total sugars by the phenol-sulfuric acid method. All incubation media contained 0.05% NaN_3.

RESULTS AND DISCUSSION

Pure cell walls prepared in HEPES-MES buffer containing EDTA were incubated in water at 35°C. After 20 h of incubation, the cell walls released as much as 10% of the cell wall dry weight as sugars (Fig. 1). Gel filtration chromatography using Sepharose 4B showed that the released sugars were polysaccharides with large molecular weights. Nearly half of the total sugars released was galacturonic acid, suggesting that large fragments of pectic polysaccharides were solubilized from the cell wall matrix during autolysis.

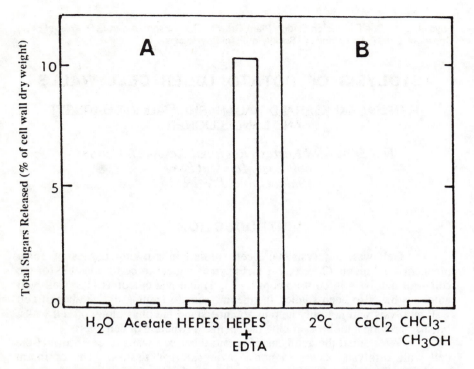

FIG. 1 (A) Effects of cell wall isolation media on the autolysis, and (B) effects of incubation conditions on the autolysis of cell walls isolated in HEPES-MES buffer containing EDTA.

Various incubation conditions prevented the autolytic process. At 2°C, only 0.2% of the cell wall dry weight was released after 24 h. Presence of 50 mM $CaCl_2$ in the incubation medium inhibited the cell wall autolysis and cell walls washed with $CHCl_3$-CH_3OH mixed solution (1:2) also showed very low autolysis (Fig. 1). At pH between 5 and 7 (50 mM sodium citrate buffer), autolysis was less than half of that in water. Water was a better incubation medium for the cell wall autolysis than the sodium citrate buffer (pH 2.5 to 7) and potassium phosphate buffer (pH 6 and 7), though the amount of sugars released in the buffer increased at lower pH than pH 5.

When the cell walls were prepared in water, sodium acetate buffer or HEPES-MES buffer without EDTA, virtually no autolysis was observed (Fig. 1). Incubation of those inactive cell walls in sodium acetate buffer (pH 4.5 and 5), sodium citrate buffer (pH 2.5 to 7) or potassium phosphate buffer (pH 6 and 7) did not greatly increase the autolysis. However, addition of 5 mM EDTA to the incubation buffers (sodium citrate and potassium phosphate) remarkably increased the sugar release, though the largest amount of the released sugars was only 40% of that in water from the cell walls prepare din HEPES-MES buffer with EDTA.

Results suggest that the potato cell wall autolysis may have been caused by EDTA chelation of Ca ions from the cell wall pectic polymers. The degree to which cell wall enzymes are involved in the autolytic process is yet to be determined.

LITERATURE CITED

1. DOPICO B, E LABRADOR, G NICOLAS 1986 Characterization and localization of the cell wall autolysis substrate in *Pisum sativum* epicotyls. Plant Sci 44: 155-161
2. HUBER DJ, DJ NEVINS 1979 Autolysis of the cell wall β-D-glucan in corn coleptiles. Plant Cell Physiol 201: 20-212
3. HUBER DJ, DJ NEVINS 1980 β–D-Glucan hydrolase activity in *Zea* coleptile cell walls. Plant Physiol 65: 768-773
4. HUBER DJ, DJ NEVINS 1980 Partial purification of endo- and exo-β-D-glucanase enzymes from *Zea mays* L. seedlings and their involvement in cell-wall autohydrolysis. Planta 151: 206-214
5. HUBER DJ, DJ NEVINS 1981 Wall-protein antibodies as inhibitors of growth and of autolytic reactions of isolated cell walls. Physiol Plant 53: 533-539
6. LABRADOR E, G NICOLAS 1982 Autolytic activities of the cell wall in rice coleptiles. Effects of nojirimycin. Physiol Plant 55: 345-350
7. LABRADOR E, G NICOLAS 1985 Autolysis of cell walls in pea epicotyls during growth. Enzymatic activities involved. Physiol Plant 64: 541-546
8. NAGAHASHI G, TS SEIBLES 1986 Purification of plant cell walls: Isoelectric focusing of $CaCl_2$ extracted enzymes. Protoplasma 134: 102-110
9. REVILLA G, MV SIERRA, I ZARRA 1986 Cell wall autolysis in *Cicer arietinum* L. epicotyls. J Plant Physiol 122: 147-157
10. RUSHING JW, DJ HUBER 1984 *In vitro* characterization of tomato fruit softening. The use of enzymatically active cell walls. Plant Physiol 75: 891-894

Physiology of Cell Expansion During Plant Growth, *D.J. Cosgrove and D.P. Knievel* Eds.,
Copyright ©1987, The American Society of Plant Physiologists

CHANGES IN WALL METABOLISM OF TOBACCO SUSPENSION-CULTURED CELLS UPON ADAPTATION TO OSMOTIC STRESS

NAIM MOHAMMAD IRAKI AND NICHOLAS C. CARPITA

*Botany and Plant Pathology Department, Purdue University,
West Lafayette, Indiana 47907*

INTRODUCTION

Cells in suspension culture adapted to osmotic stress mimicked by either polyethylene glycol or sodium chloride exhibit a reduced rate of cell expansion (6). Reduced expansion is due neither to an increase in the tensile strength of the wall (unpublished data) nor to inadequate turgor maintenance (6). Several workers have suggested that enzymes localized in the cell wall hydrolyze load-bearing glycosidic bonds of matrix polysaccharides causing wall loosening and cell expansion (8).

Wall autohydrolysis offers a convenient tool for studying enzyme and wall interrelations concerning wall-turnover events that are related to the inhibition of cell expansion. We examined changes in wall autolysis in normal tobacco cells and cells exhibiting reduced cell expansion due to adaptation to osmotic stress.

MATERIALS AND METHODS

Tobacco (*Nicotiana tabacum L.* cv. W38) cells were grown in liquid medium and were adapted to either 30% PEG or 2.5% NaCl in the medium as described in (1, 7). Cells at stationary phase of growth were suspended in 10 mM sodium phosphate, pH 6.0 containing 25 mM ascorbate and broken by release from a nitrogen pressure chamber. The broken cells were washed with 4 liters of 10 mM sodium phosphate buffer, pH 6, and with 500 ml of -20°C acetone on a coarse sintered glass funnel, followed by air drying for 10 to 15 min. Walls (3 mg dry wt./ml) were incubated at 27°C in 0.25 strength McIlvaine buffer at pH 7 for unadapted cells and pH 6 for both PEG- and NaCl-adapted cells. One drop of toluene was added to each incubated sample. Analysis of sugar composition by GLC was obtained by incubating walls in 10 mM sodium phosphate-citrate at the appropriate pH for 60 h at 27°C.

Chemical assays and sugar composition. Total sugar was determined by the phenol-sulfuric method (4) and uronic acids by the carbazol method (3) as modified by addition of sulfamate (5).

RESULTS AND DISCUSSION

Cell walls isolated from unadapted cells showed higher autolytic activity than walls isolated from either PEG-adapted or NaCl-adapted cells when incubated in the presence of 30% PEG or 2.5% NaCl, respectively (Table I).

Table I. *Amounts of total sugar and uronic acids released during 16 hours of autolysis of walls isolated from tobacco cells at stationary phase of growth.*

	Cell line					
Treatment	Unadapted		PEG-adapted		NaCl-adapted	
	T.sugar	U.acids	T.sugar	U.acids	T.sugar	U.acids
	µg glc (gal A) equiv./ mg cell wall					
Walls+Buffer	30	8	64	80	11	12
Boiled Walls+Buff.	1	0	22	27	14	19
Walls+PEG	Nd	Nd	9	0	Nd	Nd
Walls+NaCl	Nd	Nd	Nd	Nd	4	0

However, when the walls of the adapted cells were incubated in the absence of osmotica the amount of the released wall material was much higher than that released by the walls of the unadapted cells, but most of this material was released non-enzymically since heat inactivation of the walls did not decrease this process drastically (Table I). A large part of the material released by walls of adapted cells incubated in the absence of either PEG or NaCl is uronic acid-rich (Table I). The release of this material is blocked when the walls were incubated in the presence of osmotica (Table I). We found that the release of uronic acid-rich material into the medium of the adapted cells is almost totally blocked *in vivo*. Furthermore, walls of these cells contain a larger proportion of loosely bound pectin extracted by cold EDTA than do walls of the unadapted cells. These observations indicate that the metabolism of polyuronides is impaired upon reduction in cell expansion which is a consequence of adaptation to osmotic stress.

Analysis of sugar composition of polysaccharides released during wall autolysis (Table II) demonstrated that walls of adapted cells released mainly arabinose and galactose but less xylose and glucose compared to walls of unadapted cells. Preliminary results from linkage analysis of these

271

Table II. *Composition of sugars release during autolysis of walls isolated from tobacco cell cultures at stationary phase of growth.*

Culture	Rhm	Fuc	Ara	Xyl	Man	Glc	Gal
				mol%			
Unadapted cells	5	1	19	19	2	25	29
NaCl-adapted cells	8	1	27	12	6	21	25
PEG-adapted cells	6	3	24	14	4	31	18

polysaccharides showed that the majority of the xylosyl residues in all three lines of cells are terminal-, and 2-linked. Also, the glucosyl residues are mainly 4,6-linked. These xylosyl and glucosyl residues are diagnostic linkages for xyloglucans. These observations imply that during autolysis walls of adapted cells release mainly arabinogalactan but with a lower proportion of xyloglucan compared to walls of unadapted cells.

Our results suggest that PEG as well as NaCl inhibit both autolysis and non-enzymic release of uronic acid-rich material. In vivo, this effect causes the accumulation of polyuronides in the cell wall. Further, the solubilization of xyloglucan is reduced upon reduction in cell expansion.

LITERATURE CITED

1. BRESSAN RA, PM HASEGAWA, AK HANDA 1981 Resistance of cultured higher plant cells to polyethylene glycol-induced water stress. Plant Sci Lett 21: 23-30
2. CARPITA NC 1984 Cell wall development in maize coleoptiles. Plant Physiol 76: 205-212
3. DISCHE Z 1947 A new specific color reaction of hexuronic acids. J Biol Chem 167: 189-198
4. DUBOIS M, DA GILLES, JK HAMILTON, PA REBERS, F SMITH 1956 Colormetric method for determination of sugars and related substances. Anal Chem 28: 350-356
5. GALAMBOS JT 1967 The reaction of carbazole with carbohydrates. 1. Effect of borate and sulfamate on the carbazole reaction. Anal Biochem 19: 119-132
6. HANDA S, RA BRESSAN, AK HANDA, NC CARPITA, PM HASEGAWA Solutes contributing to osmotic adjustment in cultured plant cells adapted to water stress. Plant Physiol 73: 834-843
7. HASEGAWA PM, RA BRESSAN, AK HANDA 1980 Growth characteristics of NaCl selected and nonselected cells of *Nicotiana tabacum* L. Plant Cell Physiol 21: 1347-1355
8. LABAVITCH JM 1981 Cell wall turnover in plant development. Ann Rev Plant Physiol 32: 385-406

Physiology of Cell Expansion During Plant Growth, *D.J. Cosgrove and D.P. Knievel* Eds.,
Copyright ©1987, The American Society of Plant Physiologists

SOLUBILITY OF POLYSACCHARIDES SYNTHESIZED *IN VITRO* BY GOLGI APPARATUS AND PLASMA MEMBRANE OF *ZEA MAYS* COLEOPTILES

DAVID M. GIBEAUT AND NICHOLAS C. CARPITA

Botany and Plant Pathology Department, Purdue University, West Lafayette IN 47907

INTRODUCTION

Synthesis and accumulation of glucuronoarabinoxylan (GAX) and xyloglucan (XG) begin in embryonal tissues of maize coleoptiles and continue throughout cell expansion. Whereas, synthesis and accumulation of β-glucan and a highly substituted GAX occur only during rapid cell elongation (2). Previous work has demonstrated that the Golgi apparatus is the principle site of non-cellulosic polysaccharide synthesis (3). Our objective was to separate the Golgi apparatus from plasma membrane. The plasma membrane produces large amounts of callose, which complicates the identification and quantitation of polymer synthesis associated with the Golgi apparatus. Furthermore, differential solubilization coupled with chemical and enzymic analysis can then be utilized to fully characterize linkage structure. Initially we have used the solubility in water, potassium hydroxide and acetic nitric reagent to separate polymers after *in vitro* labelling with UDP-^{14}C- glucose.

MATERIALS AND METHODS

Seeds of maize (*Zea mays* L. cv LH 74 x LH 51) were soaked with aerated deionized water for 24 h at 25°C, sown in vermiculite saturated with deionized water, and incubated in darkness at 28° C for 2 to 3 d. The coleoptiles were excised from the seedling, and the etiolated leaves were removed.

Preparation of cellular membranes. All preparation was done at about 4°C. Coleoptiles were gently mashed with a mortar and pestle in 50mM HEPES-KOH (pH 7.6), 14% sucrose (w/v), 0.1% ovalbumin, 14 mM β-mercaptoethanol, 10 mM KCl, 10 mM ascorbic acid, 3 mM EDTA, and 3 mM EGTA. All gradient buffers contained 50 mM HEPES-KOH (pH 7.6), 14 mM β-mercaptoethanol, 0.5 mM EDTA, 0.5 mM EGTA, and the appropriate concentration of sucrose (w/v). The homogenate was filtered through nylon screen then centrifuged at 800 x g for 10 min. The supernatant was layered upon a 60% to 20% sucrose step gradient and centrifuged at 125,000 x g for 30 min. The mat of membranes upon the 60% sucrose step was resuspended with 8.6% sucrose. The resuspended membranes were layered on a 20 to 50% sucrose linear gradient and centrifuged at 130,000 x g for 90 min. Fractions of 1.2 ml were collected and monitored by absorbance at 254 nm.

Membrane markers. The following markers and methods were used: total protein (BioRad), endoplasmic reticulum - NADPH cytochrome c reductase (7), mitochondria - cytochrome c oxidase (5), Golgi apparatus - total detergent soluble IDPase (7), and plasma membrane - specific binding of napthylpthalamic acid (NPA) (1).

Glucan synthetase reactions. Four neighboring fractions were pooled for the NPA binding and glucan synthetase reactions. Glucan synthetase I (GS I) and II (GS II) reactions were done at 25°C with 0.2 ml of membrane and were initiated with 0.1 ml of reaction buffer. GS I reaction buffers contained 30 µM UDP-glucose 50 mM HEPES-KOH (pH 7.0), and 5 mM $MnCl_2$. GS II reaction buffers with or without 5 mM $CaCl_2$ contained 50 mM HEPES-KOH (pH 7.0) and 1 mM UDP-glucose. In addition, each reaction mixture contained UDP-^{14}C-glucose (25 or 50 nCi). GS I reaction mixtures were incubated for 15 min, GS II for 30 min. Reactions were stopped by adding 1.2 ml ethanol containing 3 mg/ml non-ionic cellulose and boiled for 5 min then chilled to -20°C. Unreacted substrate was removed by washing three times with hot 80% ethanol and chilling between washes. Lipids were removed by washing three times with chloroform : methanol (1:2,v/v) at room temperature.

Solubilization of products. Solubilization of the reaction products was done at room temperature, first with three repetitions of distilled water followed by three repetitions of 4 N KOH. The KOH extracts were neutralized with acetic acid. Solubilization by acetic nitric reagent (A/N) was done once with boiling for 1 h. Each repetition of an extraction was pooled, and dried before liquid scintillation spectroscopy.

RESULTS AND DISCUSSION

GS I activity is greatest at low UDP-glc concentration (1-10 µM) and is associated with the Golgi apparatus; whereas, GS II activity is greatest at higher UDP-glc concentration (1 mM) and is associated with the plasma membrane (5). In our gradients, the Golgi apparatus and plasma membrane were well separated as indicated by IDPase, NPA and GS activities. GS I activity was three to ten fold greater in the Golgi apparatus than plasma membrane fractions. GS II activity was three to five fold greater in the plasma membrane than Golgi apparatus fractions and was stimulated by $CaCl_2$ three to seven fold in all membrane fractions.

Polymer size, sugar composition and linkage structure may have an effect on the solubility of a polysaccharide. Small polymers, generally less than ten sugar units long, are water soluble. At GS I reaction conditions, the water soluble product was greater in the plasma membrane than Golgi apparatus fractions. Polymers larger than ten sugar units may require KOH to be soluble. The KOH soluble product was the greatest proportion of product for each reaction and membrane fraction. Polymers requiring A/N reagent for solubilization may be tightly bound to cellulose or be even larger than those requiring KOH for solubilization. A/N soluble product was a substantial proportion of the total product from GS I reaction conditions, especially in the Golgi apparatus fractions. Little A/N soluble product was formed at GS II reaction conditions regardless of the membrane fraction. No product was left in the cellulose pellets after A/N solubilization.

For good structural analysis of the polymers synthesised by Golgi apparatus, cell fractionation techniques must separate Golgi apparatus from plasma membrane as completely as possible, but some overlap of membranes and GS activities will still exist. By chelating Ca^{2+} in the grinding and gradient buffers, GS II activity can be suppressed. Furthermore, synthesis of products insoluble in 4N KOH but soluble in A/N may be a characteristic function of the Golgi apparatus not found in the plasma membrane.

LITERATURE CITED

1. BRUMMER B, RW PARISH 1983 Identification of specific proteins and glycoproteins associated with membrane fractions isolated from *Zea mays* L. coleoptiles. Planta 157: 446-453
2. CARPITA NC 1984 Cell wall development in maize coleoptiles. Plant Physiol 76: 205-212
3. FINCHER GB, BA STONE 1981 Metabolism of noncellulosic polysaccharides. *In* Encyclopedia of Plant Physiology New series, Vol 13B. Tanner W, FA Loewus eds. pp 68-132. Springer Verlag
4. HENRY RJ, BA STONE 1982 Factors influencing β-glucan synthesis by particulate enzymes from suspension-cultured *Lolium multiflorum* endosperm cells. Plant Physiol 69: 632-636
5. HODGES TK, RT LEONARD 1974 Purification of a plasmamembrane-bound adenosine triphosphatase from plant roots. Methods Enzymol 32: 392-406
6. RAY PM 1979 Maize coleoptile cellular membranes bearing different types of glucan synthetase activity. *In* Methodological Surveys in Biochemistry. E Reid, ed. pp 135-146. Horwood Chichester
7. WEINECKE K, KE SCNEPF, DG ROBINSON 1982 Organelles involved in the synthesis and transport of hydroxyproline-containing glycoproteins in carrot root discs. Planta 155: 58-63

Physiology of Cell Expansion During Plant Growth, *D.J. Cosgrove and D.P. Knievel* Eds.,
Copyright ©1987, The American Society of Plant Physiologists

CELL WALL FORMATION BY ISOLATED PROTOPLASTS AND PLASMOLYZED CELLS FROM CARROT SUSPENSION CULTURES

E. M. SHEA AND N.C. CARPITA

*Botany and Plant Pathology Department, Purdue University,
West Lafayette, IN 47907*

INTRODUCTION

Protoplasts are a convenient model system for the study of wall formation, especially for studies on the nature and assembly of wall components and their changes as the wall develops. Previous studies with cultured cells have demonstrated that incipient wall material produced by the protoplast is different from the cell walls of the parent culture (1).

We have studied wall synthesis in plasmolyzed, intact cells as an alternative to studying wall regeneration within protoplasts because cell wall degrading enzymes may be contaminated with proteolytic enzymes which could damage the plasma membrane. Other researchers (3) have shown microscopically that a new wall forms around the plasmolyzed cell within the pre-existing wall but few studies have been done on the chemical nature of the new wall. The wall formed by a plasmolyzed cell was compared to the wall produced by protoplasts as well as the parent culture.

MATERIALS AND METHODS

Plant material. Suspension cultures of carrot cells (*Daucus carota* L. cv. Danvers) were routinely grown on M & S salts supplemented with 0.1 g/l inositol, 2.0 mg/l glycine, 0.5 mg/l pyridoxine, 0.5 mg/l thiamine-HCl, 0.4 mg/l 2,4-D and 31.5 g/l glucose (pH 5.0). Cells were subcultured weekly and inoculated at 8 mg fresh weight/ml.

Protoplast isolation. Protoplasts were isolated by a modification of the method of Klein *et al.* (2). Cells were incubated for 1 hour in the presence of 0.5%/0.1% (w/v) cellulase (Worthington), 0.1 % (w/v) pectinase (Worthington) in wall-regeneration medium (carrot medium containing 10mM glucose and 450 mM sorbitol). The protoplasts were separated from broken cells and debris by flotation on a ficoll gradient. The protoplasts were washed 3 times, collected by centrifugation and resuspended to a final concentration of 2 $\times 10^6$ protoplasts per ml in wall-regeneration medium.

Radiolabeling of regenerating walls of protoplasts. Two ml of the protoplast suspension were transferred to petri dishes (6 cm dia., 1.5 cm high) and incubated without shaking at room temperature in the light. One hundred μCi of U-^{14}C-glucose was added to each of four dishes. After three hours the contents of each dish were transferred to a conical centrifuge tube. Two of the tubes were centrifuged (3 min, 120g) and the supernatants were transferred to new tubes. Four volumes of ethanol were added to each tube and 20 mg of cellulose was added as a carrier. The tubes were mixed thoroughly and stored at -20°C. Two unlabeled dishes were sampled for determination of cell viability. Other dishes were labeled for three hours with radioactive glucose 24, 48, and 72 h after transfer. The cells and media were harvested and treated as above.

The ethanol precipitable material was collected by centrifugation (10 min, 3000g), washed three times with ice cold 80% ethanol, suspended in water and freeze-dried. Each dried wall fraction was hydrolyzed in 2N TFA, reduced with NaBH$_4$ and acetylated. The alditol acetates were separated by gas chromatography on a Hewlett-Packard 5840A.

Radiolabeling of walls of plasmolyzed cells. One gram of cells was transferred to 25 ml of medium containing 0 mM or 800 mM sorbitol. Twenty μCi of U-^{14}C-glucose were added to two flasks of each sorbitol concentration. After three hours the cultures were filtered, and the cells were frozen in liquid nitrogen. Other cultures were labeled for three hours with radioactive glucose 24, 48, and 72 h after transfer.

The cells were homogenized in ice cold 50 mM TES buffer containing 10 mM ascorbic acid, pH 7.2. Wall material was collected by centrifugation, and the pellet was washed once with ice cold 0.5 M potassium phosphate buffer, pH 7.0, twice with water, and once with methanol. The pellet was suspended in chloroform: methanol (1:1) and heated at 40C for 30 min. The pellet was washed once with methanol and once with water. It was resuspended in water and freeze-dried.

The dried wall material was suspended in DMSO and stirred overnight at room temperature. The supernatant was dialyzed overnight against deionized water, frozen and lyophilized. The pellet was suspended in 0.5% ammonium oxalate and heated in a boiling water bath for 60 min. The supernatant was dialyzed overnight against deionized water, frozen and lyophilized.

The cell wall pellet was divided in half and one half was suspended in acetic-nitric reagent [water: glacial acetic acid: nitric acid (5: 20: 2.5)] and heated in a boiling water bath for 90 min. The supernatant was assayed for radioactivity. The remaining cellulose was washed three times with water before being assayed for radioactivity.

The second half of the cell wall pellet was extracted sequentially with 0.1N KOH, 1N KOH, and 4N KOH each containing 3 mg/ml NaBH$_4$. Each extraction was carried out under N$_2$ for 55 min. at room temperature. The KOH supernatants were dialyzed against deionized water and freeze-dried.

RESULTS AND DISCUSSION

In the three hours immediately following protoplast isolation, 45% of the sugar incorporated into wall material was in the medium and 54% was associated with the pellet. After three days, only 4% of the newly synthesized material was in the medium and 96% was incorporated into the protoplast wall (Fig. 1). This is consistent with the observation that wall formation proceeds in a patchwork manner; as the wall matures more of the material synthesized by the cell is incorporated into the wall and less is lost into the medium.

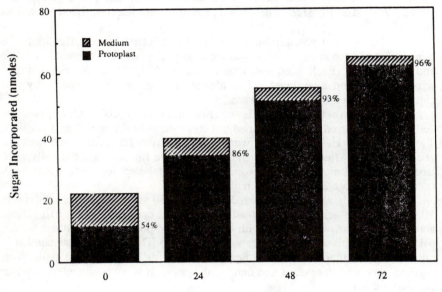

FIG. 1. Distribution of material synthesized by protoplasts during a 3 h period 0, 24, 48, or 72 hours after isolation. After three hours incubation with 100 μCi of U-^{14}C-glucose, the protoplasts were collected by centrifugation and burst in 80% ethanol. Polymers were precipitated from the medium with 80% ethanol.

When "plasmoplasts" were labeled immediately after transfer to hypertonic medium, 42% of the labeled sugar was in material extracted with DMSO. After three days, only 18% of the newly synthesized material was in DMSO extractable material. The greatest increase was in the material that could only be removed by digestion with acetic-nitric acid (Fig. 2). When plasmolyzed cells form a new wall around the protoplast within the pre-existing wall, polymers do not escape into the medium. Instead, they are trapped in the space between the plasma membrane and the pre-existing wall. Polymers that are not incorporated into a wall or are only loosely bound, are removed from the wall pellet in the first extract (DMSO).

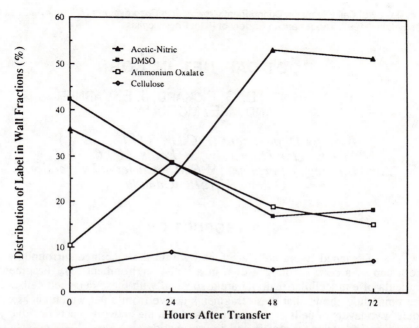

FIG. 2. Distribution of label in walls of carrot cells plasmolyzed in 800 mM sorbitol. Cells were incubated with 20 μCi U-^{14}C-glucose 1, 24, 48, or 72 h after transfer to hypertonic medium. The cells were homogenized, the walls were washed and then extracted sequentially with DMSO and hot ammonium oxalate. The remaining, non-cellulosic material was digested with acetic-nitric reagent.

LITERATURE CITED

1. BLASCHEK W, D HASS, H KOEHLER, G FRANZ 1981 Cell wall regeneration by *Nicotiana tabacum* protoplasts: chemical and biochemical aspects. Plant Sci Lett 22: 47-57
2. KLEIN AS, D MONTEZINOS, DP DELMER 1981 Cellulose and 1,3-glucan synthesis during the early stages of wall regeneration in soybean protoplasts. Planta 152: 105-114
3. ROBINSON DG, WR CUMMINS 1976 Golgi secretion in plasmolyzed *Pisum sativum* L. Protoplasma 90: 369-379

279

ROOT-CAP NET IN CORN

F. C. GUINEL, B. G. PICKARD, J. E. VARNER
AND M. E. McCULLY

Biology Department (F.C.G., B.G.P., J.E.V.),
Washington University, St. Louis, MO 63130,
and Biology Department (M.E.M.), Carleton University,
Ottawa K1S 5B6, Canada

INTRODUCTION

Detached living cells are found within the mucilage surrounding the root cap of a corn seedling grown in a moist environment (3). Fragmented reticula of extracellular material are associated with these detached cells. The current study shows that these fragments derive from a net which encases the outer two layers of cells of the root cap, and that the network reacts positively to histochemical stains for protein and polysaccharide.

MATERIALS AND METHODS

Plants. Corn (*Zea mays* L.) seeds were grown axenically on 1% agar (see 3).

Light microscopy. Four day-old seedlings were immersed in water for 3-4 min. They were removed gently to preserve the integrity of the drop of swollen mucilage at the root cap. Some of the drops were then wiped off with a brush onto a slide, mounted in distilled water, and viewed with phase-contrast or Nomarski optics. Other drops were smeared onto coverslips coated with cytochrome c, treated with various stains (see Table I), well rinsed, mounted upside down in a drop of water, and viewed with bright-field optics.

Electron microscopy. For TEM, pieces of root-tip were freeze-substituted, sectioned and stained as described by Guinel and McCully (2). For SEM, pieces of root-tip and drops of mucilage on cytochrome c-coated coverslips were fixed with 3% glutaraldehyde in phosphate buffer, and dehydrated in an ethanol series. They were then critical-point dried and coated with gold.

RESULTS AND DISCUSSION

When cells recently detached from the corn root-cap are examined in whole mounts, they are typically seen to be attached to fragments of a geometrical reticulum which is particularly dense in phase microscopy (Fig. 1). The fragments associated with the more-or-less spherical cells detaching at the tip of the cap tend to be relatively large, often exhibiting a series of interconnected squares and hexagons or of pentagons and heptagons, whereas the

Table I. *Histochemical Properties of the Net Fragments.*

Test (Reference)	Substrate	Result
Toluidine blue, pH 4.4 (7)	carboxylated polysaccharides	+
Alcian blue, pH 2.5 (8)	carboxylated polysaccharides	+
Alcian blue, pH 0.5 (8)	sulfated polysaccharides	+
Periodic acid-Schiff's (7)	vicinal hydroxyl groups	+++
Wheat germ agglutinin (1)	N-acetylglucosamine residues	++
Ulex europaeus lectin (1)	L-fucose residues	++
Calcofluor (7)	β-linked glucans	++
Schiff's reagent (4)	endogenous aldehyde groups	-
Phloroglucinol-HCl (7)	lignin	-
Autofluorescence (7)	phenolics	++
Fast-green (4)	basic proteins	-
Silver stain (6)	proteins	+
Amido black (4)	proteins	+
Coomassie blue (4)	proteins	+
Bicinchoninic acid (10)	proteins	+

fragments associated with elongate cells detaching from the flanks of the cap are usually small, consisting of a few joined rod-like elements which remain in close contact with the cell walls.

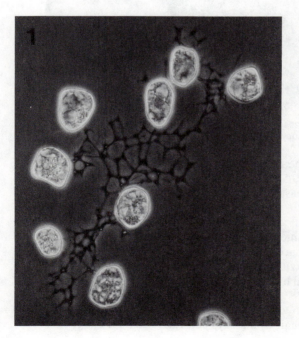

FIG. 1. A group of living axial cells suspended in root-cap mucilage. Associated with these cells is a fragment of a three-dimensional net which remains from the break-up of the middle lamella. Phase-contrast optics. (x 160)

The source of the fragments can be seen both in sections through the root cap (not shown) and in scanning electron micrographs of whole root tips (Fig. 2). The surface cells are held together by a net formed in the middle lamella where the faces of the tetrakaidecahedral cells meet; apparently, extra, reinforcing material is deposited at all these junctional edges and corners. Release of the cells into the rhizosphere is effected by partial hydrolysis of the interfacial sheets of middle lamellar material and by partial dismemberment of the net at and near the outer surface of the cap. Contrary to earlier reports (5, 9) the cell walls show no evidence of deterioration.

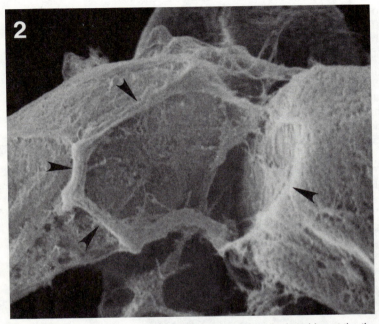

FIG. 2. Detail of net fragments associated with the spherical cells detaching at the tip of the cap. Scanning electron microscopy. (x 2300)

Susceptibility of the net, as well as its fragments (Table I), to staining with, e.g., Alcian blue (pH 0.5) has permitted determination that the net is about two cell layers thick (at least under the specified conditions of growth), and surrounds the entire cap.

Surveying the histochemical responses of fragments of reticulum associated with whole-mounted cells, it appears likely that the net may contain both protein and carbohydrate components: four reagents commonly accepted as specific for protein react with the fragments, and a test for the presence of polysaccharides with vicinal hydroxyl groups and one for β-glucans are strongly positive.

The fragments also test positive for N-acetylglucosamine residues and for L-fucose residues, although with assays considered to have lower specificity than the others.

The fragments meet two tests for the presence of carboxylated polysaccharides. Well-washed fragments also test positive with Alcian blue (pH 0.5), a stain used to test for sulfated polysaccharides (and indeed root-cap micilage gives also a positive test). Autofluorescence of fragments suggests the presence of aromatic rings.

The positive test for sulfated polysaccharides is of particular interest. Both analysis for X-ray scattering by sulfur atoms in thin sections and autoradiography of $^{35}SO_4^{2-}$ treated roots and their detached cells are being undertaken to extend the evidence. Because we are unaware of any previous report of sulfated polysaccharides in higher plants it will be important to confirm sulfation in net and mucilage, or to rule it out by extraction and chemical characterization of the polysaccharide. Because the net holds together the root cap and maintains the columella and meristem under compression, it is clear that it plays a structural role.

LITERATURE CITED

1. GUINEL FC, ME McCULLY 1985 Evaluation of the specificity of lectin binding to sections of plant tissue. Histochem 83: 265-277
2. GUINEL FC, ME McCULLY 1986 Some water-related physical properties of maize root-cap mucilage. Plant, Cell and Environment 9: 657-666
3. GUINEL FC, ME McCULLY 1987 The cells shed by the root cap of *Zea*: their origin and some structural and physiological properties. Plant, Cell and Environment. 10:565-578
4. JENSEN WA 1962 Botanical histochemistry. Principles and practice. Ed: Freeman WH and Co., San Francisco
5. JUNIPER BE, RM ROBERTS 1966 Polysaccharide synthesis and the fine structure of root cells. J Roy Micr Soc 85: 63-72
6. MORRISSEY JH 1981 Silver stain for proteins in polyacrylamide gels: a modified procedure with enhanced uniform sensitivity. Anal Biochem 117: 307-310
7. O'BRIEN TP, ME McCULLY 1981 The study of plant structure. Principles and selected methods. Termarcarphi Pty Ltd, Melbourne
8. PARKER BC, AG DIBOLL 1966 Alcian stains for histochemical localization of acid and sulfated polysaccharides in algae. Phycologia 6: 37-46
9. PAULL RE, RL JONES 1976 Studies on the secretion of maize root-cap slime. V. The cell wall as a barrier to secretion. Z Pflanzenphysiol 79: 154-164
10. SMITH PK, RI KROHN, GT HERMANSON, AK MALLIA, FH GARTNER, MD PROVENZANO, EK FUJIMOTO, NM GOEKE, BJ OLSON, DC KLENK 1985 Measurement of protein using bicinchoninic acid. Anal Biochem 150: 76-85

β-GLUCAN SYNTHASE FROM RED BEET ROOT: CHAPS SOLUBILIZATION, FUNCTIONAL RECONSTITUTION AND GROUP SPECIFIC CHEMICAL MODIFICATION

THERESA L. MASON, MARGARET E. SLOAN,
PANAYOTIS RODIS, AND BRUCE P. WASSERMAN

*Department of Food Science, New Jersey
Agricultural Experiment Station,
Cook College, Rutgers University,
New Brunswick, NJ 08903*

INTRODUCTION

UDP-Glucose:(1,3)-β-D-Glucan Synthase (GS) is a plasma membrane-bound enzyme which participates in the synthesis of ß-(1,3) glucan (callose) and possibly cellulose in the expanding cell wall (3, 4, 6, 9). It has been extensively characterized *in vitro* but has not yet been purified to the point where subunits have been definitively identified. Isolation of functional GS complexes might enable the production of genetic and immunological probes which could be used to study the regulation of cell wall biosynthetic enzymes during plant growth and development. A goal of our laboratory is to purify GS from red beet root and identify its protein subunits (5). This abstract presents findings from two approaches pursued to achieve this goal: 1. Design of a non-digitonin based solubilization and purification protocol; and 2. Group-specific chemical modification.

MATERIALS AND METHODS

Membrane isolation, enzyme assays and protein determinations were conducted as previously described (1, 5, 10). 3-[3-Cholamidopropyl)-dimethylammonio]-1-propane sulfonate (CHAPS) solubilization and functional reconstitution is diagrammed in Figure 1. Buffers contained 50 mM Tris, pH 7.5.

RESULTS AND DISCUSSION

CHAPS solubilization and functional reconstitution. The detergent digitonin has been widely used for GS solubilization, however its low critical micelle concentration (CMC) has made complete purification difficult (5). Therefore, alternate detergents with higher CMC values were sought. One detergent whose use seemed applicable was CHAPS, a zwitterionic derivative of cholic acid (7). The hydrophobic cholic acid portion of CHAPS bears structural

1. **MICROSOMES**

   ↓ Gradient Centrifugation

2. **PLASMA MEMBRANES**

   ↓ 0.3% CHAPS, 5 mM Mg^{2+}
      80,000g, 20 min

3. **GLUCAN SYNTHASE ENRICHED PELLET**

   ↓ 0.6% CHAPS, 1 mM EDTA, EGTA
      80,000g, 20 min

4. **CHAPS SOLUBILIZED GLUCAN SYNTHASE**

   ↓ Sepharose 4B Gel Filtration,
      0.1% CHAPS, 0.1 mg/ml asolectin

5. **FUNCTIONALLY RECONSTITUTED GLUCAN SYNTHASE**

FIG. 1. Protocol for sequential CHAPS solubilization and functional reconsitution procedure.

similarity to the digitogenin portion of digitonin. Advantages of CHAPS include a relatively high CMC (0.6% vs. 0.02%), dialyzability, and a much lower micellar molecular weight (6,500 kD vs. 70,000 kD).

The procedure outlined in Figure 1 gives optimal conditions for the solubilization and partial purification of GS using CHAPS. Specific activities of the various fractions are summarized in Table I. A plasma membrane enriched fraction was prepared and solubilized by a sequential procedure. First, membranes were exposed to 0.3% CHAPS in the presence of Mg^{2+}. This step solubilized little activity, but approximately 50% of the protein. Mg^{2+} was critical for maintaining GS in an insoluble form. Solubilization was then achieved with CHAPS in the presence of EDTA and EGTA. It was necessary to conduct assays of the CHAPS-solubilized GS (CSGS) in the presence of 0.01% digitonin.

CSGS was then further purified by passage through a Sepharose 4B gel filtration column where GS eluted in the void volume. To recover activity, the enzyme must be functionally reconstituted by addition of phospholipid to elution buffers. When phospholipid and CHAPS were omitted only 7% of the activity was recovered. Complete activity recovery was obtained with elution buffers containing 0.1 mg/ml asolectin and 0.1% CHAPS. These results confirm GS's requirement for phospholipid (10).

Overall, this procedure increased the specific activity of GS 21-fold relative to microsomes. The protein subunit composition of this preparation is currently under investigation.

Table I. *Sequential Solubilization and functional reconstitution of plasma membrane glucan synthase.*

Fraction	Activity	Protein	Specific Activity	Purification	Yield
	(Units)	*(mg)*	*(U/mg)*	*(fold)*	*(%)*
1. Microsomes	295	5.45	54	-	100
2. Plasma Membranes	155	1.32	117	2.2	53
3. Enriched Pellet	134	.71	189	3.5	46
4. CHAPS Solubilized	117	.15	782	14.5	39
5. Functionally Reconstituted[a]	62	.055	1127	20.8	21

[a]Sum of activity and protein measurements in the two peak column fractions. The total activity recovered in all fractions was 106 U representing 91% of the solubilized activity layered on the column.

Group-specific chemical modification. Chemical modification by group-specific chemical probes is used to identify those subunits of an enzyme complex that may be involved in catalysis (2, 8). The probes function by covalently binding to specific amino acid residues. The first step is to identify amino acid residues involved in enzymatic activity by observing inactivation. Protection against inactivation by substrate and/or effectors allows subunits to be identified by electrophoretic means (2).

Four group specific reagents were tested: phenylglyoxal (PGO, arginine), N-ethylmaleimide (NEM, cysteine), formaldehyde with $NaBH_3CN$ (HCHO, lysine) and n-acetylimidazole (NAI, tyrosine). Three of these, PGO, NEM and HCHO, were inhibitory (Fig. 2). This indicates that arginine, cysteine, and lysine residues, respectively, are important for GS activity. Intact plasma membranes and CSGS were equally susceptible to inactivation.

Protection against inactivation by incubating the inhibitory reagents in the presence of substrate and cofactors was not obtained. These results suggest that arginine, cysteine, and lysine residues affected are not part of the substrate binding site or that the substrate binding site is inaccessible to these reagents.

Acknowledgments--This research was supported by NSF Grant DMB 85-02523. N.J. Agricultural Experiment Station, Publication F-10109-1-87.

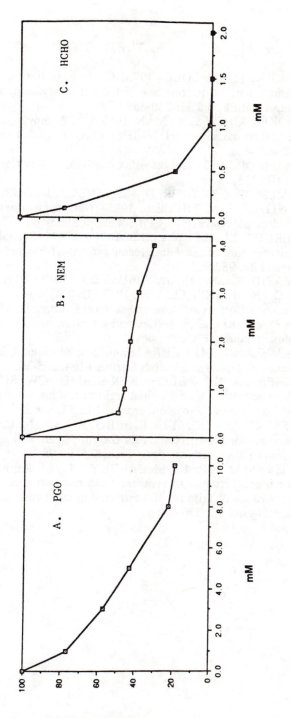

FIG. 2. Inactivation of sequentially solubilized glucan synthase by: A. PGO, B. NEM and C. HCHO. Pre-incubations were conducted in 50 mM HEPES as follows: PGO, 30°C, 15 min, pH 7.5; NEM, 0°C, 30 min, pH 8.0; HCHO + 10 mM NaBH$_3$CN, 23°C, 90 min, pH 7.5.

LITERATURE CITED

1. BRISKIN DP, RJ POOLE 1983 Characterization of a K^+-stimulated adenosine triphosphatase associated with the plasma membrane of red beet. Plant Physiol 71: 350-355

2. BROOKER RJ, CW SLAYMAN 1983 [^{14}C]N-ethylmaleimide labeling of the plasma membrane [H^+]-ATPase of *Neurospora crassa*. J Biol Chem 258: 222-226

3. DELMER DP 1983 Biosynthesis of cellulose. Adv Carb Chem Biochem 41: 105-153

4. DELMER DP, G COOPER, D ALEXANDER, J COOPER, T HAYASHI, C NITSCHE, M THELEN 1985 New approaches to the study of cellulose biosynthesis. J Cell Sci Suppl 2: 33-50

5. EIBERGER LL, BP WASSERMAN 1987 Partial purification of digitonin-solubilized beta-glucan synthase from red beet root. Plant Physiol 83: 982-987

6. HAYASHI T, SM READ, J BUSSELL, M THELEN, FC LIN, RM BROWN, Jr., DP DELMER 1987 UDP-Glucose: (1,3)-beta-glucan synthases from mung bean and cotton. Differential effects of Ca^{2+} and Mg^{2+} on enzyme properties and on macromolecular structure of the glucan product. Plant Physiol 83: 1054-1062

7. HJELMELAND LM, A CHRAMBACH 1984 Solubilization of functional membrane proteins. Methods Enzymol 104: 305-308

8. KASHER JS, KE ALLEN, K KASAMO, CW SLAYMAN 1986 Characterization of an essential arginine residue in the plasma membrane [H^+]-ATPase of *Neurospora crassa*. J Biol Chem 261: 10808-10813

9. WASSERMAN BP, LL EIBERGER, KJ McCARTHY 1986 Biotechnological approaches for controlled cell wall glucan biosynthesis in fruits and vegetables. Food Tech 40,5: 90-98

10. WASSERMAN BP, KJ McCARTHY 1986 Regulation of plasma membrane beta-glucan synthase from red beet root by phospholipids. Reactivation of Triton X-100 extracted glucan synthase by phospholipids. Plant Physiol 82: 396-400

Physiology of Cell Expansion During Plant Growth, *D.J. Cosgrove and D.P. Knievel* Eds.,
Copyright ©1987, The American Society of Plant Physiologists

PRODUCTION OF MONOCLONAL ANTIBODIES TO CELL WALL PEROXIDASE AND CELLULASE

SUNG-HA KIM, M. E. TERRY AND S. J. ROUX

*Dept. of Botany, Univ. of Texas at Austin
Austin, TX 78713*

INTRODUCTION

Among the numerous enzymes that have been reported in the cell wall, peroxidases (EC1, 11, 1, 7) are of importance in a variety of wall functions and wall peroxidase isoenzymes have been identified and partially purified from many different plant families. The possibility that the enzyme peroxidase is involved in the control of cell elongation has been proposed because there is an inverse relation between wall peroxidase levels and the growth rate. Another wall enzyme potentially involved in the regulation of wall growth is cellulase (EC 3.2.1.4). This enzyme refers to a group of enzymes which, acting together, hydrolyze cellulose.

The availability of monoclonal antibodies against them would greatly aid efforts to quantitate and biochemically characterize wall peroxidase and cellulase. Here we describe the production and characterization of a library of monoclonal antibodies against cell wall proteins, including antibodies against a wall peroxidase and a wall cellulase.

MATERIALS AND METHODS

Coleoptile sections (1.5 cm) were excised from 3 mm below the coleoptile tips of 7-day-old etiolated corn seedlings. Wall proteins were isolated from these coleoptiles by a slight modification of the procedure of Terry and Bonner (4). Immunization, hybridoma screening and cloning procedures employed were basically the same as those described in Silberman et al. (3).

RESULTS AND DISCUSSION

Monoclonal antibodies have been produced against a preparation of soluble wall proteins from the coleoptiles of etiolated corn seedlings. Among 81 hybridomas cloned, 22 that secreted the highest titers of antibody against a wall antigen, as estimated by both ELISA and Western blot analyses, were selected for subcloning. All 22 monoclonal antibodies were demonstrated to be directed to some antigenic site on various cell wall antigens. Our Western blot analysis combined with minislab gel electrophoresis to screen either hybridoma supernatants or ascitic fluids has proven to be a rapid and useful way of determining the protein specificities of monoclonal antibodies. Subcloning yielded two monoclonal antibody preparations, designated mWP3 and mWP19, which both effectively immunoprecipitated peroxidase activity from crude and partially purified preparations of wall peroxidases. When mWP3 and mWP19 were immobilized on nitrocellulose membrane and overlaid with crude preparations of wall proteins in a dot blot assay, they both bound enzymes with peroxidase activity. The molecular weight of the wall peroxidase recognized by these antibodies was estimated to be 96,000 by Western blot analysis (Fig. 1).

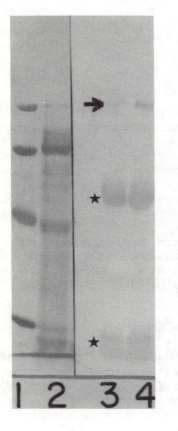

FIG. 1. Western blot analysis (1,2) of immunoprecipitate pellets of mWP3. The lanes were as follows: [1] Molecular weight markers stained with 0.1% amido black (94,000, 67,000, 43,000 and 30,000 daltons from top to botton); [2] Total extracted wall proteins stained with 0.1% amido black; [3] Protein precipitated by mWP3 from a solution eluted from a peroxidase-containing band that was electrophoretically separated on a non-denaturing gel. The immunoprecipitated protein was electroblotted and visualized using mWP3 as the primary antibody; and [4] Same as lane 3 except total extracted wall proteins were used for immunoprecipitation. Notice that * in lanes 3 and 4 show the heavy (top) and light (bottom) chain of mouse IgG.

Two of the clones produced also yielded monoclonal antibodies, mWP14 and mWP18, which show cross-reactivity with a protein near 65,000 daltons that is immunoprecipitated from corn wall proteins by polyclonal antibodies that were raised against a known extracellular, high p1(9.5) cellulase (Fig. 2).

These monoclonal antibodies can be used to study the enzymatic regulation of cell wall extension by testing their effectiveness in inhibiting the activities of wall enzyme *in situ*. They will also be employed as a probe to study the site of synthesis of peroxidase and cellulase in the cell.

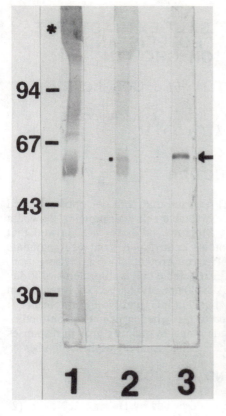

FIG. 2. Western blot identification (1,2) of proteins immunoprecipitated from corn wall proteins by the polyclonal rabbit anti-9.5 cellulase antibodies. After electroblotting the immunoprecipitated protein, visualization was carried out using the following primary antibodies: lane 2, mWP14; lane 3, mWP18. Lane 1 shows a protein stained with 0.1% amido black. Notice that * in lane 1 shows the 150,000 daltons non-reduced form of rabbit IgG (heavy chain plus light chain), and numbers represent the scale of the molecular weights (x 10^3 daltons).

LITERATURE CITED

1. JOHNSON DA, JW GAUTSCH, JR SPORTSMAN, JH ELDER 1984 Gene Anal Tech 1: 3-8
2. LAEMMLI UK 1970 Nature (London) 227: 680-695
3. SILBERMAN LG, N DATTA, P HOOPS, SJ ROUX 1985 Arch Bioch Bioph 236(1): 150-158
4. TERRY ME, BA BONNER 1980 Plant Physiol 66: 321-325

Physiology of Cell Expansion During Plant Growth, *D.J. Cosgrove and D.P. Knievel* Eds.,
Copyright ©1987, The American Society of Plant Physiologists

ANTIBODIES AGAINST SALT-EXTRACTABLE PEA CELL WALL PROTEINS AND THEIR EFFECT ON GROWTH

MELISSA A. MELAN AND DANIEL J. COSGROVE

*Biology Department, Pennsylvania State University
University Park, PA 16802*

INTRODUCTION

In 1981, Huber and Nevins (2) reported that corn coleoptiles treated with antibodies raised against 3M LiCl extractable cell wall proteins showed a 35% supression of growth. Recently, Morrow and Jones (4) showed that antibodies raised against cell wall proteins centrifuged from pea segments infiltrated with 50 mM $CaCl_2$ had no effect on growth. In this investigation, antibodies were raised against 3 M LiCl extractable cell wall proteins of 7 day old etiolated pea (*Pisum sativum* L. cv. Alaska) epicotyl tissue. Dialyzed serum (2), serum globulin fraction (2), affinity purified IgG, and $F(a,b)_2$ fragment (7) preparations of immune serum were used to test whether or not the antibodies would affect growth. This investigation attempts to quantify growth responses as well as demonstrate and characterize the immunoreactivity of the antibodies.

MATERIALS AND METHODS

Growth assays were conducted as described by Huber and Nevins (2) with minor modifications. Antibody pre-treatments were carried out for 90 min and each growth assay contained buffer and pre-immune control treatments. Dialyzed serum treatment consisted of a 1:4 dilution of serum previously dialyzed overnight against 1 mM Na Citrate, pH 6.0. Serum globulin fraction, IgG and $F(a,b)_2$ fragment treatments were made at concentrations of 1 mg/ml. Growth was followed at 1 hour intervals for 5 hours. Immunofluorescence microscopy, SDS-PAGE, Western blot analysis, affinity chromatography and metalloimmunoassay procedures were carried out according to established protocols.

RESULTS AND DISCUSSION

Growth assay results indicate that antibody pre-treatments have no effect on growth. Relative growth rates of antibody treated segments show no significant difference from pre-immune (Fig. 1) or buffer treatments. Immunofluorescence microscopy of tissue sections to which 1 mg/ml IgG was applied directly shows that the antibodies bind to cell walls.

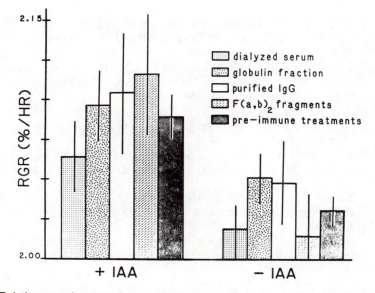

FIG 1. Relative growth rates of pea segments treated with immune or pre-immune preparations ± 95% confidence intervals. Data from a total of 8 growth assays with 10 segments per assay.

Immunofluorescence of segments floated in 1 mg/ml IgG for 90 min shows that the antibodies penetrate into the tissue. Affinity chromatography with Sepharose-bound IgG and Western blot analysis indicate that the antibodies recognize and bind to most if not all proteins extracted from the cell walls. Metalloimmunoassay results give a value of 75.5% precipitable protein.

Results from this investigation are consistent with those of Morrow and Jones (4) but contrary to those of Huber and Nevins (2). This finding may be due to the compositional and conformational differences between monocot and dicot cell walls. Indications are that the protein(s) responsible for expansive growth in peas may not be extractable with salts. Recently, Nagahashi and Siebles (5) reported that greater than 80% of α-mannosidase activity remained bound to potato tuber cell walls after 3 M LiCl treatment.

An alternative explanation for the observed results may be that the growth associated protein(s) may not be accessable to antibodies *in situ*. Although immunofluorescence results show that the antibodies penetrate the tissue, they may only be binding to proteins located on the surfaces of the cell wall. X-ray crystallography of IgG indicates an arm to arm distance of 146 Å (3) and $F(a,b)_2$ dimensions of 30 x 40 x 50 Å (6). Cell wall pore sizes have been estimated to be 35 - 52 Å (1). Thus, it may be very difficult for the antibodies to reach the protein(s) if they are intercalated into the wall.

LITERATURE CITED

1. CARPITA N, D SABULARSE, D MONTEZINOS, D DELMER 1979 Determination of the pore size of cell walls of living plant cells. Science 205: 1144-1147

2. HUBER DJ, DJ NEVINS 1981 Wall-protein antibodies as inhibitors of growth and of autolytic reactions of isolated wall. Physiol Plant 53: 533-539

3. HUBER R, J DIESENHOFER, PM COLMAN, M MATSUSHIMA, W PALM 1976 Crystallographic structure of an IgG molecule and an F_c fragment. Nature (London) 264: 415-420

4. MORROW DL, RL JONES 1986 Localization and partial characterization of the extracellular proteins centrifuged from pea internodes. Physiol Plant 67: 397-407

5. NAGAHASHI G, TS SIEBLES 1986 Purification of plant cell walls: isoelectric focusing of $CaCl_2$ extracted enzymes. Protoplasma 134: 102-110

6. POLJAK RJ, LM AMZEL, HP AVEY, LN BECKA, A NISSONOFF 1972 Structure of Fab' New at 6Å resolution. Nature New Biol 235: 137-140

7. PORTER RR 1959 The hydrolysis of rabbit gamma-globulin and antibodies with crystalline papain. Biochem J 73: 119-203.

PHYSIOLOGICAL EVIDENCE FOR AUXIN COMPARTMENTATION

DAVID J. PARRISH

*Department of Agronomy, Virginia Tech,
Blacksburg, VA 24061.*

INTRODUCTION

Freshly excised stem segments elongate at relatively high, presumably auxin-induced, rates. Within a few hours, however, their extension slows to a much lower "basal" or "non-induced" rate (2). Such segments are still quite responsive to exogenous IAA and are generally considered merely auxin-depleted. But are they truly auxin-impoverished? Data presented here suggest that auxin in a growth active form is still present in segments long after they have been excised and have slowed to basal elongation rates. The auxin that is present, however, is sequestered by aerobic processes in some intracellular compartment, isolating it from growth-active sites.

MATERIALS AND METHODS

One-cm segments of hypocotyls or mesocotyls were excised from 5- to 7-day-old seedlings of *Cucumis sativus, Glycine max, Helianthus annuus, Pisum sativum*, and *Zea mays*. Following a 2- to 3-hr aging period (2), segments were stacked in an auxanometer and exposed to 15- to 30-min pulses of anoxia, induced by bubbling their buffered medium with N_2. In another study, pea stem segments were placed in 4 μM ^{14}C-IAA for 3 hr. After a brief wash, the segments were stacked in a tube through which was pumped fresh, IAA-free buffer. Elution of ^{14}C under aerobic and anaerobic conditions was monitored by scintillation counting of 5-min fractions (3).

RESULTS AND DISCUSSION

Stem segments of each of the species tested responded to anoxia by ceasing to elongate, but restoration of O_2 resulted in a short-lived (about 1 hr) burst of growth that exceeded stored growth and whose kinetics were seemingly identical to auxin's: *viz.*, a rapid acceleration, a brief decline, and a second increase (4). Parrish and Davies (3) demonstrated and named this "emergent growth" effect in peas. The observation of emergent growth now extends to a diverse group of angiosperms ranging from grasses (Poaceae) to composites (Asteraceae).

Pea stem segments loaded with ^{14}C-IAA eluted auxin (or some auxin product) at a rapidly declining rate under aerobic conditions (Fig. 1). Efflux approached a non-zero, "basal" rate by the fifth hour. Anoxia greatly promoted loss of ^{14}C (Fig. 1 and Ref. 3). Activity remaining in stem segments reflected the time under anaerobiosis (Table I). Cyanide has a similar effect on elution (1) and also causes emergent growth when administered in 15- to 30-min pulses (unpublished results).

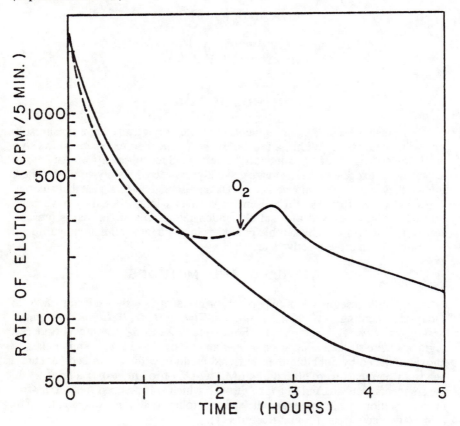

FIG. 1. Elution of ^{14}C from pea stem segments after floating for 3 hr in 4 μM ^{14}C-1AA. The dashed line represents elution of one set of segments while under anoxia.

These data provide indirect, physiological evidence for the sequestering of auxin in some compartment where it cannot interact with putative growth-active sites. The low rate of efflux (auxin in transport?) from this pool may account for the "basal" elongation rates observed in aged segments (2). The sequestering of auxin is dependent upon aerobic respiration; under anaerobiosis or cyanide, auxin escapes the compartment, making it more accessible to growth-active sites and causing a burst of growth when O_2 is restored. Restoration of aerobic respiration reestablishes polar transport and the sequestered pool, so that the auxin-induced burst of elongation is short-lived. In intact stems, the

Table I. *Radioactivity in pairs of 1-cm pea stem segments fed ^{14}C-IAA for 3 hr and measured immediately or after 6 hr of elution under aerobic or anaerobic conditions.*

Elution Treatment	Activity		Activity Lost
	No elution	6-hr elution	
(hr/gas)	-------DPM (x1000)-------		%
6/O$_2$	16.3±1.7	13.3±0.8	19
2/O$_2$+2/N$_2$+2/O$_2$	18.6±1.3	12.1±1.9	35
6/N$_2$	13.5±0.7	5.7±0.4	57

compartmentation or sequestering of auxin might play an important role in modulation of auxin-mediated events such as tropisms or other differential growth responses.

LITERATURE CITED

1. DAVIES, PJ 1974 The uptake and elution of indoleacetic acid by pea stem sections in relation to auxin induced growth. *In* Y. Sumiki, ed, Plant Growth Substances, 1973 Hirokawa Publishing Co, Tokyo, pp 767-779
2. PARRISH, DJ AND PJ DAVIES 1975 The influence of aging conditions on the short term growth of green pea stem segments. Plant Physiol 55: 586-588
3. PARRISH, DJ AND PJ DAVIES 1977 Emergent growth: an auxin-mediated response. Plant Physiol 59: 745-749
4. VANDERHOEF, LN AND CA STAHL 1975 Separation of two responses to auxin by means of cytokinin inhibition. Proc Nat Acad Sci, USA 72: 1822-1825

Physiology of Cell Expansion During Plant Growth, *D.J. Cosgrove and D.P. Knievel* Eds.,
Copyright ©1987, The American Society of Plant Physiologists

CALCIUM ANTAGONIST INHIBITS AUXIN-STIMULATED GROWTH AND WALL SYNTHESIS

D. A. BRUMMELL AND G. A. MACLACHLAN

*Biology Department, McGill University,
1205 Avenue Dr. Penfield,
Montreal, Quebec,
Canada H3A 1B1*

INTRODUCTION

In order to promote growth, auxin must first bind to a receptor. If the auxin receptor is found only on the plasma membrane and faces outwards, as is suggested by immunocytochemical studies (5), then some second messenger is necessary in order to transmit the signal from the plasma membrane to the internal machinery of the cell. In animal cells, transient increases in cytosolic Ca^{2+} levels act as a second messenger in the regulation of many processes, but the evidence for a similar mechanism occurring in plant cells is scantier (4). However, there is evidence that Ca^{2+} is necessary for the growth-regulating action of auxin in roots (3), although in roots auxin causes an inhibition rather than a promotion of growth. The present work examines the dependence on calcium of auxin-stimulated growth in segments of pea stem using 8-(N,N-diethylamino)octyl-3,4,5-trimethoxybenzoate HCl (TMB-8), a purported intracellular calcium antagonist.

MATERIALS AND METHODS

Abraded stem segments from the growing region of the third internode of etiolated *Pisum sativum*, cv. Alaska, seedlings were pretreated for 2 h in 200 µM TMB-8. They were then incubated in 50 mM glucose containing [^{3}H]glucose and 10 mM mes-NaOH pH 7.2 for a further 2 h, in the presence of 10 µM IAA, 200 µM TMB-8 and 1 mM $CaCl_2$ where indicated. Segments were washed in ice-cold H_2O x 4, ground in a pestle and mortar, centrifuged and the supernatant removed. Cell wall debris was washed in ice-cold H_2O x 2 then ice-cold 80% ethanol x 2. Radioactivity in cell walls and in the soluble phase was determined by liquid scintillation counting.

RESULTS AND DISCUSSION

Auxin-stimulated growth at 2 h was inhibited approximately 50% by the TMB-8 treatment. Since growth is dependent upon continued cell wall synthesis (1), the effect of TMB-8 on the incorporation of labeled glucose into cell walls was examined. In these experiments growth was prevented by the neutral buffer, thus avoiding the consequent large increases in water and glucose uptake due to cell expansion in the auxin-treated segments.

When IAA was not present in the incubation medium, TMB-8 had little, if any, effect on relative cell wall incorporation (i.e., incorporation into cell walls as a proportion of total label uptake) (Table I). In contrast, in the presence of IAA, TMB-8 inhibited the relative incorporation of label into cell walls by about 60% (Table I). This was the result of a 45% inhibition of cell wall incorporation and a 40% promotion of total uptake. Why TMB-8 should promote uptake of glucose is not clear, but may indicate that the glucose carrier is regulated or affected by cytosolic Ca^{2+} concentrations.

Table I. *Effect of 200 μM TMB-8 on incorporation of [^{3}H]glucose into cell walls relative to total uptake into pea segments in the absence of calcium.* Data are means of at least 10 replicates from 5 experiments.

	[^{3}H]glucose			
	H_2O	H_2O + TMB-8	IAA	IAA+ TMB-8
	(cpm x 10^{-3}/16 segs ± SD)			
Cell wall	4.0	3.7	4.8	2.6
Total uptake	103	102	105	142
Relative cell wall incorporation (x 10^{-3})	39.1 ±2.1	35.8 ±3.2	45.9 ±3.7	18.5 ±4.6

Since TMB-8 presumably brings about the above effects by antagonizing intracellular calcium, an attempt was made to reverse the inhibitions by adding extracellular Ca^{2+}. In the presence of 1 mM Ca^{2+} and IAA, the effects of TMB-8 on both cell wall incorporation and total uptake were almost completely restored to the control level (Table II). Ca^{2+} caused a small (but not statistically significant) promotion of relative cell wall incorporation in the other treatments (compare Table II with Table I). The presence of Ca^{2+} did not reverse the TMB-8 inhibition of growth, since extracellular Ca^{2+} is itself growth-inhibitory (2).

The results are consistent with the view that the actions of auxin on cell wall synthesis, and thus potentially on growth, require or involve Ca^{2+}, at least in part.

Table II. *Effect of 200 μM TMB-8 on incorporation of [³H]glucose into cell walls relative to total uptake into pea segments in the presence of 1 mM calcium.*
Data are means of at least 10 replicates from 5 experiments.

	H_2O	[³H]glucose H_2O + TMB-8	IAA	IAA+ TMB-8
		(cpm x 10^{-3}/16 segs ±SD)		
Cell wall	3.8	3.7	4.9	4.3
Total uptake	94	92	97	97
Relative cell wall incorporation (x 10^{-3})	40.5 ±4.4	40.6 ±2.3	50.5 ±6.7	44.9 ±3.3

Acknowledgments--This work was supported by a grant from the Natural Sciences and Engineering Research Council of Canada.

LITERATURE CITED

1. BRUMMELL DA, JL HALL 1985 The role of cell wall synthesis in sustained auxin-induced growth. Physiol Plant 63: 406-412
2. CLELAND RE, DL RAYLE 1977 Reevaluation of the effect of calcium ions on auxin-induced elongation. Plant Physiol 60: 709-712
3. HASENSTEIN KH, ML EVANS 1986 Calcium dependence of rapid auxin action in maize roots. Plant Physiol 81: 439-443
4. HEPLER PK, RO WAYNE 1985 Calcium and plant development. Ann Rev Plant Physiol 36: 397-439
5. LOBLER M, D KLAMBT 1985 Auxin-binding protein from coleoptile membranes of corn (*Zea mays* L.). II. Localization of a putative auxin receptor. J Biol Chem 260: 9854-9859

Physiology of Cell Expansion During Plant Growth, *D.J. Cosgrove and D.P. Knievel* Eds.,
Copyright ©1987, The American Society of Plant Physiologists

THE RELATIONSHIP BETWEEN ACCUMULATION OF AUXIN INTO, AND GROWTH OF, COLEOPTILE CELLS

M.J. VESPER, C.L. KUSS, AND J.M. MAXSON

*Department of Biology, University of Dayton,
Dayton, OH 45469*

INTRODUCTION

A persistent question in the study of plant cell growth concerns the relationship between the amount of auxin in growing plant cells and how much these cells grow. Coleoptiles from etiolated grass seedlings have become a model system for the study of many aspects of plant cell growth. Our experiments were designed to measure the growth of maize coleoptile sections over a 2-h period of exposure to an exogenous solution of IAA, and to estimate the free IAA accumulated in these sections during this period.

Since we are trying to determine if there is a direct relationship between IAA content in cells and their growth, we altered responsiveness of coleoptile sections to auxin in two ways: 1) By inclusion of the auxin transport inhibitor, naphthylphthalamic acid (NPA), in the IAA solution, an increase in growth of sections responding to IAA was accomplished. 2) We used the different levels of responsiveness to auxin that maize coleoptile sections express as a function of time after their excision (2) in another test of the relationship between auxin content and cell growth.

MATERIALS AND METHODS

Zea mays L. (B73 x MO17) seeds were grown as described previously (3). Coleoptile sections (1 cm) were cut beginning 3 mm below the tip. Growth of sections submerged in aerated solution was measured continuously by use of a position-sensing transducer (3). Coleoptile sections were given IAA at two different times after their excision. Those given IAA beginning 0-30 min after excision are called low sensitivity (LS) sections, and those given IAA beginning 120 min after excision are called high sensitivity (HS) sections (2). IAA concentration was 0.3 µM.

Uptake of externally applied [^{3}H]IAA (522 dpm/pmole) was monitored in the same sections in which growth was measured. At the end of 60 or 120 min in the [^{3}H]IAA solution, sections were placed in ice-cold water for 15 min, blotted dry, their fresh weights determined, and then placed in LSC cocktail overnight before being counted. Percent of the radioactivity in the sections which was in the free IAA form at the 60- and 120-min periods was determined by TLC performed on methanolic extracts from a separate batch of sections.

301

RESULTS AND DISCUSSION

Change in the responsiveness of maize coleoptile sections to exogenously applied IAA is a function of time from excision and is apparent both in the presence and absence of NPA. Yet NPA causes a general enhancement of growth in both LS and HS sections. This enhancement is detected as soon as the growth response to IAA begins and it is sustained. However, NPA does not appear to alter the latent period to a response to 0.3 μM IAA. A comparison of -NPA and +NPA values in column A of Table I illustrates the influence of NPA on growth.

Table I. *Relationship between growth of coleoptile sections and accumulation of free IAA from an exogenous source in the presence and absence of 10 μM NPA.*

Growth and uptake were measured in LS and HS sections for the first 60 min (LS-60, HS-60) or 120 min (LS-120, HS-120) of incubation in 0.3 μM [^{3}H]IAA (522 dpm/pmole).

	A GROWTH		B IAA		C A/B	
	-NPA	+NPA	-NPA	+NPA	-NPA	+NPA
	Δμm		*pmole/g f.w.*[1]			
LS-60	440	650	82.6	148.7	5.33	4.37
HS-60	2220	3250	118.1	191.3	18.8	16.9
LS-120	2610	3935	94.2	188.4	27.7	20.9
HS-120	6945	8550	151.6	313.9	45.8	27.2

[1]Calculated using % of radiolabel in the form of free IAA at appropriate time period, as determined by TLC of methanolic extracts of coleoptile sections.

The presence of NPA caused an increase, in all cases, in the amount of radiolabel accumulated in the sections. In the -NPA sections, only 38% of the dpm in LS sections co-chromatographed with IAA and only 50% of dpm in HS sections did so following a 120-min incubation in [^{3}H]IAA. A similar result was found by Nonhebel et al. (1). However, the presence of NPA caused a general increase in the % of label co-chromatographing with IAA. Therefore, the pmole of free IAA/g f.w. accumulated in the +NPA treatment is shown to increase by 40-50% over the -NPA treatment (Table I, column B).

Does growth increase proportionately with this increase in free IAA in NPA-treated sections? Column C in Table I gives the ratio of growth:free IAA content in the cells + and -NPA. Comparison of the paired samples in column C using Student's t test shows no significant differences between the NPA-treated sections and untreated sections, except for HS-120 (t = 0.95).

When comparing the values within the -NPA and +NPA parts of column C in Table I, it is apparent that the ratio of growth:IAA content differs (t = 0.95) between the samples taken after 60 min incubation and those taken after 120 min of incubation for either LS or HS sections. In the case of LS sections this is entirely expected since LS sections are accumulating auxin during the first 60 min of their incubation but are still within the latent period of the growth response (2). That HS sections show a similar discontinuity between the 60-min and 120-min samples, yet have a latent period of only 15 min, indicates that there may be a need for auxin to accumulate above a certain level before growth is enhanced.

The data in Table I reinforce the idea that different states of sensitivity to auxin can exist in maize coleoptile sections. LS-60 sections accumulate less IAA than HS-60 sections in the absence of NPA (82.6 vs 118.1 pmole IAA/g f.w.). One might conclude that the difference in growth between LS and HS sections during the first 60 min of an IAA response is due to a mere failure of LS sections to accumulate sufficient IAA during this time. However, LS-60 sections treated with NPA accumulate 148.7 pmole IAA/g f.w., yet no enhancement of their growth is observed during that time. Clearly, LS-60 tissue is incapable of responding to a significant IAA signal. However, during the second 60-min period in IAA (-NPA), LS sections begin to grow while IAA has accumulated to only about 94 pmole/g f.w.

We can conclude from this work that the presence of NPA causes greater accumulation of label from an external [^{3}H]IAA solution and increases the percent of label remaining in the free IAA form. Comparing sections incubated + and -NPA, a direct relationship between growth and IAA content in the tissue is suggested. However, such a direct relationship is not clearly seen when comparing sections having natural variations in responsiveness to IAA. LS sections appear to be initially limited in their response to IAA by some factor other than internal concentration of IAA.

Acknowledgments--Supported by a grant from the National Science Foundation, DBM 8515925.

LITERATURE CITED

1. NONHEBEL HM, JR HILLMAN, A CROZIER, MB WILKINS 1985 Metabolism, of (14C)-indole-3-acetic acid by coleoptiles of *Zea mays* L. J Exp Bot 36: 99-109
2. VESPER MJ, ML EVANS 1978 Time-dependent changes in the auxin sensitivity of coleoptile segments. Plant Physiol 61: 204-208
3. VESPER MJ 1985 Use of a pH-response curve for growth to predict apparent wall pH in elongating segments of maize coleoptiles and sunflower hypocotyls. Planta 166: 96-104

Physiology of Cell Expansion During Plant Growth, *D.J. Cosgrove and D.P. Knievel* Eds.,
Copyright ©1987, The American Society of Plant Physiologists

GROWTH REGULATION IN AMPHIBIOUS BUTTERCUPS

R. F. HORTON, C. BRIAND AND S. HUBLER

Department of Botany, University of Guelph, Guelph, Ontario, Canada N1G 2W1

INTRODUCTION

Amphibious plants—those that can grow successfully terrestrially or as aquatics—are subject to a range of dramatic anatomical, morphological and physiological changes on submergence. A number of *Ranunculus* species (water buttercups or crowfoots) exhibit a rapid alteration of the elongation rate of petioles ('accommodation growth') or have different leaf forms underwater ('heterophylly'). Petiole growth is related to enhanced C_2H_4 levels in underwater tissues (2). Changes in the structure of leaves underwater can be prevented by treatment with ABA (9). Here we introduce the range of effects of submergence and regulator treatments in these species, and differentiate between the effects of IAA and C_2H_4 in enhancing elongation growth in petiole cells of *R. sceleratus*.

MATERIALS AND METHODS

Plants and seeds (achenes) of *R. sceleratus* L., *R. flabellaris* Raf. and *R. Gmelini* DC were collected locally. Achenes of *R. pygmaeus* Wahl. were collected in the North West Territories. All plants were grown in soil in controlled environments and then subjected to submergence.

RESULTS AND DISCUSSION

Petioles of *R. sceleratus* increase their growth rate within 20 min after submergence (2). This growth acceleration is related to an increase in tissue C_2H_4 (6) and can be prevented by treatment with aminoethoxyvinylglycine (2) Co^{++} (7) or Ag^+ (3) which prevent C_2H_4 synthesis or action. Bladed petioles in air will respond to both IAA and C_2H_4.-a low level of IAA (10^{-7} M) is a prerequisite for sensitivity to C_2H_4 in isolated petiole segments (5). Segments incubated in 5 mg.l^{-1} ethephon + 10^{-7} M IAA elongate to a similar extent over an 18 h period to those in 10^{-4} M IAA alone. However, the nature of the growth is different (Table I). C_2H_4 treatment in the presence of a low level of IAA (like that occurring when bladed petioles are submerged) allows elongation growth with a minimal concurrence of radial expansion. In contrast, elongation growth after high IAA incubation is accompanied by marked radial expansion. The response of these tissues to IAA and C_2H_4 differs from that of pea epicotyls where radial expansion after high IAA treatments can be ascribed to IAA-induced C_2H_4 (1).

Table I. *Growth of petiole segments of Ranunculus sceleratus.*
2-cm segments were excised from 6-wk-old plants and floated on 5 ml buffered treatment solutions in the light at 25° with shaking for 18 h (5 segments in each treatment, triplicate). Initial weight of bladed (submerged) petiole taken as mean of initial weight of all treatments.

Treatment	% increase in length	% increase in weight
Control (buffer)	4.8	8.3
10^{-7} M IAA	6.1	13.4
5 mg l^{-1} ethephon	11.2	14.7
5 mg l^{-1} ethephon + 10^{-7} M IAA	71.3	88.0
10^{-4} M IAA	70.1	136.2
Submerged (in water with blade)	84.4	97.6

Petiole cells of *R. sceleratus* grow in response to submergence (or C_2H_4) by elongation. Petioles of *R. pygmaeus* (like some other amphibious plants (4)) exhibit both a changed rate of cell division and cell expansion on submergence.

R. flabellaris and *R. Gmelini*, submergence causes a change in shape of subsequently expanded leaves. The underwater leaves are finely dissected, have a different anatomy and far fewer stomata (8). However, submergence in 2.5 x 10^{-5} M ABA allows leaf production to continue but the leaves formed are anatomically and morphologically very similar to terrestrial leaves (9) and have stomata.

Acknowledgments--These studies were funded by operating grants from NSERC Canada.

LITERATURE CITED

1. EISINGER W 1983 Regulation of pea internode expansion by ethylene. Annu Rev Plant Physiol 34: 225-240
2. HORTON RF 1987 Ethylene and the growth of amphibious plants. *In* D. Klambt, ed, Plant Hormone Receptors, Nato ASI Series, Springer Verlag, Berlin (in press)
3. HORTON RF, AB SAMARAKOON 1982 Petiole growth in the celery-leaved crowfoot (*Ranunculus sceleratus* L.): effects of auxin transport inhibitors. Aquat Bot 13: 97-104
4. RIDGE I 1985 Ethylene and petiole development in amphibious plants. *In* JA Roberts, GA Tucker, eds, Ethylene and Plant Development, Butterworths, London, pp 229-240

5. SAMARAKOON AB, RF HORTON 1983 Petiole growth in *Ranunculus sceleratus* L.: the role of growth regulators and the leaf blade. Can J Bot 61: 3326-3331

6. SAMARAKOON AB, RF HORTON 1983 Petiole growth in *Ranunculus sceleratus* L.: ethylene synthesis and submergence. Ann Bot 54: 263-270

7. SAMARAKOON AB, L WOODROW, RF HORTON 1985 Ethylene- and submergence-promoted growth in *Ranunculus sceleratus* L. petioles: the effects of cobalt ions. Aquat Bot 21: 33-41

8. YOUNG J, NG DENGLER, RF HORTON 1987 Heterophylly in *Ranunculus flabellaris*: the effect of abscisic acid on leaf anatomy. Ann Bot 60: 117-125

9. YOUNG J, RF HORTON 1985 Heterophylly in *Ranunculus flabellaris* Raf.: the effect of abscisic acid. Ann Bot 55: 899-902

Physiology of Cell Expansion During Plant Growth, *D.J. Cosgrove and D.P. Knievel* Eds.,
Copyright ©1987, The American Society of Plant Physiologists

PHOTOCONTROL OF LEAF UNROLLING

NEIL VINER AND HARRY SMITH

*Department of Botany, University of Leicester
Leicester, UK.*

INTRODUCTION

The leaves of cereal seedlings grown in total darkness are tightly rolled about their long axes. Exposure to light, even briefly, causes the leaves to unroll, and this phenomenon may be studied using isolated leaf segments. The driving force for leaf unrolling is the isodiametric swelling of the mesophyll cells, and is thus a manifestation of cell growth (1). Leaf unrolling is a classical phytochrome-mediated reponse, being induced by brief red, and reversed by brief subsequent far-red, light (3). Here we present preliminary data on the possible role of calcium in phytochrome-mediated unrolling of barley leaves.

MATERIALS AND METHODS

Segments (2 cm) were excised 1 cm from the tips of 6-day dark-grown leaves of barley (*Hordeum vulgare* cv. Golden Promise) and floated on MES/NaOH buffer (35 mM, pH 6.5) with or without EGTA (2 mM). After 45 min dark incubation various light and calcium treatments were given relative to this time point (i.e. t_0). Treatments consisted of: red light, 5 min, 36.6 µmol $m^{-2}s^{-1}$; far-red light, 5 min, 186.0 µmol $m^{-2}s^{-1}$; $CaCl_2$ or $MgCl_2$, 0.1-10.0 mM. After treatment, samples were dark-incubated for 24h at 25C, and final leaf width measured using a dissection microscope.

RESULTS AND DISCUSSION

Calcium-chelation prevents phytochrome response. Figure 1a shows that leaf segments not pretreated with EGTA exhibited the standard unrolling response to 5 min red light followed by darkness; this response was readily reversed by far-red given immediately after the red light. The unrolling response is saturated at relatively low fluences. Smith (2) found that 90% of the maximum response was produced with a red light fluence of 1 mmol m^{-2}; total fluence in the experiments outlined here was more than 11 mmol m^{-2}. Pre-incubation for 45 min in buffer containing 2 mM EGTA caused leaf segments to become insensitive to red light, but the phytochrome control of leaf unrolling could be restored by the addition of Ca^{2+} (Fig. 1b). Threshold Ca^{2+} concentration for unrolling under these conditions was between 0.1 and 1.0 mM, corresponding to an activity range in the EGTA buffer of 10-40 µM. The requirement for divalent cation could not be satisfied by Mg^{2+}.

307

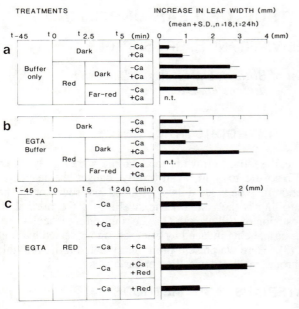

TREATMENTS   INCREASE IN LEAF WIDTH (mm)

(mean+S.D.,n=18,t=24h)

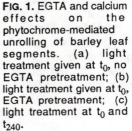

FIG. 1. EGTA and calcium effects on the phytochrome-mediated unrolling of barley leaf segments. (a) light treatment given at t_0, no EGTA pretreatment; (b) light treatment given at t_0, EGTA pretreatment; (c) light treatment at t_0 and t_{240}.

Escape from phytochrome and calcium control. After red light, in the absence of EGTA, cell expansion normally progresses beyond phytochrome control within about 2h, as evidenced by the escape from far-red reversibility (Fig. 2; see also Smith, 1975). Similarly, with leaves pretreated with EGTA, the addition of Ca^{2+} failed to restore the leaf unrolling response if delayed until about 2h after the red light treatment; indeed, significant loss of sensitivity to added Ca^{2+} occurred within 1h of the light treatment (Fig. 2).

Since the Pfr form of phytochrome is unstable in etiolated tissues, the possibility was considered that the observed kinetics (Fig. 2) may be due to the loss of Pfr after red light. In the absence of EGTA, Pfr was lost progressively, reaching about 40% of the t_0 value within 4h; EGTA treatment increased the rate of loss of Pfr, such that by 4h approximately 25% of the t_0 level remained. However, both far-red reversibility in the controls, and the capacity of Ca^{2+} to restore the leaf unrolling response in the EGTA-pretreated seedlings, were lost more rapidly than the loss of measurable Pfr. The fact that, after EGTA incubation, the original light effect could not be rescued by Ca^{2+} given more than 2h after the red light, even though saturating amounts of Pfr were still present, probably indicates that calcium does not interact with Pfr itself, but with a short-lived component of the primary transduction chain of Pfr action.

The situation is, however, more complicated. If, 4h after the original red, a further red-light treatment was given *together* with added Ca^{2+}, then leaf unrolling was induced; witholding Ca^{2+} again prevented the response (Fig. 1c). These data suggest that Pfr remaining in the cells 4h after the first red treatment

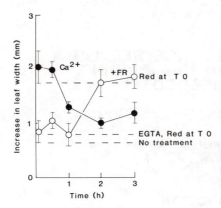

FIG. 2. Time courses for escape from far-red photoreversibility and for the restoration of the response to red light by calcium.

is not competent to induce cell expansion even if Ca^{2+} is provided, whereas Pfr formed anew by photoconversion is fully capable of initiating the transduction chain. Such apparently paradoxical findings can only be reconciled by assuming that the functional, but not the spectral, properties of Pfr change at some point after first being formed by photoconversion. It is generally regarded that the first step in phytochrome action is for Pfr, upon photoconversion from Pr, to "couple" with its primary reaction partner. If Ca^{2+} were necessary for the subsequent steps in the transduction chain, and if its absence allowed the permanent uncoupling of any existing Pfr molecules, then these data would, in principle, be understandable; other speculations are, no doubt, equally feasible.

Acknowledgments--N.V. is a recipient of a SERC Research Studentship.

LITERATURE CITED

1. BURSTROM H 1942 Uber Entfeltung und Einrollen eines mesophillen Grasblattes. Bot Notis 7: 351-4622
2. SMITH H 1975 Phytochrome and Photomorphogenesis. McGraw-Hill, Maidenhead. 235pp
3. VIRGIN H 1962 Light-induced unfolding of the grass leaf. Physiol Plant 15: 380-389

Physiology of Cell Expansion During Plant Growth, *D.J. Cosgrove and D.P. Knievel* Eds.,
Copyright ©1987, The American Society of Plant Physiologists

EFFECT OF LOW FLUENCE RED LIGHT ON GROWTH AND NAA BINDING IN MAIZE MESOCOTYL

ALAN M. JONES AND SHUBO ZHOU

Dept. of Biology, University of North Carolina, Chapel Hill, NC 27514

INTRODUCTION

Linear growth of the mesocotyl is affected by fluences of red light (R) spanning ten orders of magnitude (2, present study). Low fluences (1 mmol m^{-2}) cause a 50% decrease in the elongation rate of this tissue fully apparent at 2 hours with a 40-minute lag phase (5). This R-induced inhibition is reversed by far-red light (FR) indicating the involvement of phytochrome (4, 5). High fluences (1 mol m^{-2}) induce an 80% decrease in the rate of cell elongation as well as many other developmental responses. Exogenous IAA added simultaneously with red-light treatment to excised tissue prevents most of the red-light induced inhibition of growth (4). The current working model explaining the above observations is that phytochrome, in some way, modulates the flux (2) or pool size (1) of auxin within the elongating cells of the mesocotyl. This would account for most of the decrease in growth rate, assuming that auxin is limiting in the unirradiated cells. The remaining long-term effect of light is on the capacity for cells to respond to auxin and, at least for high fluences, this may be due to changes in the number of auxin receptors (6). Here, we show: 1) maize mesocotyl has a saturable Very Low Fluence (VLF) response, 2) that the capacity to respond to auxin is decreased by red light via phytochrome action, 3) that the total number, but not necessarily the availability, of 'site I' auxin binding sites is unaltered by low fluences, and 4) that the change in the capacity to respond to auxin is probably mediated by auxin itself.

MATERIALS AND METHODS

Maize caryopses (B73 x Mo17, Ohio Seed Foundation) were grown in complete darkness. Mesocotyl was visualized using an infrared scope and marked with ink then irradiated with the indicated fluences of R or FR. NAA-binding was performed by the method of Jones et al. (3). The apical 1-cm section of mesocotyl was used for all experiments.

RESULTS AND DISCUSSION

Three distinct phases of R-induced inhibition of mesocotyl growth are observed over ten orders of fluence (data not shown). The VLF response, previously shown to be unsaturable (2), saturates here at 10^{-7} mol m^{-2}. The High Irradiance Response is also shown to exist in this maize hybrid. Seedlings were given a low fluence of R and/or FR, returned to darkness for 10 h, then excised mesocotyl was assayed for NAA-induced elongation. R-treated seedlings grew 50% less at the optimal NAA dosage but the dosage of NAA which induced 50% of the maximum growth was essentially unchanged (Fig. 1). FR reversed the R effect. Low fluence R had no effect on NAA binding in this tissue (data not shown). The capacity to elongate in response to NAA is also reduced in

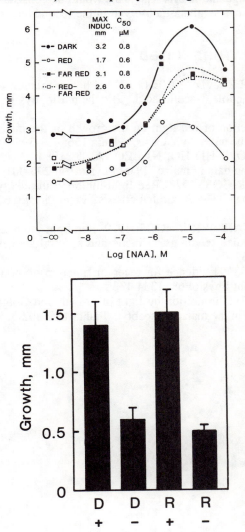

FIG. 1. NAA dose response curves for mesocotyl growth. Mesocotyls were excised from seedlings pretreated with R (1 mmol m^{-2}) and/or FR (100 mmol m^{-2}) then 10 h of darkness.

FIG. 2. The effect of an NAA pretreatment on the capacity for NAA-induced growth of mesocotyl from dark (D) and R-treated (R) seedlings. Seedlings were irradiated then tissue was excised and placed in buffer with (+) or without (-) 10 µM NAA for 4 h then washed in buffer without NAA for 2 h. After this, NAA-induced growth for all pretreated tissue was determined.

311

excised mesocotyl incubated in auxin-free medium regardless of light treatment (Fig. 2). Pretreatment of this tissue with NAA prevents the reduction of growth capacity (Fig. 2) suggesting that a decrease in IAA itself mediates the observed reduction in growth capacity and that R operates by reducing IAA levels.

The equilibrium constant of auxin binding, equals the C_{50} of auxin-induced growth in etiolated seedlings suggesting that it is binding which is the growth limiting step. If this remains to be true for R-irradiated plants, then in view of the current data, one may conclude that the availability of receptors and not the pool size decreases after R. However, this is not necessarily the case. Some other step downstream of binding such as the production of the cell wall matrix may become limiting after R. If this were true, then the total pool size of the enzyme involved must change and not its K_m. This would be consistent with the R-induced changes in glucan synthetase observed in the second but not the first cm of maize mesocotyl (7).

LITERATURE CITED

1. BANDURSKI RS, A SCHULZE, JD COHEN 1977 Photoregulation of the ratio of ester to free indole-3-acetic acid. Biochem Biophys Res Comm 79: 1219-1223
2. IINO M 1982 Inhibitory action of red light on the growth of the maize mesocotyl: Evaluation of the auxin hypothesis. Planta 156: 388-395
3. JONES AM, LL MELHADO, T-HD HO, NJ LEONARD 1984 Azido auxins: Quantitative binding data in maize. Plant Physiol 74: 295-301
4. VANDERHOEF LN, WR BRIGGS 1978 Red light-inhibited mesocotyl elongation in maize seedlings. The Auxin hypothesis. Plant Physiol 61: 534-537
5. VANDERHOEF LN, PH QUAIL, WR BRIGGS 1979 Red light-inhibited mesocotyl elongation in maize seedlings. Kinetic and spectral studies. Plant Physiol 63: 1062-1067
6. WALTON JD, PM RAY 1981 Evidence for receptor function of auxin binding sites in maize. Plant Physiol 68: 1334-1338
7. WALTON JD, PM RAY 1982 Inhibition by light of growth and Golgi-localized glucan synthetase in the maize mesocotyl. Planta 156: 302-308

Physiology of Cell Expansion During Plant Growth, *D.J. Cosgrove and D.P. Knievel* Eds.,
Copyright ©1987, The American Society of Plant Physiologists

THE EFFECT OF S-3307 ON THE IAA LEVELS IN ELONGATING STEM OF ALASKA PEA

R. H. HAMILTON AND D. M. LAW

*Biology Department, Pennsylvania State University,
University Park, PA 16802*

INTRODUCTION

A number of reports have shown GA_3 treated plant tissues can have high auxin levels. We (4) have confirmed reports of Muir (2) that GA_3 treated Little Marvel pea stems had high IAA levels. The small decreases in indoleacetylaspartate did not appear to account for the large increases in IAA levels observed. The objective of this work was to determine if a low GA level would also be correlated with a low IAA level in elongating pea stems.

MATERIALS AND METHODS

Eight-day-old Alaska peas (*Pisum sativum* L.) were treated with the GA biosynthesis inhibitor S-3307 (1). The resulting growth inhibition was later (14 day old) reversed by application of GA_3 for 3 days. The length of the upper two internodes and their IAA content was determined. The IAA determinations were by isotope dilution HPLC with a fluorescence detector after small column "cleanup" (3).

RESULTS AND DISCUSSION

The dwarfing of Alaska pea was apparent both in a greenhouse and a growth chamber experiment (Table I). In each experiment GA_3 gave partial to complete reversal. The IAA content seemed closely related to the length of the upper two internodes and was much higher in the greenhouse experiment (Table I). In both experiments the IAA level was reduced by S-3307 and the inhibition was partly to almost completely reversed by GA_3.

313

Table I. *The growth and IAA content of the upper two internodes of 17 day old Alaska pea treated with S-3307 for 9 days or S-3307 (6 days) followed by GA₃ for 3 days.*[a]

Treatment Greenhouse Exp.	Length[b]	IAA per stem segment[c]
	mm	*pmol*
Control	78	34
S-3307	31	14
S-3307 + GA₃	56	25
Growth Chamber Exp.		
Control	48	10
S3307	32	4
S3307 + GA₃	47	9

[a]The chemical dwarfing agent S-3307 [RS (E)-1-(4-chlorophenyl-4,4-dimethyl-2-(1,2,4-triazol-1-yl)-1-pentene-3-ol] was applied at 2 mg/box (18 x 25 x 6 cm) containing 30-35 plants growing in vermiculite. The boxes were sprayed with 15 ml of 5×10^{-4} GA₃.
[b]Average of two boxes (9-14 plants per box).
[c]Average of two boxes. Recovery of ^{14}C-IAA was from 34 to 66%. HPLC sensitivity was about 1 pmol lower limit.

LITERATURE CITED

1. IZUMI, K., Y. KAMIYA, A. SAKURAI, H. SHIO, N. TAKAHASHI. 1985. Studies on the sites of action of a new plant growth retardant (E-1-(4-chlorophenyl)-4, 4-dimethyl-2-(1,2,4-triazol-1-yl)-1-pentene-3-ol (S3307) and comparative effects of its stereoisomers in a cell-free system from cucurbita maxima. Plant Cell Physiol. 26: 827-828

2. LANTICAN, B. P., R. M. MUIR. 1969. Auxin physiology of dwarfism in *Pisum sativum*. Physiol. Plant. 22: 412-423.

3. LAW, D. M., R. H. HAMILTON. 1982. A rapid isotope dilution method for analysis of indole-3-acetic acid and indoleacetylaspartic acid from small amounts of plant tissue. Biochem. Biophys. Res. Commun. 106: 1035-1041.

4. LAW, D. M., R. H. HAMILTON. 1984. Effects of gibberellic acid on endogenous indole-3-acetic acid and indoleacetylaspartic acid levels in a dwarf pea. Plant Physiol. 75: 255-256.

INDEX

α–amanitin, 135
abscisic acid, 219
ACC synthase, 234-235
ACC, 230
acid growth hypothesis, 101, 133, 219-222
    (see also wall acidification)
acid phosphatase, 265-266
acid-induced wall extension, 8-9, 22-26
actinomycetes, 78
Allium cepa, 58
aminocyclopropane-1-carboxylic acid (ACC), 230
aminoethoxyvinylglycine (AVG), 230, 233, 304
aminooxyacetic acid (AOA), 230, 233
ampicillin, 82
amylase, 236
amylolytic activity, 235-236
amyloplasts, 211
Ancylobacter aquaticus, 80-82
anoxia, 295
antibiotics, 76
antibodies
    against gibberellins, 166, 167
    against NTPase, 150
    against wall proteins, 122-130, 289-291, 292-293
    penetration into tissue, 292-293
apoplastic solutes, 97, 239-240
apposition, 1
Arabidopsis thaliana, 39, 173-176
arabinogalactan, 271
arabinoxylan, 31-34, 49
ascorbate, 50, 270, 273
Aspergillus, 31
ATP/ADP, 103
ATPase, 265
    Ca^{2+}-dependent, 206-207
    in pea nuclei, 146
    in plasma membrane, 102-103, 205, 263
    in tonoplast, 102
autolysins, 76-77
autolysis of walls, 267-268, 270-271
    suppression by antibodies, 125
auxin, 8, 34, 92, 101, 125, 205-209, 211-213, 198, 215-224, 260, 262, 298-299, 301-303, 304, 310, 313-314
    and gene activiation, 12, 13-15, 133-143
    binding, 310-312
    compartmentation, 295-297
    sensitivity, 301-303

Avena, see oat

Bacillus subtilis, 36, 37, 77-78
bacteria, 74-85
    cell wall, 74-75-79
    osmotic presssure, 79-81
    turgor pressure, 79-84
barley leaf growth, 241-242
bean
    cotyledons, 249-250
    leaves, 92, 194-196
    seeds, gibberellin isolation, 156
beet, 102, 103, 284
Beta vulgaris, see beet
bio-assay of gibberellins, 162, 163
birch leaf growth, 197

Ca^{2+}, 29-31, 33, 145-154, 265, 268, 298-300, 307
    -ATPase, 206-207
calcimedin, 149
calcium, 104
    and root growth, 208-209
    and wall properties, 24, 266, 268
    antagonist, 206-207, 211, 298
    as second messenger, 202-213
    binding proteins, 146-151
Calcofluor, 63-66
Callitriche platycarpa, 229
calmodulin, 146-153, 203-207, 210-212
    antagonists, 206-207, 211
Capsicum, 49
carbon dioxide, 232
carrot cell walls, 52, 276-279
castor bean leaves, 92
cDNA clones to auxin-regulated mRNA, 133-135
cell division, 1, 194, 199, 228
cell enlargement, cessation, 198
cellulase, 32, 289
cellulose, 5, 29-31, 36, 58-59, 274, 277, 284
    microfibril orientation, 5, 29-31, 58-70, 216
    synthesis, inhibition of, 25
    synthesizing complexes, 58-64
cerulenin, 136-137
chemical creep, 22
chlorophyll, 194
colchicine, 59
corn, 39, 103
    coleoptiles, 124, 206, 215, 218-220, 258-260, 273, 289-291, 301-303
    leaf growth, 181-185
    mesocotyl, 295, 310

roots, 188-190, 280-283
    stem growth, 156
cotyledon excision, 252-254
creep, 19
cucumber hypocotyl, 54, 92, 295
cucumber, long-hypocotyl mutant, 176-178
cuticles, 54
cutin, 54
cyanide, 20-21, 296
cycloheximide, 15, 136-137
cytokinin, 198

dictyosomes, 222
diferulate, 48-49, (see also ferulic acid)
diffuse growth, 1
dilution and growth, 97-98, 180
directionality of growth, 5
dithiothreitol, 50
dwarf mutants, 38-40, 172-174, 156-160

EDTA, 29, 210, 267-268, 271, 273
EGTA, 29, 63, 208-209, 273, 307
endoglucanase, 38
endoglycosidases, 31
epidermis as growth controlling layer, 215-224, 54
Escherichia coli, 78
ethylene
    and growth directionality, 5
    and growth of aquatic plants, 229, 304
    and growth of deepwater rice, 229-232
    biosynthesis, 232
exoglucanase, 38, 125, 128
extensibility, see wall extensibility
extensin, 33, 37, 46-48, 50, 53, 123
extrusion pores in cell walls, 249-251

ferulic acid, 31 (see also diferulate)
freeze fracture microscopy, 58-64
fungi, 78
fusicoccin, 13, 102-104, 136-141, 220

galactan, 29, 49
gas vesicles, 79-82
gene activation hypothesis for auxin action, 13-15, 133
genetic dwarfism, 53, 158
Gibberella fujikuroi, 156-157, 233
gibberellin, 173-175, 198
    and stem growth, 8, 38-40, 53, 156-167, 233-234, 313-314
    biosynthesis inhibitor, 313
    levels, 161, 163-164
    primary action of, 167
    synthesis, 157-167
β–glucan, 36-38
glucan synthase, 274-275, 284-286

glucuronoarabinoxylan, 273
Glycine max, see soybean
Golgi apparatus, 8, 31, 102, 262, 273-275
grasses, 48
gravitropism, 59-64, 209-213

H⁺ secretion, 135-143
H⁺-ATPase, 12, 101, 141-143
hemicelluloses, 8, 29-32, 34, 49, 262
Hevea brasiliensis, 102
Hordeum vulgare, 241, 307
hydraulic conductance, 2-4, 95-97, 115-117, 245, 256
hyphae, 78

indole-3-acetic acid, (also see auxin) 102, 104, 208-209, 215, 260, 295, 301, 310
inositol phospholipid turnover, 202-204
inositol triphosphate, 202-204
Instron technique, 19, 195, 197
interference microscopy, 259-261
internode growth of rice, 227-237
intramembranous particles, 62-64
intussusception, 1
isodityrosine, 33, 46-49, 52

kinase, 145-154, 203-204

leaf expansion, 193-199, 230-231, 239-240, 241-242
leaf unrolling, 307-308
lectins, 123,
light, 145-146
    blue, 53-54, 176-177, 260
    far-red, 146, 307-309, 310
    red, 145-146, 176-178, 194, 307-309, 310-312
    white, 194
lignin, 50
lysozymes, 76

maize, see corn
    coleoptile walls, 49
mannitol, 136
membrane
    potential, 103
    vesicles, 101-103
metabolic control analysis, 105
monensin, 11, 223
mRNA, 13, 133-143, 166
mucilage, 280
multinet theory, 77
mutants,
    gibberellin-deficient, 172-174, 156-160
    photomophogenetic, 172, 174-178
myo-inositol, 222-223

β-N-acetylglucosaminidase, 265
NAA, 310
naphthylphthalamic acid (NPA), 301-303
nephelometer, 80
Nitella, 7, 185, 188, 258
Norflurazon, 177
2,5-norbonadiene, 233
NTPase, 145-151
nuclear enzymes, 145-154

oat coleoptiles, 22-24, 59-70, 104, 217
onion, see Allium
osmotic adjustment, 4-5, 112, 181-182,
        255-256
osmotic potential or pressure, 97-98,
        115-116, 239, 243-244, 255-
        256, 252-254
osmotic stress, 270

pea,
    cotyledons, 249
    dwarf mutants, 39
    plumules, 145-148
    shoots, 163-166, 313
    stems, 25, 90-94, 156, 215, 220-223,
        252-254, 255-257, 258-260,
        262-263, 292, 295, 299
    cDNA clones from, 134
pearl millet internodes, 54
pectins, 8, 29-31, 48, 50, 267, 271
peptidoglycan, 8, 75-77
peroxidase, 33, 46, 49, 50-55, 289-291
pH, 84, 101-104, 51
    and microfibrils, 66-70
Phaseolus vulgaris, see bean
phenolic cross-linking, 46-55
phenylpropanoids, 49
phospholipase, 206
photoreceptor mutants, 175
phototropism, 260
phytochrome, 145-146, 150-151, 174-
        178, 194-195, 307
Pisum, see pea
polyamines, 145, 151-153, 235
    biosynthesis, 235
polyethylene glycol, 270-271
potato, 50, 52, 265-266, 267-268
pressure probe, 90, 92, 109, 188, 255-
        257
pressure-block technique, 92-94
protein kinase, see kinase
protein phosphorylation, 150-151,
proton fluxes and growth, 104 (see also
        acid growth)
proton pump, 104
proton secretion and mRNA induction,
        133, 135-143
protoplasts, 276-279
psychrometry, 109, 243

Q_{10} for growth, 20-21
quercetin, 146
Quercus, 245-248
quinone-methides, 49-50

radish, 104
Ranunculus, 304-305
rhamnogalacturonans, 29-31
rhubarb stalk, 54
rice endosperm walls, 49
rice, deepwater, 227-237
Robertson Mutator, 160
root
    cap, 280-283
    elongation, 188-190, 208
    pressurization technique, 241-242
rye coleoptiles, 91

S-adenosylmethionine (SAM)
        decarboxylase, 235
Saccharum, 239-240
salinity, 198, 255-256, 270
secondary messengers, 202-213
secretory vesicles, 141, 262
solute
    absorption, 88, 252-254
    accumulation, see osmotic
        adjustment
    transport and growth, 97-98
soybean
    hypocotyl, 24, 92, 110, 133,
    leaf, 243

spinach, 49
starch breakdown, 236
strain-hardening, 6
Streptococcus faecium, 77-78
stress relaxation in vivo, see wall
        relaxation
stress relaxation, 2, 19, 130
    (see also wall relaxation)
submergence, 227
sugar beet, 48, 50
sunflower
    hypocotyl, 206, 295
    leaf growth, 196
surface stress theory, 75
surface tension, 74
sycamore leaf growth, 196

tentoxin, 194, 196
terminal complexes, 58-70
tetcyclacis, 233
tissue tension, 215-217
tobacco, 270-271
tomato, 103
    GA-deficient mutants, 173
    phytochrome-deficient mutant, 175
    suspension cultures, 47, 50
transcription, 133-140, 150, 166

transpiration, 181, 190
transposon line of maize, 160
Tricoderma, 31
turgor
    maintenence, 109
    pressure, 2-5, 33, 75-76, 79, 84,
        88-92, 109-118, 182-184,
        188-189, 195, 196 241-242,
        243-244, 252-254, 255-257,
        252-254
tyrosine trimers, 48

UDP-glucose 274
    and cell elongation, 105

vacuole, 110
vanadate, 102, 136-137, 146-147
vesicle fusion, 63 (see also secretory
        vesicles)
Vigna angularis, 25
viscoelastic deformation of wall, 19-20,
        218
viscoplastic flow, 6
volumetric elastic modulus, 90, 256

wall
    acidification, 8-9, 13, 198, 136-143,
        219-222
    (see also acid-growth hypothesis)
    composition, 28-40
    enzymes or proteins, 112, 123-128,
        265, 270
    expansion, 2, 88, 123
    extensibility, 18-26, 90-91, 95, 111,
        115-117, 183-188, 197-198,
        217-219, 255-257
    relaxation, 5, 88-94, 115, 130, 183,
        184, 256-257
    stress, 88-89, 198
synthesis, 8, 10-12, 222-223, 258-261,
        262-263, 273, 276, 298
    yield coefficient, see wall
        extensibility
    yielding, 88-91, 186, 255-257
water deficits & cell expansion, 109-118,
        180-191, 241-242, 255-257
water potential
    gradient, 2-3, 93, 95-97, 239
    of growing corn leaf, 181-182
    of growing sugarcane leaf, 239-240
    of growing soybean leaf, 243
    errors, 89, 183, 239
water stress, 109-118, 196, 241-242,
        243-244, 255-257, 270
water uptake, 1, 88, 95-97, 110
wheat shoot apex, 181
wood, 245-248

xylem, 109-110, 118, 245-248
xyloglucan, 31-33, 262-263, 272, 273

yield threshold, 90-92, 95, 109, 183-188,
        243-244, 255-257
Young's modulus, 83

zucchini ER vesicles, 206-207
zucchini hypocotyls, 92